ANIMAL PRODUCTION AND REPRODUCTION

ANIMAL PRODUCTION AND REPRODUCTION

Editor

Prasenjit Nagarkar

scitus academics

Animal Production and Reproduction
Edited by **Prasenjit Nagarkar**

This Book is Printed on Acid-Free Paper.

ISBN: 9781681179223
Library of Congress Control Number: 2017948951

SCITUS Academics LLC,
616, Corporate Way, Suite 2, 4766,
Valley Cottage, NY 10989

www.scitusacademics.com

Notice

Printed in United States of America.

Preface

To meet projected demand for animal products, the industry must streamline production and advance capacity to prevent, detect, diagnose, and treat animal diseases. The health and quality of livestock is influenced by genetics, care, nutrition, and environment. The quality of animal products is affected by production methods and marketing practices. Maintaining the diversity of animal genetic resources is essential to satisfy basic human needs for food and livelihood security. Genetic diversity ensures that different species and breeds are able to adapt to extreme conditions of drought, humidity, cold, and heat, making it possible for humans to obtain livelihoods in inhospitable areas. The genetic improvement has been quite successful in many instances; however, selective inbreeding and use of highly productive genetic lines and breeds while neglecting or blanket upgrading indigenous animals with exotic breeds is leading to deterioration in genetic diversity. Efficient reproduction is of the utmost importance for the sustainable improvement of animal productivity and is a critical factor influencing the economic viability of livestock farmers. Low fertility is often identified as one of the primary constraints hindering the effectiveness of livestock production systems in developing countries. Decreasing the age at first parturition, the interval from parturition to conception, interval between parturitions, and peripartal mortality, and increasing conception rate, litter size, and productive life are the key targets and crucial factors in improving livestock productivity. Increased numbers of animals also enables more intense selection of animals of superior quality to serve as parents of the next generation. Healthy and well-fed animals, properly reared and genetically selected will grow faster and produce more and better quality yields.

'Animal Production and Reproduction' aims to review developments and indicate which reproductive technologies can be used commercially or in research. It begins by discussing artificial insemination and how this has recently been refined in semen sexing technology. This volume also addresses in detail both livestock's role in climate change and the impacts of climate change on livestock production and reproduction. The objectives of this volume is to improve livestock production and reproduction of breeds adapted to the local environments, with emphasis on the characterization of livestock genetic resources, the identification of genes controlling productive and economic traits, and the establishment or the strengthening of artificial insemination programs.

Table of Contents

CHAPTER 1

Environmental Stress Effect on Animal Reproduction

Córdova-Izquierdo Alejandro[1], Villa-Mancera Abel[2], Olivares Pérez Jaime[3], Sánchez-Aparicio Pedro[4, 5]

[1]*Departamento de Producción Agrícola y Animal, Universidad Autónoma Metropolitana Unidad Xochimilco, México DF, México*

[2]*Laboratorio de Genética y Reproducción, Facultad de Medicina Veterinaria y Zootecnia, Benemérita Universidad Autónoma de Puebla, México DF, México*

[3]*Universidad Autónoma de Guerrero-Unidad Académica de Medicina Veterinaria y Zootecnia, México DF, México*

[4]*Facultad de Medicina Veterinaria y Zootecnia, Universidad Autónoma del Estado de México, Toluca, México*

[5]*Bioterio del Hospital Veterinario para Pequeñas Especies UAEM, Toluca, México*

ABSTRACT

The aim of this review is to present some aspects of environmental stress effect on reproduction of farm animals. It describes the following: neuroendocrinology of stress, types of stress, stress characteristics in the reproductive process and recommendations for control environmental stress. Stress is the result of confinement and an ambitious vision and uncaring by the man, who is in the interest of improving production, has participated unconsciously, by manipulating animal production in its way, even trying to tame new species, impossible to adapt to our environment healthy. At present, it has been concluded that stress is one of the environmental factors that affect handling and decrease the production of livestock, it is considered that the main factor which should be controlled in units animal production, since it is closely related to pathogens and infectious agents that may try to the health of animals. Finally, some recommendations are outlined to lessen stress for heat, by handling and by feeding.

Keywords: Environmental Stress; Animal Reproduction; Caloric Stress; Caloric Handling; Caloric Nutrition

1. INTRODUCTION

At present, the need to produce more efficiently, the need to improve reproductive and productive parameters so that new technologies have been adopted which are becoming more sophisticated, all in order to increase the production in animals.

Despite the advances in technology are applied to animal production, few producers or farmers who have shown concern for improving the general environmental conditions of the animal production units, not realizing that these conditions are largely, are responsible for the health of animals, altering thereof and being significantly impacting on the presence of such problems reproductive and productive, significantly increasing production costs.

Stress is the result of confinement and an ambitious vision and uncaring by the man, who is in the interest of improving production, has participated unconsciously, by manipulating animal production in its way, even trying to tame new species, impossible to adapt to our environment healthy.

At present, it has been concluded that stress is one of the environmental factors that affect handling and decrease the production of livestock, it is considered that the main factor which should be controlled in units animal production, since it is closely related to pathogens and infectious agents that may try to the health of animals.

The man plays a very important role in creating the stress. It can provide the means to alleviate, remove or modify any potential stressor [1]. Some of the criteria that have been used to indicate the presence of stress in animals are the thyroid hormones such as corticosteroids plasma or urinary catecholamine's and/or its metabolites, lactic acid, serum enzymes, blood glucose, alterations in the frequency respiratory or cardiac and somatic cell count in milk.

2. STRESS NEUROENDOCRINOLOGY

Mammals, respond to environmental changes by adaptation mechanisms, regulated by the release of hormones such as: Adrenocorticotropin (ACTH), glucocorticoids (GH) and catecholamine (CA), the amount of release depends on the type and intensity of the stress that comes along.

Under stress, there is a negative feedback effect of progesterone on luteinizing hormone (LH) increasing the first. Are elevated cortisol levels with subsequent depression, there are changes in the release of LH. The level of prostaglandin F2 alpha (PGF2 ∞) and ACTH increase. The levels of epinephrine and norepinephrine, also increase [1]. All this adversely affects reproductive function in mammals [2].

3. TYPES OF STRESS

3.1. Caloric

Solar radiation, wind speed, air temperature and humidity are all factors which are concerned with the presentation of heat stress. An early reproductive responses to heat stress are the decrease in the intensity of estrus and consequently low fertility [3]. It is known that mammalian females are more sensitive to heat stress, during the 12 days before estrus, increasing this sensitivity on day 2 before [4].

Studies by [2] found that the absence of high temperatures cause libido, as well as affecting the quantity and quality of semen. They also reduce heat filing and when present, are less intense and of shorter duration. If the female is pregnant, the possibility of embryonic mortality is high, factors lead to a decrease in the fertility of domestic mammalian females [5].

Recent studies conducted by [6], mentioned for the first time, evidence of the delayed effect of heat stress on follicular steroidogenesis revealed that exposure of female domestic heat stress led to a decrease in steroidogenesis, 20 and 26 days later, in mid-sized preovulatory follicles. Furthermore, the theca cells more susceptible found that granulose cells and they expressed low estrogen production in medium-sized follicles and low viability in preovulatory follicles. In terms of production of steroids, theca cells appeared to be consistently susceptible to heat stress and expressed transport effects on androgen production in both types of follicles.

When the body temperature exceeds 40°C, the developed follicles can suffer severe damage and become nonviable [7], are secreted higher levels of progesterone, which inhibits the release of LH during estrus, interrupting the process of Ovulation [8].

A study by [9] has revealed that heat stress reduces dominant follicle size of the first and second follicular wave. In another study, [9], indicate that exposing domestic female mammals during heat stress is altered follicular dynamics and lowers the follicular dominance in the next cycle. In 1997, Wolfensohn et al., Argued that the preovulatory follicle is a key component in the reproductive system and the deterioration of their function during heat stress may trigger other reproductive effects, such as distortion in gonadotropin secretion, underdevelopment of the body luteum and embryo, resulting in a low fertility.

Studies by under in vitro conditions, suggesting that exposure of pregnant domestic females at high ambient temperatures and humidity, which often occur during the summer months, can disrupt the balance between the biochemical and endometrial factors responsible for the maintenance of pregnancy. [9] indicated that the lipid composition of membranes of oocytes are more deteriorated in summer than in winter. An important aspect to counteract the effect of heat stress on the quality of the oocyte and embryo development is the potential use of antioxidants to increase the strength of both the oocyte and embryo.

In 1994, [10] found that the cooled for 2 to 3 to 5 to 6 days after the pregnancy, pregnancy rates improved as compared with females under heat stress. Moreover, [8] found that thyroxine and triiodothyroxine can reduce heat stress.

It has been shown that during periods of severe heat stress, only 10% to 20% of inseminations can result in normal pregnancies [7]. In Israel, conception rates during the winter months reach more than 50%, while during the summer can lose up to 20% [11]. In cattle, it has been found that the Jersey breed is more tolerant to heat stress that Holstein.

Work performed in the United States, indicate that Holstein and Jersey cows set in hot, express signs of estrus for only 12 to 13 hours, which means a difference of 5 - 6 hours less than the normal duration of estrus animals warm places [11].

In the 2000 [13] indicated that dairy cows with black layer absorb more heat from sunlight, cows of white layer. In this regard, it is known that the temperature from which affected reproductive efficiency in cattle, is in a range ranging from 21.1°C to 32.2°C, so that indicated [13], that the cows need to be cooled as much as possible throughout the summer, in order to raise fertility.

In other studies in Florida, by [14] showed that in summer with heat and humidity conditions, pregnancy rates are down to 10%.

The best way to control the effects of heat stress are: 1) physical modification of the environment, 2) breeding races less sensitive to heat stress and, 3) good nutritional management [11].

3.2. Nutrition

In 1998, Zapiola said that in cattle, stress during transportation, has a detrimental effect on the physiology of the animal that the stress caused by lack of food and drink, for a period of equal length, which is because the blood flow is diverted from the internal organs, to the peripheral tissues in an attempt to reduce the body temperature by increasing heat losses. This mechanism leads to a reduction of blood flow, aimed at internal organs such as the uterus, fallopian tubes and ovaries, decreased blood supply to these organs also implies low nutrient availability and therefore low functional capacity [12].

Studies by [8], reflected the corpus luteum of nutrition-stressed females, is lighter and also indicated by chemical constituents of allantoic fluid, the underfed female embryos, degenerate.

Females consume less food during warm seasons, consequently, there may be insufficient nutrients are available, after the production of milk and therefore the ovaries do not function during the first 6 weeks of lactation [14].

Vitamin E may decrease moderately, the effects of heat stress during hot months, does not necessarily increase fertility [10]. A proactive factor could be the adequate food supply in quantity and quality, which have direct effects on reproduction and on the alterations caused by the indirect effects of nutritional stress [14].

3.3. By Handling

The interactions between employees and animals influence the response of cattle to stress management known as [14].

In 2000, [13], indicated that the way to handle livestock, animals may produce an unnecessary handling stress, which may affect the biological mechanisms of reproduction and the immune response. On the other hand, electrical bites, immobilization and other stress factors management weaken the female reproductive functioning.

There are many driving situations can cause stress seriously affecting the reproductive process of the species, such as excessive mobilization of inseminated females before, the use of aggressive or driving the separation of animals, mobilization of animals on the sleeve of management for different purposes, etc. [14].

The time when stress management, represents greater adverse effects on reproduction, it is just no time to estrus [13].

The temperament of cattle can be classified as follows: 1) stand still while subject, 2) is restless, 3) shaken vigorously driving sleeve and 4) shakes violently management sleeve and tries to escape. Overall, in cattle, the Bos indicus have a more aggressive and nervous temperament than Bos taurus [14].

4. CHARACTERISTICS OF STRESS IN REPRODUCTIVE PROCESS

In 2000 [13] indicated that the process of reproduction is an important physiological system for the development of species, linked to stress, that [1] has classified into two groups: environmental stress and handling. Environmental stress includes room temperature, cold and/or hot and cold, wind and moisture. The stress management includes animal density, to handling procedures, the flow of animals, to the interaction between animals of the same or different species and the existing social condition, as nonspecific psychological distress, noise, physical trauma, etc. The combination of both types of stress, act as stressors, which compromises the homeostasis of the animal.

Moreover, [2] mentioned the following types of stress: physical, which includes the transport of animals and physical damage, psychological, which provides for the isolation of animals, physiological, in which considered to hypoglycemia and changes in blood pressure, all aspects related to the reproductive process of mammals.

Cattle, is one of the most adaptable and homeocinéticos have mechanisms to maintain body functions at the expense of critical changes in other physiological functions, such as reproduction, so that the reproductive function is determined largely by the environment [11].

Constant maintenance of a normal body temperature is known as homeotermia and is vital to the functioning of the brain tissue. In cattle, the normal diurnal variation in body temperature should be from 0.6 to 1.2°C. A further increase to 1.2°C is a sign of illness or poor adaptation to temperature elevation. This temperature range is known as a zone of comfort, varies according to the species and animal breeds. When animals have a body temperature below or above their level of comfort, decreases or increases metabolic rate than normal, appearing in imbalance in animal welfare. Heat or body temperature can be eliminated by the following mechanisms: radiation, conduction, convection and evaporation [11].

4.1. In the Male

4.1.1 Semen Quality

Environmental stress can cause low sperm quality, which is closely related to low fertility in females, probably due to a combination of low fertilization rates and increased embryonic mortality [1] [14].

Direct exposure of the testis at high temperatures, causes changes in certain critical stages of spermatogenic cycle, which is also directly related to the quality of the ejaculate.

Recent studies conducted by [13] suggested that the effect of stress on sperm quality, can be improved by implementation of the seminal freezing technology, however, the uterus of female stress may represent heat to sperm.

4.2. In the Female

Sexual behavior and fertility rate are the main indicators of the mammalian female reproduction that are negatively affected by environmental stress. So that the programs undertaken in order to increase the fertility of domestic females, are less successful in hot weather than in temperate [11].

A temperature increase of 0.5°C uterine during hot days, caused a decrease in the rate of fertilization. In cattle, heifers exposure to 32°C for 72 hours after insemination, inhibit embryonic development, however, it is known that 48% of the females held 21°C, can become pregnant without any problem. If heat stress is presented after 10 days after the service, fertility is not affected.

4.2.1. On Fertility

Fertility in lactating cows, varies by season. In the winter, decreases about 50%, 20% in the summer and fall is lower than in the winter. A few years earlier, Gilad et al., (1993) reported that conception rates in Israel fell from 52% in winter to 24% in the summer.

In summer, 80% of estrus may be undetectable [14]. Also indicated, Ryan et al. (1992) when the rectal temperature of the animals increased from 38.5°C to 40°C in 72 hours after insemination service, pregnancy rates can decrease up to 50%.

Heifers and cows studies have indicated that the decline in oocyte quality in the early postpartum period is associated with negative energy balance and low body condition of the animals, which is expressed in developing embryos increased and abnormal having embryos resulting in loss of the hottest months of the year [13].

4.2.2. On Embryos

Several studies [10] [14] have indicated that in cattle, embryonic development is highly sensitive to high temperatures, in the top three to 11 days after service; acquiring more heat tolerance as the gestation period progresses. It is known that the embryos obtained by in vitro fertilization (IVF) are more susceptible to heat stress than those obtained under natural conditions. In this regard, [14] indicated that the greatest loss of bovine embryos from IVF, occur before 42 days, when females are under heat stress.

5. RECOMMENDATIONS FOR CONTROLLING ENVIRONMENTAL STRESS

5.1. For Heat

- Provide a ventilation system that controls body temperature.
- Implement water sprinklers.
- Animal protection against solar radiation, directly and indirectly, through shadows or appropriate ceilings.
- Provide shade in feeding and drinking, to increase feed intake in animals with heat stress.
- Provide water sprays.
- Provide animals, spray baths, in the hottest part of the day.
- Try to have animals with white fur, as they are the most easily absorb heat and therefore are less sensitive to heat stress.
- Develop genetically adapted in animals, as they may be less sensitive to heat stress.
- Implementation of frozen embryos.
- Insemination with frozen semen in less hot time.

5.2. By Handling

- Provide free areas in the Production Unit and shaded.
- Provide the required area per animal, for comfort.
- Bathing females before service and 3 to 5 days.
- Breed or service in less hot periods.
- Implement estrus synchronization programs, to schedule inseminations or services.
- Do not isolate females long before artificial insemination or service.

5.3. By Feeding

- Balancing diets properly, providing the energy needed to offset declining intake presented.
- Reduce your intake of fiber and protein and increase energy.

6. CONCLUSION

Córdova-Izquierdo Alejandro, Villa-Mancera Abel, Olivares Pérez Jaime, Sánchez-Aparicio Pedro, The effects of the environment on the reproduction of farm animals, negatively affect reproductive performance in both females and males, in females, on the percentage of fertility, declining oocyte quality and therefore the quality of the embryo; in males negatively affect the quantity and quality of sperm. Therefore, it is recommended to follow the guidelines described in this paper to reduce the effect of environmental stress on reproduction of farm animals.

REFERENCES

1. Coubrough, R.I. (1985) Stress and Fertility. A Review. Onderstepoort Journal of Veterinary Research, 52, 153-156.
2. Dobson, H. and Smith, R.F. (1995) Stress and Reproduction in Farm Animals. Journal of Reproduction and Fertility. Supplement, 49, 451-461.
3. Ullah, G., Fuquay, J.W., Keawkhong, T., Clark, B.L., Pogue, D.E. and Murphey, E.J. (1996) Effect of GonadotropinReleasing Hormone at Estrus on Subsequent Luteal Function and Fertility in Lactating Holsteins during Heat Stress. Journal of Dairy Science, 79, 19501953. http://dx.doi.org/10.3168/jds.S0022-0302(96)76565-7
4. Gilad, E., Meidan, R., Berman, A., Graber, Y. and Wolfenson, D. (1993) Effect of Heat Stress on Tonic and GnRHInduced Gonadotrophin Secretion in Relation to Concentration of

Oestradiol in Plasma of Cyclic Cows. Journal of Reproduction and Fertility, 99, 315-321. http://dx.doi.org/10.1530/jrf.0.0990315

5. Roth, Z., Arav, A., Bor, A., Zeron, Y., Braw-Tal, R. and Wolfenson, D. (2001) Improvement of Quality of Oocytes Collected in the Autumn by Enhanced Removal of Impaired Follicles from Previously Heat-Stressed Cows. Reproduction, 122, 737-744.http://dx.doi.org/10.1530/rep.0.1220737
6. Lee, C.N. (1993) Environmental Stress Effects on Bovine Reproduction. Veterinary Clinics of North America: Food Animal Practice, 9, 263-273.
7. Roth, Z., Meidan, R., Braw-Tal, R. and Wolfenson, D. (2000) Immediate and Delayed Effects of Heat Stress on Follicular Development and Its Association with Plasma FSH and Inhibin Concentration in Cows. Journal of Reproduction and Infertility, 120, 83-90.
8. Ealy, A.D., Arechiga, E.F., Bray, D.R., Risco, C.A. and Hansen, P.J. (1994) Effectiveness of Short-Term Cooling and Vitamin E for Alleviation of Infertility Induced by Heat Stress in Dairy Cows. Journal of Dairy Science, 77, 3601- 3607.http://dx.doi.org/10.3168/jds.S0022-0302(94)77304-5
9. Wolfenson, D., Lew, B.J., Thatcher, W.W., Graber, Y. and Meidan, R. (1997) Seasonal and Acute Heat Stress Effects on Steroid Production by Dominant Follicles in Cows. Animal Reproduction Science, 47, 9-19. http://dx.doi.org/10.1016/S0378-4320(96)01638-7
10. Ambrose, J.D., Drost, M., Monson, R.L., Uutledge, J.J., Leibfried-Rutredge, M.L., Thatcher, M.-J., Kassa, T., Binelli, M., Hansen, P.J., Chenoweth, P.J. and Thatcher, W.W. (1999) Efficacy of Timed Embryo Transfer with Fresh and Frozen in Vitro Produced Embryos to Increase Pregnancy Rates in Heat-Stresed Dairy Cattle. Journal of Dairy Science, 82, 2369-2376. http://dx.doi.org/10.3168/jds.S0022-0302(99)75487-1
11. Dobson, H., Tebble, J.E., Smith, R.F. and Ward, W.R. (2001) Is Stress Really All That Important? Theriogenology, 55, 65-73. http://dx.doi.org/10.1016/S0093-691X(00)00446-5
12. Zapiola, M. (1998) La reducción del estrés del manejo mejora la productividad y el bienestar animal. http://www.grandin.com/spanish/reduccion.estres.manejo.html
13. Dobson, H. and Smith, R.F. (2000) What Is Stress, and How Does It Affect Reproduction? Animal Reproduction Science, 2, 743-752. http://dx.doi.org/10.1016/S0378-4320(00)00080-4
14. Rutledge, J.J. (2001) Use of Embryo Transfer and IVF to Bypass Effects of Heat Stress. Theriogenology, 55, 105-111. http://dx.doi.org/10.1016/S0093-691X(00)00449-0

CHAPTER 2

Alleviating Heat Stress Leads to Improved Cow Reproductive Performance

Siriwat Suadsong

[1] Department of Obstetrics Gynaecology and Reproduction, Faculty of Veterinary Science, Chulalongkorn University, Bangkok, Thailand

1. INTRODUCTION

Thailand, a country situated in the south eastern part of Asia, is situated in between 15° 00' North latitude and 100° 00' East longitude and located in a tropical area with high temperature and humidity. Crossbred Holstein dairy cattle are popular because, during times of high environmental thermal stress, their milk production and reproductive efficiency is not depressed as it is with purebred Holstein cattle. However, these crossbred cattle have been inseminated with purebred Holstein frozen semen to improve milk production. Although the genetic potential for milk yield has improved, the predominant dairy breed has now become Holstein and the impact of heat stress on production and reproduction has increased.

This chapter will be showed the finding of our studied that show the impact of heat stress on postpartum reproductive performance and milk production and evaluate the effects of utilizing an evaporative cooling system for improving cow comfort, energy balance, postpartum reproductive performance and milk production of early lactating dairy cows under hot and humid climatic conditions.

Climatic conditions in the tropics are such that the hot season is relatively long, there is intense radiant energy for an extended period of time, and there is generally high relative humidity. Thus heat stress is chronic in nature, there is often little relief from the heat during the night, and intense bursts of combined heat and humidity depress performance. Lactating dairy cows create a large quantity of metabolic heat and accumulate additional heat from radiant energy. Heat production and heat accumulation, coupled with a compromised cooling capability, because of environmental conditions, causes heat load in the cows to increase to a point when body temperature rises, feed intake declines and ultimately the cow's productivity falls.

During period of elevated temperature, animals show less physical activity and seek shelter to decrease radiant heat exposure. Elevated body temperatures will rapidly trigger adaptive mechanisms to restore body temperature to normal. These adaptations, including panting, sweating, reduced feed intake and lowered metabolism, may be necessary for survival, but they are not generally favorable to milk production [1]. Moreover, because of their relative size and their high metabolic rate, associated with milk production, dairy cows are particularly susceptible to the effects of heat stress.

Heat stress has a significant impact on dairy cattle in hot and humid climates. Environmental factors, which contribute to heat stress, include high ambient temperatures, radiant energy, and high humidity, all of which compromise the cow's ability to dissipate body heat. When the cow cannot dissipate sufficient heat to maintain thermal balance, body temperature rises and heat stress occurs. Ambient temperature is a major component of heat stress, however humidity must also be considered because evaporative heat loss is more effective when humidity is low. The temperature-humidity index (THI) combines these two factors into an indicator of cow comfort. Cows are beginning to be stressed when the THI exceeds 72 [2].

Dairy cows have several mechanisms to help dissipate body heat and maintain body temperature, such as; conduction, convection, radiation and evaporation. Conduction, convection, and radiation depend on a relatively large differential between the body and the environmental temperature, and evaporation works bests at a low relative humidity. When the environmental temperature nears the cow's body temperature and is coupled with high relative humidity, all the cow's cooling mechanisms are impaired. As a result the cow's body temperature rises and the cows exhibits physiological responses to hot weather. Cows in hot climates generally produce additional heat, compared to those in cool climates, because of greater physical activity (such as panting) which is necessary to enhance cooling in hot conditions. In addition lactating dairy cows produce large amounts of heat from both ruminal fermentation and metabolic processes. As production increases, the total amount of heat produced increases. In order to maintain body temperature within the normal range, dairy cows must exchange this heat with the environment. The most noticeable response to heat stress is reduced feed intake, reduced milk yield, reduced activity, and increased respiration rate and water intake.

2. EFFECTS OF HEAT STRESS ON PRODUCTION

Temperature and humidity combine to decrease dry matter intake (DMI) in dairy cows as a physiological means of regulating internal body temperature. This is accomplished by decreasing rumen fermentation and the metabolic rate [3,4]. A reduction in DMI decreases the nutrients available for milk synthesis, milk production declines and many lactation parameters are affected [5-7]. High environmental temperatures also increase the respiration rate and the water intake, which consequently reduces DMI due to gut full [8]. Because of many dairy cows in hot weather are unable to consume enough feed to meet energy demands during early lactation, they typically mobilize body reserves to maintain their milk production until the intake of feed can match or exceed nutritional requirements [9,10] thus, entering a state of negative energy balance (NEB).

In heat stressed dairy cows there is a reduction in DMI [11], which prolongs the period of negative energy balance. Negative energy balance leads to decreased plasma concentration of insulin, glucose and insulin-like growth factor-I (IGF-I), and increased plasma concentrations of growth hormone (GH) and non-esterified fatty acid (NEFA) [12,13]. All of these metabolic hormones can affect reproduction. Metabolic hormones acting on the hypothalamo-pituitary axis and the ovary probably mediate the inhibitory effects of negative energy balance on postpartum fertility.

3. EFFECTS OF HEAT STRESS ON REPRODUCTION

Heat stress affects reproductive performance both by direct action on reproduction and by indirect actions mediated through alterations in energy balance. There is an interaction between DMI, stage of lactation, milk production, energy balance and heat stress, that results in reduced luteinizing hormone (LH) secretion and a decreased diameter of the dominant follicle in the postpartum period [14], this results in reduced oestradiol secretion from the dominant follicle, leading to poor expression of oestrus. The postpartum anovulatory interval of dairy cow, is characterized by a variable period of negative energy balance that is reported to modulate the recrudescence of ovarian cyclicity [9,15,16]. In heat stressed cows, motor activity and other manifestations of oestrus are reduced [17] and the incidence of anoestrus and silent ovulation are increased [18].

There is a decrease in fertility in lactating dairy cows during summer in hot climate [19]. The magnitude of the depression depends on the geographical location and the milk yield [20-22]. In tropical climates, high ambient temperatures and humidity are important determinants of reproductive performance. Heat stress decreases the intensity and duration of oestrus, which in turn reduces both the number of inseminations and the pregnancy rates [23]. Heat stress alters the concentration of circulating hormones by increasing the circulating concentration of corticosteroids [24] and by reducing progesterone concentration [25]. The viability of pre-fixation embryos is reduced [26], and the uterine environment is altered by a decreased blood flow [27] and increased uterine temperature [28]. These changes are associated with increased early embryonic loss and a reduced proportion of successful inseminations. Cows exposed to heat stress have a high incidence of early embryonic mortality [26,29], and some of this effect is due to the direct effect of elevated temperature on the embryo [30].

Lactating dairy cows are susceptible to heat stress because of the elevated internal heat production which is associated with lactation. During periods of heat stress, milk production, feed intake, and physical activity are decreased [11]. At the same time, reproductive ability is compromised [31,32]. The exposure of lactating cows to heat stress has been shown to cause a decrease in follicular growth and to reduce serum estradiol [33],which also concluded that decreased follicular size or decreased dominant follicle function occurred in lactating cows that were exposed to heat stress[34,35]. Some of the reproductive losses, in heat stressed cattle, are associated with decreased expression of oestrus caused by anoestrus and silent ovulation [36,37]. Heat stress delays follicle selection and lengthens the follicular wave having potentially adverse effects on oocytes quality [34] and follicular steroidogenesis [35].

Heat stress cause infertility and represented a major source of economic loss in dairy cows under tropical conditions. In recent study, the conception rate of dairy cows in Thailand decrease 20-30 % in hot season. The conception rate was lower in April and May when compared to other month, which lowest in summer and highest in winter (Figure 1). The effects of heat stress can be directly related to the increase in body temperature of dairy cow, which affects the reproductive function and embryonic development.

The detrimental effects of heat stress on the reproductive performance of dairy cows have been well documented. These include a suppressed intensity of oestrus, a reduced preovulatory LH surge and decreased secretion of luteal progesterone [25], altered ovarian follicular development [33], decreased embryo development [38] and lower fertility

[39]. In an attempt to minimize these effects, modifications to dairy cattle housing environments have been implemented to alleviate thermal stressors and improve cow comfort, through the use of shade, fans, sprinklers, and evaporative cooling [11,37,40]. These methods can enhance pregnancy rates significantly [24,41].

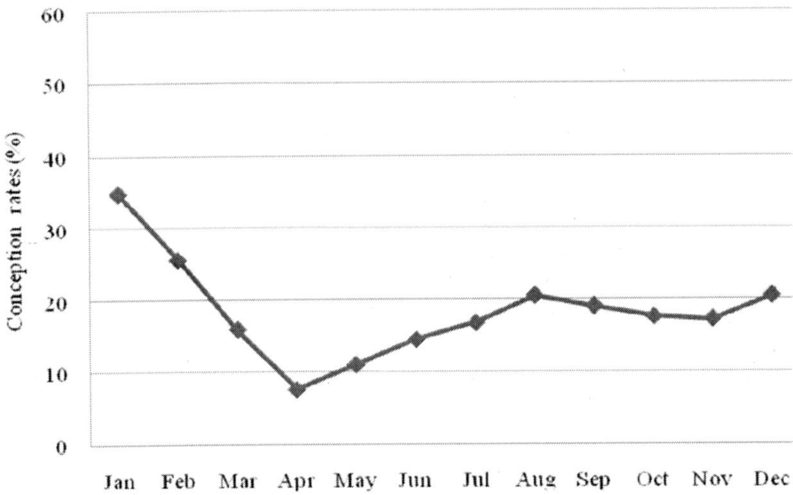

Figure 1. Conception rates of dairy cows in commercial farm located in the central part of Thailand.

4. EFFECT OF ENVIRONMENTAL MODIFICATION FOR DAIRY COWS UNDER TROPICAL CONDITIONS

Reducing heat stress in dairy cattle requires a multi-disciplinary approach. It involves breeding for improved heat tolerance, improved nutrition and improved reproductive management for the animals, and improved the structural design and the environmental control of their housing.

Heat stress reduces milk production and reproductive efficiency. In an attempt to minimize these effects, modifications to dairy cattle housing environments have been implemented to alleviate thermal stressors and improve cow comfort [11,37,40,42]. The major objective of any cooling system is to keep the cow's body temperature as close to normal for as much of the day as possible. An acceptable range in rectal temperature is 38.5-39.3 C. There are two general approaches to cooling dairy cows. One is to modify the environment to prevent heat stress or to utilize methods that increase heat dissipation from the skin surface of cattle. The easiest and most obvious way to help heat-stressed cows is to provide shade. Direct sunlight adds a tremendous heat load to the cow and can be blocked by shades, but shade alone is inadequate to reduce the effect of heat stress. A more economical method to reduce the effect of heat stress is by evaporative cooling. Evaporative cooling can be accomplished by two approaches; 1) direct evaporation from the skin surface of the cows (fan and sprinkler combinations) and 2) indirect evaporation which involve cooling the micro-environment of the cows, with cooling pads and fans, in an enclosed barn. When water evaporates it absorbs heat, there by reducing the temperature and increasing heat dissipation from the skin of cattle. When water evaporates it also increases the relative humidity, due to the increased level of water vapor present. In hot and humid regions, evaporative cooling always requires the use of forced ventilation.

A number of studies have shown that housing systems in hot climates can be modified by the use of evaporative cooling to improve both milk production and reproductive efficiency of dairy cows [43-46]. There is a great potential to reduce temperature and THI. However, as relative humidity increases and or temperature decreases, the potential for evaporative cooling to modify the environment decreases. In hot and humid climates, high relative humidity reduces the potential of evaporative cooling. Therefore, there are questions regarding the effectiveness of evaporative systems in climates with high relative humidity. This chapter showed the impact of heat stress on postpartum cow performances and evaluated the effects of utilizing an evaporative cooling system for improved cow comfort and cow performance of early lactating dairy cows in a hot and humid climate.

5. EFFECT OF EVAPORATIVE COOLING SYSTEM ON THE ENVIRONMENTAL CONDITION AND PHYSIOLOGICAL RESPONSE OF DAIRY COWS

The upper critical temperature for heat stress to begin was between 25 and 26C [47]. Climatic conditions in the present study are such that the hot season is relatively long, and generally accompanied by high relative humidity. Thus heat stress is chronic in nature and there is little relief from the heat during the evening through to the morning, and also includes intense bursts of combined heat and humidity which further depresses performance.

Environmental modifications by evaporative cooling system equipped with tunnel ventilation in this study led to a decrease in the ambient temperature and an increase in the relative humidity, during the day. The air temperature in the tunnel barn was up to 6 C cooler ($P<0.05$) during the daytime than that in the outside barn, while the relative humidity increased by up to 16%. As air was drawn through the wet cooling pads, water was evaporated into the air causing the temperature to be reduced while increasing the air moisture level. The amount of initial moisture in the outside air will directly impact on how much reduction in air temperature might be expected. Therefore, when the outside air is initially very high humid (relative humidity greater than 70%), the reduction in air temperature will be minimal (less than 2.8 to 5.6C).

The average daily minimum and maximum temperatures were 24.2±0.1 and 28.4±0.1C, respectively in the evaporative, cooled tunnel, ventilated barn and 24.7±0.1 and 34.4±0.1C, respectively in the outside barn, indicating less variation and more consistency in this cooling system. This evaporative, cooled tunnel, ventilated barn reduced daily fluctuation in the ambient temperature, relative humidity and THI during hot and humid climatic conditions. Although, this system reduced ($P<0.05$) afternoon barn temperature, but relative humidity increased ($P<0.05$) when compared to the barn without a supplemental cooling system.

Although, high environmental relative humidity reduced the cooling capacity of the evaporative cooling, but on the days when animals were observed it reduced heat stress of the lactating dairy cows. This study shows that evaporative cooling with tunnel ventilation reduced the severity of afternoon heat stress in dairy cows. THI in the tunnel ventilated barn was decreased when compared to the outside barn. There was no thermoneutal zone (THI72) during the study but exposure to conditions of moderate heat stress (THI79) that occurred outside was decreased by utilizing the evaporative cooling system. This difference in environmental conditions had a dramatic effect on the physiological response of the cows, as THI was highly correlated with both rectal temperature and respiration rate. The average rectal temperature and respiration rate in the cooled cows was lower ($P<0.05$) than in the uncooled cows. Changing in cow body temperatures was most sensitive to same day climatic factors [48]. These results suggest that evaporative cooling and tunnel ventilation has the potential to decrease exposure to heat stress and alleviate the symptoms of heat stress.

The average daily THI in the morning and afternoon in the evaporative, cooled tunnel, ventilated barn were lower ($P<0.05$) than that in the outside barn. However, the mean THI exceeded the critical point of 72 at daytime and nighttime, suggesting that the cows were exposed continuously to conditions conducive to mild heat stress for the cooled cows and moderate heat stress for the uncooled cows. In hot, arid conditions this system would work well and evaporative cooling has already been used very successfully to cool such dairy cows [46], but in high humidity locations its effectiveness would be limited by the evaporation potential. In this study the air temperature increases during the day and decreases in the evening until the next morning. As the air temperature rises during the day, the relative humidity will decrease. Accordingly during the hottest portion of the day, the outside relative humidity dropped to a level that allowed for maximum evaporation potential, making the system effective for reducing the severity of heat stress.

Table 1. The environmental measurement and physiological responses of dairy cows in the experimental housing. (MeanS.E.M.)

	Cooled (inside)	Uncooled (outside)	P-value
Number of animals (n)	17	17	
Environmental measurement :			
Daily minimum temperature (°C)	24.2±0.1	24.7±0.1	*
Daily maximum temperature (°C)	28.4±0.1	34.4±0.1	*
AM milking			
Daily mean temperature (°C)	25.3±0.1	25.6±0.1	*
Daily mean relative humidity (%)	88.5±0.3	91.6±0.2	*
Daily mean THI	76.0±0.2	76.7±0.2	*
Rectal temperature (°C)	38.4±0.0	39.2±0.1	*
Respiratory rate (breaths/min)	42.7±1.4	54.5±1.4	*
PM milking			
Daily mean temperature (°C)	26.5±0.1	31.4±0.1	*
Daily mean relative humidity (%)	86.5±0.3	74.6±0.8	*
Daily mean THI	78.0±0.2	83.1±0.2	*
Rectal temperature (°C)	38.7±0.0	39.7±0.1	*
Respiratory rate (breaths/min)	46.3±1.8	68.3±2.2	*

* Means differ ($P<0.05$) between groups

Air temperature had the greatest impact on physiological measurement, while radiation was second in importance, followed by vapor pressure and air movement [49]. Increasing the air temperature reduces the temperature differential between the cow's body temperature and the ambient temperature and decreases the transfer of heat to the environment. As ambient temperatures increase in the presence of low or high relative humidity, the cooling mechanisms employed by the cows shifted from a nonevaporative processes (convective, conductive, and radiation) to evaporative (sweating and panting) [48], this demonstrate that the percentage of cooling originating from the non evaporative processes declines as ambient temperature increases, while the evaporative process increase. As a result the uncooled cows had greater (P<0.05) rectal temperatures and respiration rates than the cooled cows.

6. EFFECT OF EVAPORATIVE COOLING SYSTEM ON DRY MATTER INTAKE AND DAILY MILK YIELD OF DAIRY COWS

Dry matter intake, expressed as kilograms per day (kg of DM/day) was greater (P<0.05) in cooled (12.0 0.2 kg/d) than uncooled cows (9.1 0.2 kg/d). DMI increased (P<0.05) from wk 1 to 22 of lactation in both groups of cows but treatment x week postpartum did not affect (P>0.05) it (Figure 2) [50].

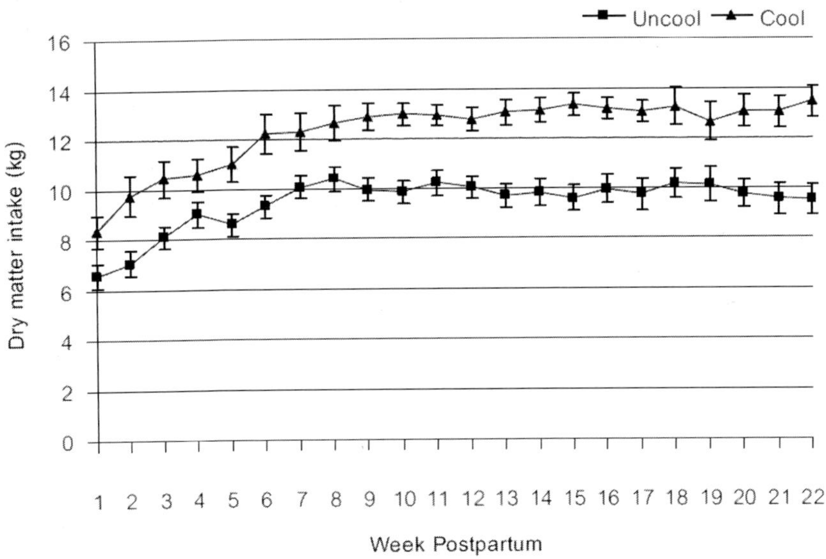

Figure 2. Weekly changes in average daily DMI for the cooled and uncooled cows during the first 22 week postpartum.

Daily milk yield was greater (P<0.001) in cooled (16.9 0.3 kg/d) than uncooled (12.6 0.2 kg/d) cows. Daily milk yield increased (P<0.001) from wk 1 to wk 22 of lactation in both groups of cows but treatment x week postpartum did not affect (P>0.50) daily milk yield (Figure 3). The 4% FCM also differ (P<0.001) between cooled and uncooled cows. Cooled cows had more persistent milk production than uncooled cows. Milk composition did not differ (P>0.50) between the groups of cows over the 12 week study (Table 2) [50].

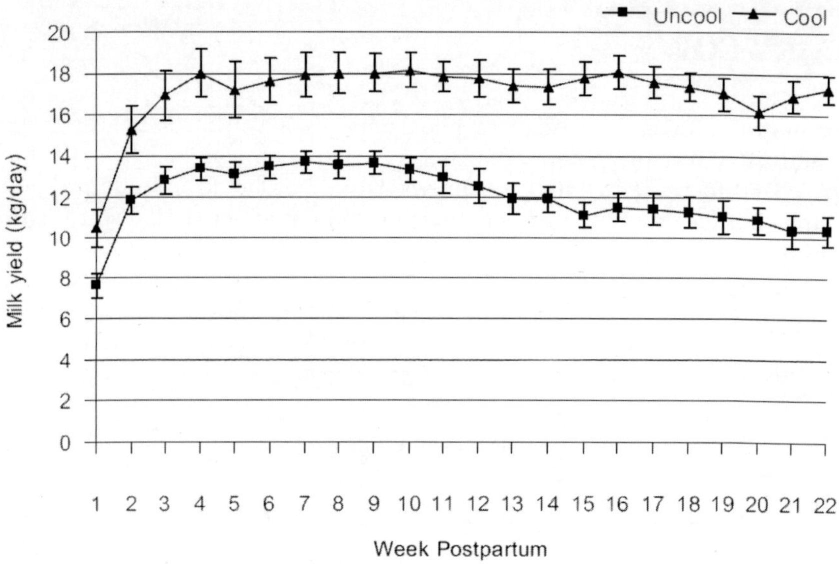

Figure 3. Weekly changes in average daily milk production for the cooled and uncooled cows during the first 22 week postpartum.

Table 2. Dry matter intake, milk production and milk composition of cooled and uncooled cows during the 12 week study.

	Cooled (inside)	Uncooled (outside)	P-value
Dry matter intake (kg/d)	12.0 ± 0.2	9.1 ± 0.2	0.001
Milk yield (kg/d)	16.9 ± 0.3	12.6 ± 0.2	0.001
Fat (%)	3.3 ± 0.6	3.4 ± 0.6	0.810
Protein (mg/ml)	3.2 ± 0.3	3.1 ± 0.3	0.650
Lactose (mg/ml)	5.0 ± 0.2	5.0 ± 0.2	0.571
Solid not fat (mg/ml)	8.9± 0.4	8.8 ± 0.4	0.450

In this study, rectal temperature was positively correlated with air temperature and THI but negatively correlated with DMI. Dry matter intake was positively correlated with milk production. Milk yield and DMI exhibited a significant decline when maximum THI reached 77 [51]. There is a significant negative correlation between THI and DMI [52,53], and the effect of THI is probably mediated through the effects of increasing body temperature on cow performance.

Mean air temperature had the greatest influence on milk yield for Holstein cows under hot conditions [48]. The mean daily ambient temperature was highly correlated with the p.m. rectal temperature and milk yield was highly correlated with the cows p.m. rectal temperature. The elevated p.m. rectal temperatures were associated with concomitant reductions in DMI and milk yield [54]. Uncooled cows, in this study, had rectal temperatures exceeded 39C at both the a.m. and the p.m. milking, when cow temperatures should have been near their lowest and highest points, respectively. Such consistently elevated rectal temperatures, result in a significant decline in DMI and milk yield when compare with the cooled cows which had a lower rectal temperature.

The negative effects of heat stress on milk production could be explained by decreased nutrient intake and decreased nutrient uptake by the portal drained viscera of the cow [55]. Blood flow which moves to peripheral tissues for cooling purposes may alter nutrient metabolism and contribute to lower milk yields during hot weather. Providing cows with supplemental shade and cooling to mitigate heat stress has been evaluated in economic terms using the increase in milk production data [2]. Adding sprinklers and fan cooling increased the feed intake (7.1 to 9.2%) and milk production (7.1 to 15.8%), and decreased rectal temperature (-0.4 to -0.5 C) and respiration rate (17.6 to 40.6 %) [56].

In this study the cooling system improved cow comfort and milk production. The cooling system decreased ambient temperatures and THI. The cows with this cooling system had decreased rectal temperatures and respiration rates, and increased feed intake (30.9 %) and milk production (42.5%). In addition, an analysis of the economics in the this study showed that the rate of return on investment in cooling equipment and additional feed plus electric costs of this cooling system, showed it was profitable in hot and humid conditions.

7. ECONOMIC ANALYSIS OF THE UTILIZING EVAPORATIVE COOLING SYSTEM

There are questions that arise regarding the cost effectiveness of an evaporative cooling system over a period of years. In this study, an economic analysis of the evaporative cooling system showed that cows in the tunnel ventilation barn produced 5.1 kg/cow/day more than the cows housed outside. The milk price was 11.50 baht/kg. Therefore, the use of tunnel ventilation cooling increased revenue by 58.65 baht/cow/day and increased feed costs by 22.17 baht/cow/day. Thus, income over feed cost was 36.48 baht/cow/day.

The cost of the fan and water pump operation was 6.88 baht/cow/day. The total equipment and supply cost to build the tunnel barn facility was 125,000 baht (for 18 cows), which, when depreciated over the expected life of the equipment (including maintenance costs of 25,000 baht over 5 years), was 4.11 baht/cow/day. Thus, the cows housed in the tunnel barn earned 25.49 baht/day or 3823.50 baht more than the cows housed outside over the 22-week study (Table 3).

Table 3. The economic analysis of the evaporative cooling system over the first 22 week of lactation in dairy cows.

Item	Difference	Income-Cost (Baht/cow/day)
Milk yield (kg/cow/day)	5.1	58.65
DMI (kg/cow/day)	2.9	-22.17
Electric cost (fan & water pump operation)		-6.88
Equipment and maintenance cost (5 years)		-4.11
Total		25.49

8. EFFECT OF EVAPORATIVE COOLING SYSTEM ON ENERGY BALANCE AND BODY WEIGHT OF DAIRY COWS

Responses of dairy cow to heat stress include panting and sweating. If these are not successful in alleviating the heat load, body temperature will rise. Increased body temperature will result in reduced feed intake, higher maintenance requirements (panting) and less efficient productive ability. The higher maintenance requirements dictates that cows need to increase feed intake to maintain milk production. However, this not possible as feed intake declines when ambient temperatures exceeded 26C. For uncooled cows in this study, as a result of this increase in requirement and a decrease in intake, milk production may decline as much as 25-30%, and typically they mobilize body reserves and lose body weight to maintain milk production until the intake of feed can match or exceed nutritional requirements [9,10], thus entering a state of negative energy balance.

The energy balance and body weight of dairy cows is presented in Figure 4 and Figure 5. During 12 wk of lactation, average body weight of postpartum cows was greater in cooled than uncooled cows. In both groups of cows, body weight decreased between wk 1 and 4, and increased between wk 5 and 22. Week postpartum ($P<0.001$) and treatment ($P<0.001$) affected EB but treatment x week postpartum was not significant. Cows in both groups entered into NEB immediately after calving. Averaged EB was greater ($P<0.001$) in cooled (0.916 0.194 Mcal/day) than uncooled cows (-0.268 0.195 Mcal/day). During the 12 week study, the week of EB nadir was at wk 2 in both groups and the degree of EB nadir did not differ significantly ($P>0.50$) between the groups, although the average was lower in uncooled than cooled cows. After reaching the EB nadir, uncooled cows required more days to reach a positive energy balance than the cooled cows. The first week that EB was greater than zero was at wk 5 in cooled cows and at wk 7 in uncooled cows. Because of dry matter intake in uncooled cows was lower ($P<0.05$) than in cooled cows. Resulting in uncooled cows having a prolonged period of negative energy balance and postpartum body weight in these cows were lesser ($P<0.05$) than in cooled cows [50]. The negative energy balance is directly related to the postpartum interval to first ovulation, and follicle size was adversely affected by negative energy balance in early postpartum dairy cows [9]. The average EB during the first 4 week of lactation was negatively correlated to the postpartum interval to first ovulation [57].

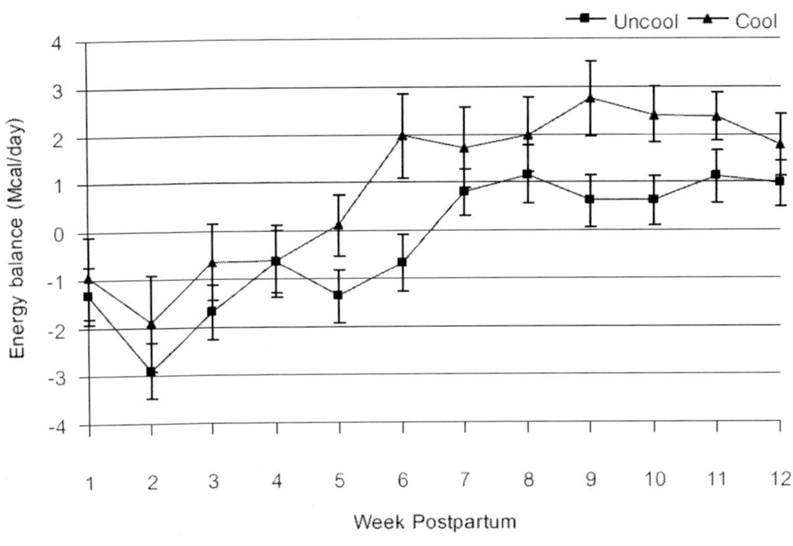

Figure 4. Weekly changes in energy balance for the cooled and uncooled cows during the first 12 week postpartum

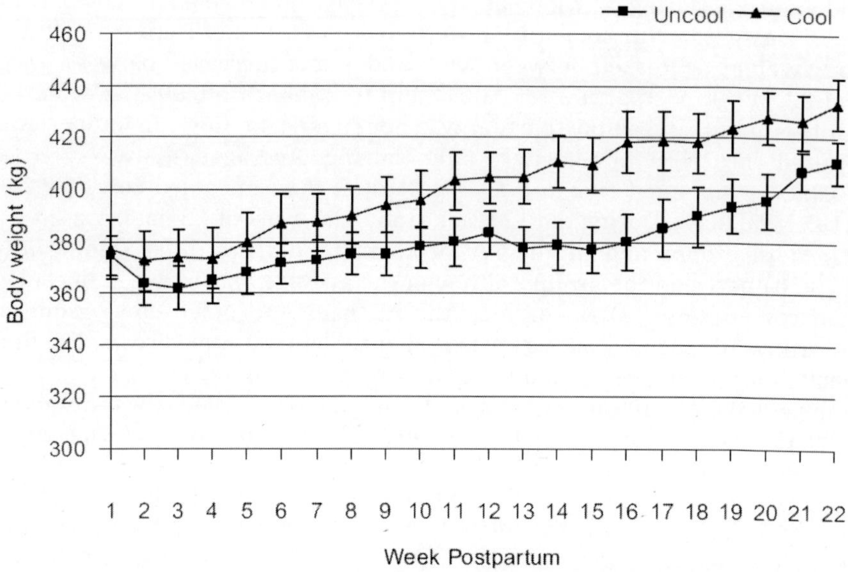

Figure 5. Weekly changes in average body weight for the cooled and uncooled cows during the first 22 week postpartum.

9. EFFECT OF EVAPORATIVE COOLING SYSTEM ON SYNCHRONIZATION OF FOLLICULAR DEVELOPMENT AND OVULATION

The exposure of lactating cows to heat stress has been shown to cause a decrease in follicular growth and to reduce concentrations of serum estradiol [33]. In this study the cooled cows had a greater ($P < 0.05$) average diameter of the largest ovulatory follicle than the uncooled cows. Heat stress inhibited follicular growth and dominance during the preovulatory period. Abnormal ovarian function in heat stressed cows was manifested as a decrease in the proestrus rise in estradiol, and the smaller size of the second wave dominant follicle [33]. Circulating estradiol concentration during the preovulatory period is necessary to produce an LH surge and ovulation. In addition, a reduced estradiol peak may also alter aspects of the LH surge that could account for some types of anovulation in lactating cows [58]. A reduction of the endogenous LH surge by heat stress was reported in heifers [59]. It has been suggested that these differences are related to preovulatory estradiol levels because the amplitude of tonic LH pulses and GnRH-induced preovulatory plasma LH surges are decreased in cows with low plasma concentrations of estradiol but not in cows with high plasma concentrations of estradiol [60]. Therefore, the synchronization rate in the uncooled cows tended to lower than in the cooled cows. In addition, the ovulation rate in response to a second injection GnRH of Ovsynch was reported to be between 87% [61] and 91% [62] in cycling cows. Therefore a second GnRH injection after the PGF_2treatment might be used to improve the ovulation rate in dairy cows under heat stress.

Table 4. The effect of evaporative cooling and tunnel ventilation system on follicular development, time of ovulation and the response rates of synchronized cows to GnRH and PGF2. Results are expressed as mean±SEM.

	Cooled	Uncooled	P-value
Number of animal (n)	17	17	
Synchronization rate (%)	82.4(14/17)	52.9(9/17)	0.08
Interval to new follicular wave emergence (day)	2.2±0.1	2.3±0.2	0.55
Size of the largest ovulatory follicle at $PGF_{2\alpha}$ (mm)	11.5±0.6	10.2±0.5	0.12
Maximal size of the largest ovulatory follicle (mm)	14.6±0.5	14.2±0.4	0.57
Growth rate of the largest ovulatory follicle after $PGF_{2\alpha}$ (mm/d)	0.9±0.1	1.1±0.1	0.17
Interval from $PGF_{2\alpha}$ treatment to ovulation (h)	83.6±5.1	88.0±6.9	0.63

10. EFFECT OF EVAPORATIVE COOLING SYSTEM ON EMBRYONIC DEVELOPMENT AND CONCEPTION RATE OF DAIRY COWS

Heat stress can compromise reproductive performance by decreasing the expression of oestrus behavior, altering follicular development, affecting oocyte competence, inhibiting embryonic development due to a reduced synchronization of ovulation response, lowering fertilization rates, and reducing embryo quality. In this study the modification of the barn environment and fixed TAI in lactating dairy cows in a hot and humid climate resulted in higher initial conception rates compared to cows housed in a barn without a supplemental cooling system (43.8 and 23.5%, respectively) (Figure 6). The results indicate that this method has the potential to attenuate some of the detrimental effects of heat stress on embryo survival during this period. Although, initial conception rates in cooled cows was higher than uncooled cows, it was still compromised by heat stress as the conception rate decreased further in both groups, indicating that embryo mortality may still have occurred after that. This was probably due to environmental heat stress [63], which causes maternal body temperature to rise leading to the impairment of embryo survival, or this cooling was not sufficiently cool to protect embryos from direct effect of high temperatures, or early embryos during this period might be sensitive to elevated temperature.

Climatic factors that may influence the degree of heat stress include: temperature, humidity, radiation and wind [32]. The upper critical temperature for heat stress begins between 25 and 26C [47]. When environment temperatures exceed 30C the day after insemination, pregnancy rates were adversely affected in lactating dairy cows than in heifers [64]. Maternal hyperthermia is detrimental to embryonic development and survival [65-67]. The oocyte and early cleavage stage embryo are the most sensitive to heat stress, while embryos that are 3 d or older are more tolerant [26,29]. Conception rates decline from 61 to 45% when rectal temperature 12 h, post breeding, increased by 1C [68]. Furthermore, cattle with a rectal temperature of 40C, as a result of exposure to 32.2 C ambient temperatures for 72 h after inseminating, had conception rates of 0%, compared with a conception rate of 48% when rectal temperature was 38.5C, for cows in an ambient temperature of 21.1 C [68]. Given that exposure to conditions of high environmental temperature and humidity has been shown to elevate rectal and uterine temperatures [28].

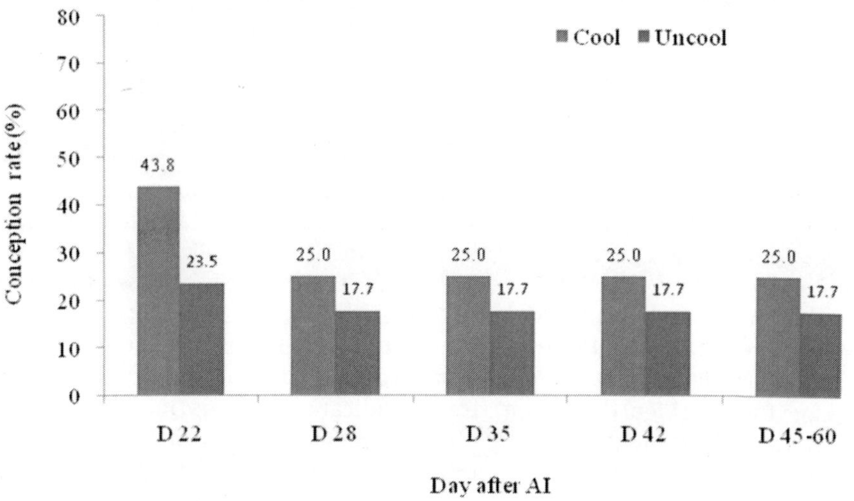

Figure 6. The conception rate after the synchronization of ovulation and fixed time AI of cooled and uncooled cows.

The modified environment had been used to reduce the effect of heat stress, however, this approach has not eliminated all problems. Timed AI might be particularly effective during heat stress periods because of the decreased incidence of missed oestrus, but heat stress has also a direct effect on the development of the embryo. Conception rates have ranged between 31 to 42 % in cows following the Ovsynch protocol [69-73]. However, conception rates in heat-stressed cows following the Ovsynch protocol were lower than in non stressed cows.

Timed AI programs based on follicular recruitment by synchronized ovulation have been developed [74]. The submission rates and pregnancy rates between d 27 and 30 were enhanced when a TAI program was used. However, the advantage was lost between d 40 and 50 due to increased embryonic mortality in the cows bred using TAI [75]. These results may indicate that cows were successfully induced to ovulate and subsequently conceive but had a reduced ability to maintain pregnancy. Previous studies also showed high (9.3 to 16.8 %) pregnancy losses between 28 and 56 d after AI [76,77].

In recent study, reported that dairy cow had a cumulative hours of body temperature greater than 38.5C more than 4-8 h per day, significantly ($P<0.05$) decline (Figure 7) in conception rate [78]. This study indicated that long period of hyperthermia (body temperature > 38.5C) had an adverse effect on dairy cows and suggested that dairy cows in the tropical area need additional cooling system to completely eliminate heat stress result in close to normal fertility of dairy cows depend on the severity of the local environment conditions. In addition, *in vitro* heat stress during the critical stage of early embryo development significantly increases the incidence of early embryonic mortality [79]. These results indicate that increased maternal body temperatures adversely affect embryo quality and the conception rate. An increase in maternal body temperature may result in an increase in the ambient temperature of oocytes, zygotes and embryos in the oviduct or the uterus. At temperature increase of 0.5C above the basal body temperature has been associated with a decreased pregnancy rate [28]. These studies indicated that elevated environmental temperature lead to hyperthermia in lactating cows. Intensive cooling cows have the potential to eliminate the decline in conception rate of dairy cows under tropical conditions.

Infertility of dairy cows under tropical conditions is primarily caused by elevated body temperature. Therefore, cooling cows should improve conception rates. A variety of cooling systems are available for heat-stressed cows. The intensive cooling cows with the combination of sprinklers and fans were used for improved reproductive performance. Cows were cooled in the holding area for a total 6-8 cooling periods and 4-6 cumulative h/day. Each cooling period combined cycles of sprinkling (0.5

min) and forced ventilation (4.5 min). Intensive cooling had significantly affected on decreased body temperature and allowed cows to maintain normal body temperature throughout the day [78]. In the same study, uncooled cows had high body temperature, daytime significant portion of the day and returned to normal body temperature during the late night and early morning (Figure 8).

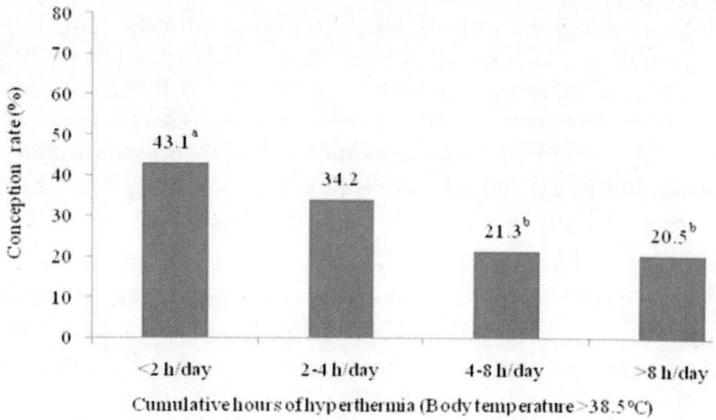

a, b Means with different superscripts differ (P<0.05)

Figure 7. This chart show conception rates of dairy cows. It also illustrated that cow had cumulative hours of hyperthermia greater than 4-8 h per day that conception rate decline significantly (P<0.05).

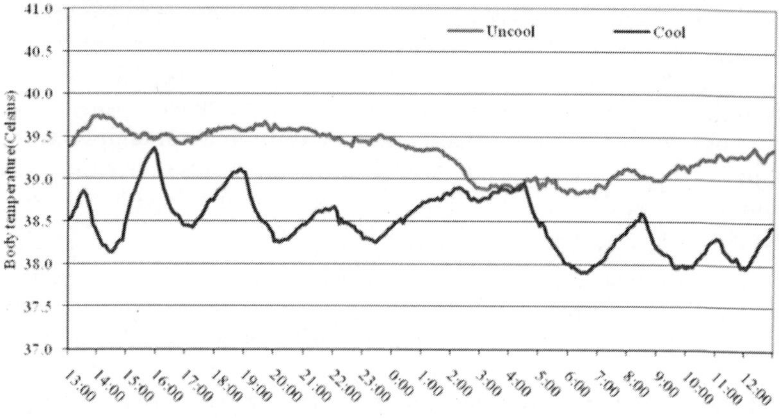

Figure 8. This chart show changes in mean vaginal temperature of cooled and uncooled cows throughout the day.

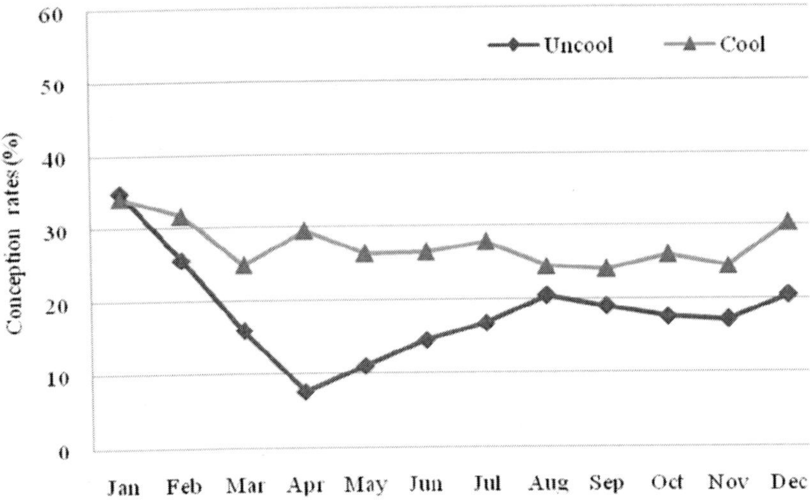

Figure 9. This chart show conception rates obtained in uncooled cows during 2004-2007 and cooled cows during 2008-2011 on commercial dairy farm in the central part of Thailand.

For large scale survey was carried out during a 8 yr period (2004-2011) in commercial dairy farm. The conception rate of intensively cooled cows with sprinkler and forced ventilation was significantly higher than that of uncooled cows. Conception rates obtained in intensively cooled cows in this study were similar to those obtained in that same winter in uncooled cows (Figure 9). This current finding confirmed this adverse effect of heat stress. In hot climates there is a large decrease in the fertility of dairy cows during summer months. In addition, intensively cooling cows has the potential to eliminate the decline in conception rate of dairy cows under tropical conditions. Therefore, intensive cooling is essential in dairy cows under tropical conditions to prevent hyperthermia and its harmful effects on those cows.

11. SUMMARY

The exposure of dairy cows to elevate temperatures have a variety of effects, including decreased fertility, depressed appetite, and decreased milk production, all of which contribute to the goal of decreasing the production of metabolic heat in order to maintain thermo-neutrality. This chapter showed significant advantages for the evaporative cooled barn can be used to reduce heat stress of dairy cows housed in hot and humid climates. The combined effect of higher milk production and increasing lactation persistency, with minimal costs could improve the financial status of dairy operations. The benefit demonstrated increased income over costs.

In addition, respiration rates and rectal temperatures, which affect both milk production and reproduction, were reduced by this environmental modification. These finding suggest that the evaporative cooling and tunnel ventilation system has the potential to decrease the exposure to heat stress, alleviate the symptoms of heat stress and improve milk production and metabolic efficiency during early lactation. This modification of the barn environment can reduce some of the detrimental effects of heat stress on follicular development and can improve the response rate to the synchronization of ovulation in dairy cows in hot and humid climates.

The implementation of evaporative cooling systems for dairy cows in hot and humid climates increased the percentage of cows that initially established a pregnancy and increased successful pregnancy while decreasing early embryonic mortality, if these cows were sufficiently cooled after breeding. It appears that modification of environment need to be developed further to improve reproductive performance of dairy cows in hot and humid climates. The finding that the cooling of cows did not alleviate all the effects of heat stress on pregnancy rates suggests that the degree of cooling was not sufficient to prevent the adverse effects of heat stress. It is also possible that the cooling of dairy cows needs to be done not only in the housing unit but also in the holding and milking areas to improve pregnancy rates. Dairy cows in hot and humid climatic condition need to be intensively cooled to completely eliminate heat stress to improve production and fertility close to normal. Therefore, additional research is needed to determine the effects of environmental modification or improved cooling system on postpartum reproductive performance and production when compared to conventional methods of cow cooling in the tropical area. However, cooling intensification should be combined with reproductive management, hormonal application and nutritional management to minimize the decline in cow performances under hot and humid climatic conditions.

ACKNOWLEDGEMENT

This study was supported by the Thailand Research Fund and Faculty of Veterinary Science, Chulalongkorn University.

REFERENCES

1. Stott GH.What is animal stress and how is it measured. Journal of Animal Science 198152150153
2. D. Armstrong, Heat stress interactions with shade and cooling. Journal of Dairy Science 19947720442050
3. Moody EG, Van Soest PJ, McDowell RE, Ford GL.Effect of high temperature and dietary fat on milk fatty acids. Journal of Dairy Science 19715414571460
4. Richardson CW, Johnson HD, Gehrke CW, Goerlitz DF.1961Effects of environmental temperature and humidity on the fatty acid composition of milk fat. Journal of Dairy Science 1961;4419371940
5. Bauman DE, Currie WB.Partitioning of nutrients during pregnancy and lactation: A review of mechanisms involving homeostasis and homeorhesis. Journal of Dairy Science 19806315141529
6. Collier RJ, Beede DK, Thatcher WW, Israel LA, Wilcox CJ.1982Influences of environment and its modification on dairy animal health and production. Journal of Dairy Science 1982;6522132227
7. W. A. Smith, B. Harris, H. H. Van Horn Jr, C. J. Wilcox, Effect of forage type on production of dairy cows supplemented with whole cottonseed, tallow, and yeast. Journal of Dairy Science 1993
8. Mallonee PG, Beede DK, Collier RJ, Wilcox CJ.Production and physiological responses of dairy cows to varying dietary potassium during heat stress. Journal of Dairy Science 19856814791487
9. W. R. Butler, R. D. Smith, 1989Interrelationships between energy balance and postpartum reproductive function in dairy cattle. Journal of Dairy Science 1989;72767783
10. Nebel RL, McGilliard ML.Interactions of high milk yield and reproductive performance in dairy cows. Journal of Dairy Science 19937632573268
11. Fuquay JW.Heat stress as it affects animal production. Journal of Animal Science 198152164174
12. M. C. Lucy, Badinga. L. Savio, R. L. de la Sota, W. W. Thatcher, Factors that affect ovarian follicular dynamics in cattle. Journal of Animal Science 19927036153626
13. P. D. Jolly, S. Mc Dougall, L. A. Fitzpatrick, K. L. Macmillan, K. Enwhitsle, Physiological effects of under nutrition on postpartum anoestrous in cows. Journal of Reproduction and Fertility 1995Suppl.);49477492
14. N. N. Jonsson, M. R. Mc Gowan, K. Mc Guigan, T. M. Davison, A. M. Hussain, M. Kafi, Relationship among calving season, heat load, energy balance and postpartum ovulation of dairy cows in a subtropical environment. Animal Reproduction Science 199747315326
15. Butler WR, Everett RW, Coppock CE.The relationships between energy balance, milk production and ovulation in postpartum Holstein cows. Journal of Animal Science. 198153742748
16. Staples CR, Thatcher WW, Clark JH.Relationship between ovarian activity and energy status during the early postpartum period of high producing dairy cows. Journal of Dairy Science 199073938947
17. Nebel R.L, Jobst SM, Dransfield MBG, Pandolfi SM, Balley TL.Use of radio frequency data communication system, HeatWatch, to describe behavioural estrus in dairy cattle. Journal of Dairy Science 1997Abst.)
18. Gwazdauskas FC, Thatcher WW, Kiddy CA, Pape MJ, Wilcox CJ.Hormonal pattern during heat stress following PGF2alpha-tham salt induced luteal regression in heifers. Theriogenology 198116271285
19. Hansen PJ.Effects of environment on bovine reproduction. In: Current therapy in large animal theriogenology. Philadelphia: WB Saunders; 1997403415
20. L. Badinga, R. J. Collier, C. J. Wilcox, W. W. Thatcher, Interrelationships of milk yield, body weight and reproductive performance. Journal of Dairy Science 19856818281831

21. L. Badinga, R. J. Collier, W. W. Thatcher, C. J. Wilcox, Effects of climatic and management factors on conception rate of dairy cattle in subtropical environment. Journal of Dairy Science 1985687885
22. Al-Katanani YM, Webb DW, Hansen PJ.Factors affecting seasonal variation in nonreturn rate to first service in lactating Holstein cows in a hot climate. Journal of Dairy Science 19998226112616
23. Thompson JA, Magee DD, Tomaszewski MA, Wilks DL, Fourdraine RH.1996Management of summer infertility in Texas Holstein dairy cattle. Theriogenology 1996;46547558
24. H. Roman-Ponce, W. W. Thatcher, D. E. Buffington, C. J. Wilcox, H. H. Van Horn, Physiological and production responses of dairy cattle to a shade structure in a subtropical environment. Journal of Dairy Science 197766424430
25. Howell JL, Fuquay JW, Smith AE.Corpus luteum growth and function in lactating Holstein cows during spring and summer. Journal of Dairy Science 199477735739
26. D. Ealy, M. Drost, P. J. Hansen, Developmental changes in embryonic resistance to adverse effects of maternal heat stress in cows. Journal of Dairy Science 19937628992905
27. H. Roman-Ponce, W. W. Thatcher, D. Caton, D. H. Barron, C. J. Wilcox, Thermal stress effects on uterine blood flow in dairy cows. Journal of Animal Science 197846175180
28. F. C. Gwazdauskas, W. W. Thatcher, C. J. Wilcox, 1973Physiological, environmental and hormonal factors at insemination which may affect conception. Journal of Dairy Science 1973;56873877
29. D. J. Putney, M. Drost, W. W. Thatcher, Embryonic development in superovulated dairy cows exposed to elevated ambient temperatures between Days 1 to 7 post insemination. Theriogenology 198830195209
30. Rivera RM, Hansen PJ.Development of cultured bovine embryos after exposure to high temperatures in the physiological range. Reproduction 2001121107115
31. G. Francos, E. Macer, Observations on some emvironmental factors connected with fertility in heat- stressed cows. Theriogenology 198319625634
32. F. C. Gwazdauskas, Effects of climatic on reproduction in cattle. Journal of Dairy Science 19856815681578
33. Wilson SJ, Marion RS, Spain JN, Spiers DE, Keisler DH, Lucy MC.Effects of a controlled heat stress on ovarian function of dairy cattle. 1. Lactating cows. Journal of Dairy Science 19988121242131
34. L. Badinga, W. W. Thatcher, T. Diaz, M. Drost, D. Wolfenson, Effect of environmental heat stress on follicular development and steroidogenesis in lactating Holstein cows. Theriogenology 199339797810
35. D. Wolfenson, W. W. Thatcher, L. Badinga, Meidan. R. Savio, B. J. Lew, R. Braw-Tal, A. Berman, Effect of heat stress on follicular development during the estrous cycle in lactating dairy cattle. Biology of Reproduction 19955211061113
36. E. Her, D. Wolfenson, I. Flamenbaum, Y. Folman, M. Kaim, A. Berman, productive. Thermal, responses. reproductive, high. of, cows. yielding, to. exposed, cooling. short-term, summer. in, Journal of Dairy Science 19887110851092
37. M. Younas, J. W. Fuquay, A. E. Smith, A. B. Moore, Estrous and endocrine responses of lactating Holsteins to forced ventilation during summer. Journal of Dairy Science 199376430436
38. Hansen PJ, Arechiga CF.1999Strategies for managing reproduction in the heat-stressed dairy cow. Journal of Animal Science 1999;773650
39. Gwazdauskas FC, Wilcox CJ, Thatcher WW.Environmental and management factors affecting conception rate in a subtropical environment. Journal of Dairy Science 1975588892
40. H. Roman-Ponce, W. W. Thatcher, C. J. Wilcox, Hormonal relationships and physiological responses of lactating dairy cows to a shade management system in a subtropical environment. Theriogenology 198116139154
41. W. W. Thatcher, F. C. Gwazdauskas, C. J. Wilcox, J. Toms, H. H. Head, D. E. Buffington, W. B. Frederickson, 1974Milking performance and reproductive efficiency of dairy cows in an environmentally controlled structure. Journal of Dairy Science 1974;57304307

42. Flamenbaum, N. Galon, Management of heat stress to improve fertility in dairy cows in Israel. Journal of Reproduction and Development 2010S36S41.
43. D. Armstrong, M. Wise, M. Torabi, F. Weirsma, R. Hunter, K. Kopel, Effect of different cooling systems on milk production of late lactation Holstein cows during high ambient temperatures. Journal of Dairy Science 1988Suppl. 1):212. (Abstr.)
44. D. Armstrong, S. De Nise, F. Delfino, E. Hayes, P. Grundy, S. Montgomery, M. Correa, Comparing three different dairy cattle cooling systems during high environmental temperatures. Journal of Dairy Science 1993Suppl. 1):24. (Abstr.)
45. I Flamenbaum, D. Wolfenson, A. Mamen, A. Berman, Cooling cattle by a combination of sprinkling and forced ventilation and its implementation in the shelter system. Journal of Dairy Science 19866931403147
46. D. P. Ryan, M. Scland, E. Kopel, D. Armstrong, L. Munyakazi, G. Gorlke, R. Ingergam, Evaluating two different evaporative cooling management systems for dairy cows in hot dry climate. Journal of Dairy Science 1992Suppl. 1):240. (Abstr.)
47. Berman, Y. Folman, M. Kaim, M. Marnen, Z. Herz, D. Wolfensen, A. Arieli, Y. Graber, Upper critical temperature and forced ventilation effects for high-yielding daily cows in a subtropical climate. Journal of Dairy Science 19856814881495
48. West JW, Mullinix BG, Bernard JK. Effects of hot, humid weather on milk temperature, dry matter intake, and milk yield of lactating dairy cows.Journal of Dairy Science 200386232242
49. Legates JE, Farthing BR, Casady RB, Barrada MS.Body temperature and respiratory rate of lactating dairy cattle under field and chamber conditions. Journal of Dairy Science 19917424912500
50. S. Suadsong, J. Suwimonteerabutr, P. Virakul, S. Chanpongsang, A. Kunavongkrit, Effect of improved cooling system on reproduction and lactation in dairy cows under tropical conditions. Asian-Australasian Journal of Animal Sciences 200821555560
51. West JW.Effects of heat-stress on production in dairy cattle. Journal of Dairy Science 20038621312144
52. Holter JB, West JW, McGilliard ML, Pell AN.Predicting ad libitum dry matter intake and yields of Jersey cows. Journal of Dairy Science 199679912921
53. Holter JB, West JW, McGilliard ML.Predicting ad libitum dry matter intake and yield of Holstein cows. Journal of Dairy Science 19978021882199
54. Maust LE, McDowell RE, Hooven NW.Effect of summer weather on performance of Holstein cows in three stages of lactation. Journal of Dairy Science 19725511331139
55. McGuire MA, Beede DK, DeLorenzo MA, Wilcox CJ, Huntington GB, Reynolds CK, Collier RJ.Effects of thermal stress and level of feed intake on portal plasma flow and net fluxes of metabolites in lactating Holstein cows. Journal of Animal Science 19896710501060
56. Bucklin RA, Turner LW, Beede DK, Bray DR, Hemken RW.Methods to relieve heat stress for dairy cows in hot, humid climates. Applied Engineering Agriculture 19917241247
57. Spicer LJ, Vernon RK, Tucker WB, Wettemann RP, Hogue JF, Adams GD.Effects of inert fat on energy balance, plasma concentrations of hormones and reproduction in dairy cows. Journal of Dairy Science 19937626642673
58. M. C. Wiltbank, A. Gumen, R. Sartori, Physiological classification of anovulatory conditions in cattle. Theriogenology 2002572152
59. Madan ML, Johnson HD.Environmental heat effects on bovine luteinizing hormone. Journal of Dairy Science 19735614201423
60. E. Gilad, R. Meidan, A. Berman, Y. Graber, D. Wolfenson, Effect of heat stress on tonic and GnRH-induced gonadotrophin secretion in relation to concentration of estradiol in plasma of cyclic cows. Journal of Reproduction and Fertility 199399315321
61. Vasconcelos JLM, Silcox RW, Rosa GLM, Pursley JR, Wiltbank MC. Synchronization rate, size of the ovulatory follicle, and pregnancy rate after synchronization of ovulation beginning on different days of the estrus cycle in lactating dairy cows.Theriogenology 19995210671078
62. F. Moreira, C. Orlandi, Mattos. R. Risco, F. Lopes, W. W. Thatcher, Effects of presynchronization and bovine somatotropin on pregnancy rates to a timed artificial insemination protocol in lactating dairy cows. Journal of Dairy Science 20018416461659

63. D. J. Putney, M. Drost, W. W. Thatcher, Influence of heat stress on pregnancy rates of lactating dairy cattle following embryo transfer or artificial insemination. Theriogenology 198931765778
64. Thatcher WW, Collier RJ.Effect of climate on bovine reproduction. In: Current Therapy in Theriogenology, Morrow, D.A. (ed). Philadelphia: WB Saunders Co, 1986301309
65. Gordon, M. P. Boland, H. Mc Govern, G. Lynn, Effect of season on superovulating responses and embryo quality in Holstein cattle in Saudi Arabia. Theriogenology 1987Abstr.)
66. D. E. Monty, C. Racowsky, In vitro evaluation of early embryo viability and development in summer heat-stressed, superovulated dairy cows. Theriogenology 198728451465
67. Ulberg LD, Sheenan LA.Early development of mammalian embryos in elevated temperature. Journal of Reproduction and Fertility 197319155161
68. Ulberg LD, Burfening PJ.1967Embryo death resulting from adverse environment on spermatozoa or ova. Journal of animal Science 1967;26571577
69. Pursley JR, Kosorok MR, Wiltbank MC.Reproductive management of lactating dairy cows using synchronization of ovulation. Journal of Dairy Science 199780301306
70. Pursley JR, Wiltbank MC, Stevenson JS, Ottobre JS, Garverick HA, Anderson LL.Pregnancy rates per artificial insemination for cows and heifers inseminated at a synchronized ovulation or synchronized estrus. Journal of Dairy Science 199780295300
71. Fricke PM, Wiltbank MC.Effect of milk production on the incidence of double ovulation in dairy cows. Theriogenology 19995211331143
72. Drillich. M. Tenhagen, W. Heuwieser, Analysis of cow factors influencing conception rates after two timed breeding protocols. Theriogenology 200156831838
73. Pursley JR, Fricke PM, Garverick HA, Kesler DJ, Ottobre JS, Stevenson JS, Wiltbank MC.Improved fertility in noncycling lactating dairy cows treated with exogenous progesterone during Ovsynch. Journal of Dairy Science 2001Suppl. 1):1563. (Abstr.)
74. Pursley JR, Mee MO, Wiltbank MC.Synchronization of ovulation in dairy cows using PGF2 and GnRH. Theriogenology 199544915923
75. Cartmill JA, El-Zarkouny SZ, Hensley BA, Lamb GC, Stevenson JS.2001Stage of cycle, incidence, and timing of ovulation, and pregnancy rates in dairy cattle after three timed breeding protocols. Journal of Dairy Science 2001;8410511059
76. Vasconcelos JLM, Silcox RW, Lacerda JA, Pursley JR, Wiltbank MC. Pregnancy rate, pregnancy loss, and response to heat stress after AI at 2 different times from ovulation in dairy cows.Biology of Reproduction 1997Suppl 1):230.
77. J. E. P. Santos, L. Thatcher, L. Pool, M. W. Overton, Effect of human chorionic gonadotropin on luteal function and reproductive performance of high-producing lactating Holstein dairy cows. Journal of Animal Science 20017928812894
78. S. Suadsong, T. Chaikhun, J. Suwimonteerabutr, Effect of improved cooling system on diurnal body temperature patterns and conception rate in dairy cows under tropical conditions 2011conference proceeding, July 26-29, 2011, Suranaree university of technology, Nakhon Ratchasima, Thailand. SAADC; 2011.
79. S. Sugiyama, M. Mc Gowan, M. Kafi, N. Phillips, M. Yong, Effects of increased ambient temperature on the development of in vitro derived bovine zygotes. Theriogenology 20036010391047

CHAPTER 3

Breeding Dairy Cows to Reduce Greenhouse Gas Emissions

M.J. Bell[1], R.J. Eckard[2] and J.E. Pryce[3]

[1] Melbourne School of Land and Environment, University of Melbourne, Australia

[2] Primary Industries Climate Challenges Centre, The University of Melbourne & Department of Primary Industries, Australia

[3] Biosciences research Division, Department of Primary Industries, Victorian AgriBiosciences Centre, Bundoora, Australia

1. INTRODUCTION

The dairy industry has made large advances in efficiencies over the past 60 years as a result of changes in breeding, nutrition and management practices [1]. To meet the growing demand for dairy products, which is projected to continue out to the year 2050 [2], milk production per cow has increased over the last thirty years primarily by genetic selection and better nutrition. Genetic selection has tended to focus on mostly production traits (kilograms milk, kilograms fat and protein) rather than fitness (lameness, mastitis, fertility and lifespan) traits, although most countries now include fitness traits in addition to production traits in modern breeding goals. The Holstein Friesian is a popular breed due to its high genetic potential to produce milk; however it is characterised by having a lower body condition score, and reduced fertility and survival compared to other breeds [3]. Even with these negative attributes, in tandem with efficiencies in production in recent decades have come reductions in greenhouse gas (GHG) emissions and resource inputs per unit product [1, 4, 5], while emissions per unit area have increased.

The main GHGs attributed to livestock systems are methane and nitrous oxide emissions [6]. Due to the variability in lifespan of gases in the atmosphere and the ability of gases to reflect and trap radiant energy, the average potential of a GHG to warm the earth's near-surface air is expressed in carbon dioxide equivalents (CO_2-eq.) emissions (its global warming potential). Methane and nitrous oxide are capable of trapping about 25 and 298 times more radiant energy respectively, over a 100-year time horizon, than one kilogram of carbon dioxide [6]. The dairy sector's total CO_2-eq. emissions are estimated to be 4% of total global GHG emissions, of which, about half are methane and a third nitrous oxide emissions [2].

The main benefits of selection to improve production efficiencies are by increased productivity and gross efficiency (i.e. the ratio of yield of milk to resource input) by firstly, diluting the maintenance cost of animals in the system and secondly, less animals are required to produce the same amount of product [1, 7]. Studies have found [8, 9] that more energy efficient animals produce less waste in the form of methane and nitrogen excretion per unit product. A study in the UK [4] calculated that the genetic improvement in dairy cows by economic and production efficiency in the last 20 years had reduced GHG emissions per unit product by 0.8% per year and would continue to reduce emissions at a rate of 0.5% per year over the next 15 years. A reduction of 0.6% per year in GHG emissions per unit product was found in the US [1] over a 63 year period. Emissions of methane and nitrous oxide per unit product were estimated to have shown large declines of about 1.3% and 1.5% per year respectively over the last 20 years in the UK, and will continue to decline over the next 15 years albeit at a slightly slower rate per year [4]. These rates of decline for methane are similar to those reported in other studies [10] for enteric methane emissions per unit product of 1.1% per year for cows selected on increased milk fat and protein production (Select line cows) and at 1.4% per year for cows selected to represent the UK average for milk fat and protein production over a similar time period.

In this review, we investigate the potential role of selective breeding in reducing GHG emissions.

1.1. Selecting. Animals for reduced emissions

Level of feed intake and its composition are important factors influencing methane and nitrogen losses. As the feed intake of an animal increases, the percentage of dietary gross energy (GE) intake lost as methane decreases by an average of 1.6% per unit of intake [11]. A higher feed intake level increases its fractional passage rate through the rumen and reduces its retention time, rumen digestion (depending on the diet) and methane production [11, 12, 13]. As rumen retention time decreases with increased feed intake the rate of nitrogen excretion increases, increasing the potential for nitrous oxide emissions [9, 14]. However, if cows are to meet their genetic potential for milk production, they need to maximise their feed intake [15, 16]. At pasture the nutrient intake can vary and impair the milk production potential of the animal, particularly during the peak of lactation [17]. To meet the genetic potential for milk production forage based diets are supplemented with high energy dense feed in the form of concentrate. Supplementing the diet of high milk yielding dairy cows at pasture with concentrate was found [17] to result in a lower rate of pasture intake substitution and a higher response in improved milk yield compared to lower milk yielding cows.

Cows with a high body weight have been found to have a greater bite weight when eating and therefore are more efficient in their use of time spent feeding [16]. The larger North American Holstein genotype has been found to produce between 8 to 11% less methane as a percentage of GE intake, on both a total mixed ration and pasture-based diet, than a small New Zealand Holstein [18], presumably due to differences in level of feed intake. However, larger cows have greater maintenance requirements. For the same level of production, a smaller cow is obviously a more efficient converter of feed into milk. This is why selection programmes in both New Zealand and Australia, in particular have focused on increasing the rate of genetic gain in traits that contribute to profitability per unit of feed eaten [19].

Selecting dairy animals for efficient feed use could bring both higher production and reduced resource requirements. In comparison to other mitigation strategies, selective breeding offers a medium to long-term approach to GHG mitigation, which can be cost effective [20]. The response from selective breeding depends on the selection intensity, genetic variation, generation interval and the economic importance of the trait, with annual rates of response typically being between 1% and 3% [7] of the mean in the trait under selection. In intensive poultry and pig production, profitability on cereal-based diets has encouraged selection for feed efficiency (ranging from 1.7 to 2.4 kg cereal feed per kg animal weight) compared to ruminant systems [21]. The improvement in feed conversion efficiency in non-ruminants has been quite remarkable, for example, it has been reported [22] that the feed conversion efficiency (kg lean meat/t of feed) in pigs has nearly doubled from 85 kg/t in the 1960s to 12 170 kg/t in 2005. It is reasonable to assume that in non-ruminant animals where selection on feed use efficiency has been made that current selection goals will account for a moderate to high proportion of any genetic variation in methane output or nitrogen use efficiency.

Average daily dry matter intake and milk yield are moderately heritable in dairy cows at about 0.30 [15, 23] compared to about 0 to 0.15 for health and fertility traits [24]. Therefore, genetic improvement in dry matter intake and milk production traits is easier to achieve than for health and fertility. Increasing the genetic potential of a cow to produce milk increases total system GHG emissions, due a higher feed intake [25]. However, the milk produced per unit of feed eaten is likely to reduce due to improvements in gross efficiency. For example, the genetic correlation between intake and milk yield has been found [15] to account for just less than half the genetic improvement in milk production being covered by an increase in dry matter intake. This implies that the apparent improvement in gross efficiency is partly due to feed conversion efficiency. The remainder could be due to increased reliance of body tissue mobilisation.

The direct selection of animals on a trait such as methane production in ruminants may be of little importance given its relationship with feed intake [26, 27], which is a more easily measured trait. The additional benefit from directly measuring GHG emissions from animals would be if selection on a measure of feed use efficiency was not possible. Whether breeding goals are able to account for all the genetic variation in methane output or nitrogen efficiency is unlikely, and therefore there may be some benefit to directly selecting on these traits if possible. In ruminants, measurements of methane output, nitrogen efficiency and overall feed efficiency are difficult and costly to obtain, which has limited the direct selection of these traits in the past. A large part of the variation in methane emissions from dairy cows has been found to be genetic, with a heritability of 0.35 for methane output and 0.58 for methane output per unit product [28], presumably due to a prediction of methane being used and its close unity correlation with feed intake. In comparison, a lower heritability of 0.13 has been found [29] in sheep for methane output. Once individual measurements for total animal methane emissions become more affordable to carry out for a large number of animals, selecting animals on methane output will become possible.

Variation in feed use efficiency and enteric methane emissions between-animals, breeds and over time means there is potential to reduce GHG emissions through genetic selection [30, 31, 32]. With a positive genetic correlation between feed efficiency and methane output, with an estimated range from 0.18 to 0.84 [28], it can be inferred that selecting cows that are more efficient will reduce methane production, possibly in the order of 1.1% to 2.6% per year. From a range of production and fitness traits, breeding studies [4, 14] found feed efficiency to have a large impact on reducing the GHG emissions from dairy systems compared to other production or fitness traits. Feed efficiency can be assessed by feed intake required per unit product (gross efficiency) or by net or metabolic efficiency commonly calculated as residual feed intake [4]. Residual feed intake is the difference between the observed and predicted feed intake; where the predicted feed intake is often calculated as energy requirements. Studies [31, 33, 34] looking at selecting beef cattle based on a lower residual feed intake (difference between actual and expected feed intake) found that growth performance was not compromised and the lower expected feed intake resulted in less methane produced. Heritability estimates for feed efficiency tend to be moderate (0.16 to 0.46; [4]). Low correlations between residual feed intake and other production traits imply that little or no genetic improvement has previously been made in residual intake in beef cattle as a result of selection on production traits [4]. In dairy cows, calculating residual feed intake accurately is difficult as changes in body tissue composition need to be fully accounted for. This is because without accounting for body composition changes, residual feed intake is mathematically equivalent to energy balance [35]. Negative energy balance (often considered to be very similar to condition score loss) in early lactation has been the subject of intense phenotypic (and genetic) investigation [36]. Mobilisation of body tissue, or low body condition score is associated with reductions in fertility [37].

Taking direct feed intake measurements can be costly due to the equipment required, therefore an indirect measure of feed efficiency may be a more appropriate option for dairy animals, but further research is required to investigate measures that might be correlated with intake. Biologically inactive markers released in the rumen (such as *n*-alkanes) have successfully been used to predict feed intake, however, as with other markers released in this way (i.e. such as SF_6) there are concerns about the consistency of the marker release rate (see [38] for a review). A new technology, known as genomic selection is especially promising for difficult or expensive to measure traits, as measurements need only to be made on a representative sample of the population. Genomic breeding values are calculated as the sum of the effects of dense genetic markers that are approximately equally spaced across the entire genome, thereby potentially capturing most of the genetic variation in a trait. Here the prediction equation is formed in a reference population with genotype and phenotype data. The prediction equation can be used to predict breeding values in animals that are genotyped but without phenotype data. The availability of SNP chips at affordable prices has made implementation of this technology commercially feasible. For example, many countries now publish genomic breeding values for bulls on a range of traits of economic importance.

Recently, there has been interest in estimating genomic breeding values for traits in the feed conversion efficiency complex. The estimated accuracy of genomic prediction of RFI calculated in a population of 1000 New Zealand and 1000 Australian non-lactating heifers was around 0.4 [39]. The accuracy of the genomic prediction could be increased further still if countries were to pool together their phenotypes that are expensive to record and their genotype resources, as these data could be used to increase the accuracy of genomic selection further still. Collaborative efforts between research organizations in the Netherlands, the UK and Australia have already demonstrated that the accuracy of genomic predictions of dry matter intake can be increased by combining datasets (de Haas submitted, 2012). The ultimate aim of these collaborative research efforts is to develop genomic breeding values for dry matter intake or a feed conversion efficiency trait that could be used in breeding programs to improve efficiency and mitigate emissions.

1.2. Breeding x feeding system

Due to the profitability of Holstein cows, Holstein genes are present in a large proportion of dairy cows globally, particularly North American. Larger North American Holstein-Friesian cows have been found to show a better response in milk yield with a higher proportion of concentrate in their diet than smaller genotypes like the New Zealand Holstein-Friesian, which have been selected for higher milk yield performance from pasture [16]. Cows which were ~88% North American Holstein and selected on increased milk fat and protein production (Select line cows) were found to grow faster and had increased kg milk per kg dry matter intake during their productive life when on a high energy dense diet, compared to cows selected to represent the UK average for milk fat and protein production on the same diet [10]. Select genetic line animals have a high genetic potential for mobilising body energy reserves for production, which has been found to have deleterious effects on health and fertility [3, 24], particularly later in life [7]. However, it was found [14] that Select line cows responded to a diet containing a low proportion of forage, rather than a high forage diet, by having a significantly shorter calving interval. Select line animals on a low forage diet also produced lower CO_2-eq. emissions per energy corrected milk compared to non-select and cows on a high forage diet over their lifetime (Figure 1).

Systems emissions can be minimised by improvements in herd health and fertility (improving longevity and productivity), and by reducing the number of replacement animals retained on the farm to reduce wastage [13, 40, 41]. Improving herd fertility in the UK back to 1995 levels could amount to a 24% reduction in methane emissions per cow by improved efficiencies of reduced herd replacements and calving interval length [40]. Cows of predominantly North American Holstein genes may be better suited to a high energy dense feeding system, typically found in the US, rather than a diet containing a high proportion of forage. In contrast, the performance of animals of New Zealand origin had higher yields of milk solids and better fertility compared to animals of North American origin when compared on a range of New Zealand grazing systems [42]. Therefore, selecting animals for an environment is important. In a study in the US [1], good health and welfare in modern high input systems (cows of 90% Holstein genes) was reported, with better production efficiency and CO_2-eq. emissions per unit product compared to the past. This may be explained by optimal nutrition being provided to these animals, which may not hold true for the same cows on a lower quality forage diet.

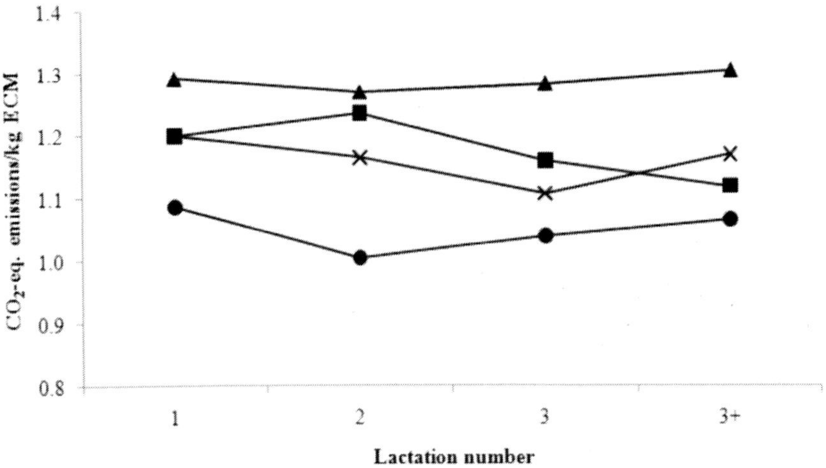

Figure 1. Carbon dioxide equivalent (CO_2-eq.) emissions per kg energy corrected milk (ECM) for cows selected for increased milk fat and protein fed a low proportion of forage (•) and a high proportion (×) of forage in their diet and cows selected to represent the average for milk fat and protein production fed a low proportion of forage (■) and a high proportion (▲) of forage in their diet (from [5]).

To bring about reductions in livestock GHG emissions, it has been suggested [21] that significant technological innovations will be required in the future, in addition to managing our consumption of meat and milk products. Technologies that can bring affordable efficiencies to production are being developed. Using genomic information, such as through genomic breeding values for feed related traits, described previously, and sexed semen [43] offer the potential for better selective breeding.

2. CONCLUSIONS

Reductions in GHG emissions by genetic selection of dairy cows in the past has been achieved largely by increased productivity and gross efficiency, whereby the maintenance cost of animals in the system has been reduced and less animals are required to produce the same amount of product. Based on current breeding goals, a similar rate of reduction in emissions intensity can be expected in the near future. Selecting dairy cows on feed efficiency, and possibly methane and nitrogen losses, will have a large impact on the environmental footprint of milk production, once implemented in breeding schemes. Further research and development of novel technologies to better understand the physiological and genetic differences between animals that lead to differences in energy and nitrogen efficiencies (or overall feed use efficiency) are still required.

Dairy farming is a highly managed system and has the potential to make reductions in GHG emissions intensity though increased efficiencies, such as optimum animal performance (less non-productive and ill animals) and reduced inputs as shown in Figure 2, whilst still maintaining productivity.

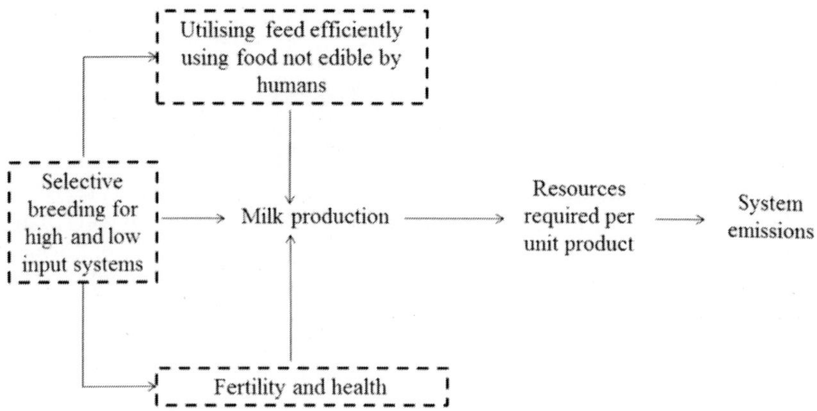

Figure 2. Production efficiencies using management (---) that can reduce GHG emissions beginning with selective breeding of a genotype for a particular system (from [5]).

Genetic improvement is a relatively cost-effective mechanism by which to achieve reductions in emissions, as the effect is cumulative and permanent. Many of the strategies for minimising emissions bring economic benefits to dairy farming through a reduction in production costs, with the added benefit of also reducing GHG emissions at little extra cost. The success of selective breeding as a mitigation strategy however, is dependent on producers being committed to its implementation.

Until direct measurements on GHG traits become available on sufficiently large numbers of animals, selecting for improved efficiency, RFI, or a measure of gross efficiency (unit of output per unit of feed eaten) offer attractive ways of reducing GHG emissions. However, the direct impact on GHG emissions is currently approximate and more research is needed to accurately assess the impact. Furthermore, selection strategies in dairy cattle need to be considered in a multi-trait framework, so that unfavourable correlated responses to selection (such as reduced fertility and excessive mobilisation of body reserves) are avoided.

ACKNOWLEDGEMENT

This work was supported by funding from Dairy Australia, Meat and Livestock Australia and the Australian Government Department of Agriculture, Fisheries and Forestry under its Australia's Farming Future Climate Change Research Program. Research on genomic breeding values for feed conversion efficiency was funded by the Dairy Future's CRC and Geoffrey Gardiner Foundation.

REFERENCES

1. Capper J.L, Cady R.A, Bauman D.E (2009) The environmental impact of dairy production: 1944 compared with 2007. J. Anim. Sci. 87:2160-2167.
2. Gerber P, Vellinga T, Opio C, Steinfeld H (2010) Greenhouse gas emissions from the dairy sector – A life cycle assessment. Food and Agriculture Organisation of the United Nations, Rome, Italy.
3. Dillon P, Berry D.P, Evans R.D, Buckley F., Horan B (2006) Consequences of genetic selection for increased milk production in European seasonal pasture based systems of milk production. Livest. Sci. 99:141-158.
4. Jones H.E, Warkup C.C, Williams A., Audsley E (2008) The effect of genetic improvement on emission from livestock systems. In: Proceedings of the European Association of Animal Production, 24-27 August, Vilnius, Lithuania, p. 28.
5. Bell M. (2011) Modelling the effects of genetic line and feeding system onmethane emissions from dairy systems. PhD Dissertation. The University of Edinburgh, UK.
6. IPCC (2007) Climate Change 2007 Series. Cambridge University Press, New York,USA.
7. Wall E, Simm G, Moran D (2010) Developing breeding schemes to assist mitigation of greenhouse gas emissions. Animal 4:366-376.
8. van de Haar M.J, St Pierre N (2006) Major advances in nutrition: relevance to the sustainability of the dairy industry. J. Dairy Sci. 89:1280-1291.
9. Chagunda M.G.G, Römer D.A.M, Roberts D.J (2009) Effect of genotype and feedingregime on enteric methane, non-milk nitrogen and performance of dairy cows during the winter feeding period. Livest. Sci. 122:323-332.
10. Bell M, Wall E, Russell G, Morgan C, Simm G (2010) Effect of breeding for milk yield,diet, and management on enteric methane emissions from dairy cows. Anim. Prod. Sci.50:817-826.

11. Johnson K.A, Johnson D.E (1995) Methane emissions from cattle. J. Anim. Sci. 73:2483-2492.
12. Ulyatt M.J, Lassey K.R, Martin R.J, Walker C.F, Shelton I.D (1997) Methane emissionfrom grazing sheep and cattle. In: Proceedings of the New Zealand Society of AnimalProduction, 57:130-133.
13. Tamminga S, Bannink A, Dijkstra J, Zom R (2007) Feeding strategies to reduce methaneloss in cattle. Animal Science Group report, Wageningen, The Netherlands.
14. Bell M, Wall E, Russell G, Simm G, Stott A (2011) The effect of improving cowproductivity, fertility, and longevity on the global warming potential of dairy systems.J. Dairy Sci. 94:3662-3678.
15. Veerkamp R.F (1998) Selection for Economic Efficiency of Dairy CattleUsing Information on Live Weight and Feed Intake: A Review. J. Dairy Sci. 81:1109-1119.
16. Dillon P (2006) Achieving high dry-matter intake from pasture with grazing dairy cows.In: Elgersma A, Dijkstra J, Tamminga S, editors. Fresh Herbage for Dairy Cattle, the Keyto a Sustainable Food Chain. Wageningen UR Frontis Series Volume 18: Springer,Dordrecht, The Netherlands. pp. 1-26.
17. Kennedy J, Dillon P, Delaby L, Faverdin P, Stakelum G, Rath M (2003) Effect of geneticmerit and concentrate supplementation on grass intake and milk production withHolstein-Friesian dairy cows. J. Dairy Sci. 86:610-621.
18. Robertson L.J, Waghorn G.C (2002) Dairy industry perspectives on methaneemissions and production from cattle fed pasture or total mixed rations in New Zealand. In: Proceedings of the New Zealand Society of Animal Production, 62:213-218.
19. Pryce J.E, Harris B.L, Montgomerie W.A (2007) Limits to selection for productionefficiency in new zealand dairy cattle. In: Proceedings of the Association for the Advancement of Animal Breeding and Genetics, 17:453-460.
20. Moran D, Barnes A, McVittie A (2007) The rationale for Defra investment in R&Dunderpinning the genetic improvement of crops and animals (IF0101). Final report toDefra. Defra, London, UK.
21. Garnett T (2009) Livestock-related greenhouse gas emissions: impacts and options forpolicy makers. Environ. Sci. Policy 12:491-503.
22. van der Steen H.A.M, Prall G.F.W, Plastow G.S (2005) Applications of genomics to thepork industry. J. Anim. Sci. 83: E1-E8.
23. Koenen E.P.C, Veerkamp R.F (1998) Genetic covariance functions for live weight,condition score, and dry-matter intake measured at different lactation stages of HolsteinFriesian heifers. Livest. Prod. Sci. 57: 67-77.
24. Pryce J.E, Nielson B.L, Veerkamp R.F, Simm G (1999) Genotype and feeding systemeffects and interactions for health and fertility traits in dairy cattle. Livest. Prod. Sci.57:193-201.
25. Lovett D.K, Shalloo L, Dillon P, O'Mara F.P (2006) A systems approach to quantifygreenhouse gas fluxes from pastoral dairy production as affected by managementregime. Agricult. Sys. 88:156-179.
26. Münger A, Kreuzer M (2008) Absence of persistent methane emission differences inthree breeds of dairy cows. Aust. J. Exp. Agric. 48:77-82.
27. Martin C, Morgavi D.P, Doreau M (2010) Methane mitigation in ruminants: frommicrobe to the farm scale. Animal 4:351-365.
28. de Haas Y, Windig J.J, Calus M.P.L, Dijkstra J, de Haan M, Bannink A, VeerkampR.F (2011) Genetic parameters for predicted methane production and potentialfor reducing enteric emissions through genomic selection. J. Dairy Sci. 94:6122-6134.
29. Hegarty R.S, McEwan J.C (2010) Genetic opportunities to reduce enteric methaneemissions from ruminant livestock. In: Proceedings of the 9th World Congress onGenetics Applied to Livestock Production, 1-6 August, Leipzig, Germany.
30. Herd R. M, Arthur P.F, Hegarty, R.S, Archer, J.A (2002) Potential to reduce greenhousegas emissions from beef production by selection for reduced residual feed intake. In:Proceedings of the 7th World Congress on Genetics Applied to Livestock Production, Montpellier, France, Communication No. 10–22.
31. Hegarty R.S, Goopy J.P, Herd R.M, McCorkell B (2007) Cattle selected for lowerresidual feed intake have reduced daily methane production. J. Anim. Sci. 85:1479-1486.

32. Yan T, Mayne C.S, Gordon F.G, Porter M.G, Agnew R.E, Patterson D.C, Ferris C.P,Kilpatrick D.J (2010) Mitigation of enteric methane emissions through improvingefficiency of energy utilization and productivity in lactating dairy cows. J. Dairy Sci.93:2630-2638.
33. Okine E.K, Basarab J.A, Goonewardene L.A, Mir P, Mir Z, Price M.A, ArthurP.F, Moore S.S (2003) Residual feed intake - what is it and how does it differfrom traditional concepts of feed utilization. In: Proceedings of Canadian Society for Animal Science, Annual meeting, 10-13 June, Saskatoon, Saskatchewan, Canada.
34. Nkrumah J.D, Okine E.K, Mathison G.W, Schmid K, Li C, Basarab J.A, Price M.A, WangZ , Moore S.S (2006) Relationships of feedlot feed efficiency, performance, and feedingbehavior with metabolic rate, methane production, and energy partitioning in beefcattle. J. Anim. Sci. 84:145-153.
35. Veerkamp R.F (2002) Feed intake and energy balance in lactating animals. In:Proceedings of the 7th World Congress on Genetics Applied to Livestock Production,Montpellier, France,Communication No. 10–01.
36. Coffey M.P, Simm G, Hill W.G, Brotherstone S (2003) Genetic evaluations of dairy bullsfor energy balance profiles using linear type scores and body condition score estimatedusing random regression. J. Dairy Sci. 86:2205-2212.
37. Pryce J.E, Harris B.L (2006) Genetics of body condition score in New Zealand dairy cows. J. Dairy Sci. 89:4424-4432.
38. Lassey K.R (2007) Livestock methane emission: From the individual grazing animalsthrough national inventories to the global methane cycle. Agricult. Forest Meteorol.142:120-132.
39. Pryce J.E, Arias J, Bowman P.J, Davis S.R, Macdonald K.A, Waghorn G.C, Wales W.J,Williams Y.J, Spelman R.J, Hayes B.J (2012) Accuracy of genomic predictions of residualfeed intake and 250-day body weight in growing heifers using 625,000 single nucleotide polymorphism markers. J. Dairy Sci. (In Press).
40. Garnsworthy P.C (2004) The environmental impact of fertility in dairy cows: amodelling approach to predict methane and ammonia emissions. Anim. Feed Sci.Technol. 112:211-223.
41. O'Mara F (2004) Greenhouse gas production from dairying: Reducing methaneproduction. Advances in Dairy Technology. Department of Animal Science and Production, University College Dublin, Ireland. 16:295.
42. Macdonald K.A, McNaughton L.R, Verkerk G.A, Penno J.W, Burton L.J, Berry D.P,Gore P.J.S, Lancaster J.A.S, Holmes C.W (2007) A Comparison of three strains of Holstein-Friesian cows grazed on pasture: growth, development, and puberty. J. DairySci. 90:3993-4003.
43. Weigel K.A (2004) Exploring the role of sexed semen in dairy production systems. J.Dairy Sci. 87:E120–E130.

CHAPTER 4

Production and Reproduction Performances of Local Dairy Cattle: In the Case of Rural Community of Wolaita Zone, Southern Ethiopia

Zereu G and Lijalem T*

Department of Animal and Range Sciences, Agriculture College, Wolaita Soddo University, Ethiopia

ABSTRACT

The study was conducted to assess local dairy cattle production and reproduction performances in rural community of Wolaita Zone, Southern Ethiopia from January to May 2015. Multistage sampling method was used and it was collected from both primary and secondary sources. The collected data was analyzed statistically using SPSS (version 20) for windows. The study reported that in overall average, 1.723, 1.536, 1.59, 1.589, 1.49 and 5.704 of dairy cows, heifers, calves, oxen, bulls and total cattle per households, were found, respectively. The average values of dairy cows, calves and oxen numbers were declare significant variation across agro-ecologies, while heifers and bulls numbers were not significantly vary due to agro ecology effect in the study area. The study also showed that the average values of daily milk yield and lactation length were, respectively, 2.127 ± 0.094, 1.861 ± 0.084 and 1.989 ± 0.063 liters; 11.875 ± 0.748, 9.821 ± 0.661 and 10.802 ± 0.503 months in midland, lowland and overall in the study area and the values were declare significant between agro-ecologies. While values of 20.167 ± 1.177, 19.658 ± 1.186 and 19.913 ± 0.832 months were reported for calving interval; 4.022 ± 0.142, 3.926 ± 0.090 and 3.818 ± 0.104 years for age at first calving cow in midland, lowland and overall, respectively, however, they have no significant different in both parameters (calving interval, age at first calving across agro-ecologies (midland and lowland) in the study area.

Keywords: Wolaita zone; Agro-ecologies; Small-holder farmers; Heifers; Calves

1. INTRODUCTION

Livestock play a major role in the Ethiopian economy and its people by providing food, input for crop production, soil fertility management, raw material for industry, cash income, saving, fuel, social functions and employment. Ethiopia has the largest livestock population in Africa, estimated at 45,054,969 cattle; 20,562,832 sheep; 20,191,099 goats [1] and 2.4 million camels [2]. Nevertheless, productivity is generally lower than even in comparable African countries. The national per capita production of livestock and livestock products, export earnings from livestock and per capita consumption of food from livestock origin have declined since 1974 [3].

In many respects the dairy industry occupies a special position among the other sectors of livestock (agriculture). Milk is produced every day and gives a regular income to the numerous small producers. Milk production is highly labour-intensive and so provides a lot job opportunities [4]. In Ethiopia, the majority (99.25%) of the cattle population are local breeds, which are found in rural areas under subsistence type of farming system and the remaining are hybrid and exotic breeds that accounted for about 0.65 and 0.09%, respectively [5]. However, productive performances of cattle are very low even in comparison to other East African countries these may be due to animal diseases, unavailability of feed in terms of quality and quantity and genetic characteristics of animals [2]. The country has made great effort to improve the productivity of local breeds through artificial insemination (AI) program to crossbreed locally adapted cattle breeds with improved exotic dairy breed ones. This program addresses major parts of the country including Southern Nation, Nationalities Peoples and Regional State. However, the success of such programs is not satisfactory due to numerous factors, including substandard nutrition, poor husbandry practice and infrastructure status; more ever, such challenges are related by due to poor production and reproductive performances of animals as suggested by many dairy producers [6].

In general, the most part of Ethiopia, which are very well suited for dairying because a country is characterized by endowed with such enormous cattle resources and climatic situations conducive to cattle production as a result, it should be self-sufficiency in dairy products and should be encouraged to optimize the use of available resources to fill the gap between demand and supply. To improve dairy production, it is necessary to design appropriate and sustainable dairy development strategies based on 'felt' needs of smallholder farmers. Smallholder farmers represent about 85% of the population and are responsible for 98% of total milk production [7]. Milk is a cash crop and dairy farming is an investment option for smallholder farmers. So to ensure the success of the dairy production, regular supervision should be taken by evaluating the production and reproductive performances of the dairy cow under the prevailing management and environmental conditions since dairy cattle are important sources of food and income for smallholders, in the country Ethiopia. According to Le Blanc [8] dairy herd profitability is determined by dairy cows reproductive and production performance because it is a major determinant for overall productivity of dairy production systems by affecting the efficiency of milk production, the number of calves produced per cow and lifetime milk production [9]. As a result production and reproductive performance traits are mattered for being successful or bankruptcy of dairy farm enterprises as these traits are major importance in dairy production [10]. According to Cavestany and Galin [11] and Pursley, et al. [12] study calving interval, daily milk yield, lactation length and age at first calving are one of the major numerous measures of production and reproductive performance parameters for dairy cattle production. However, there is limited information on regarding to reproductive and productive performance of local cattle in Ethiopia in general, in particular Wolaita areas. Therefore, Having information on performances of reproductive and productive of local cows in the study area, wolaita Ethiopia would help to suggest the future genetic improvement options for this herd. The aim of this study was thus to determine the reproductive and productive performances of local cows of Ethiopia, in particular in wolaita areas.

2. METHODOLOGY

The study was conducted from January to May 2015 in Wolaita Zone, Southern Ethiopia. The study area is located in 6.4° to 7.1° north latitudes and37.4° to 38.2° east longitudes, and it covers a total area of 3, 982 km2. Its altitude ranges from 1,200 to 2,950 m above sea level (masl), and it is subdivided into three agro ecological zones, namely, kolla or lowland (35%, <1,500 masl), woina dega or intermediate highland (56%, 1,500 to 2,400 masl) and dega or highland (9%, >2,400 masl). Wolaita Zone has a bimodal type of rainfall pattern. The major and minor rain seasons usually last from June to September and March to May, respectively. The average total annual rainfall is 1,014 mm, and the mean daily temperature is 19.5°C.

In current study area, multistage sampling method was used. From lowland and midland, respectively, Humbo and Bolesoserea woredas were selected randomly. From each selected Woredas, three districts were selected randomly thereby a total of six districts were selected. Again, from each selected districts, 25 households were selected purposively farmers who have dairy cattle. Thus, a total of 150 households were selected for interviewee survey study. However, some households were not committed to response the questions requested by researchers. Even though, however, more 100 farmers committed in giving response of requested questions in some parameters.

Data was collected from both primary and secondary sources. The primary sources were obtained through a semi-structured questionnaire; it was pre-tested the questionnaire before the actual data collection was carried. The interviews were conducted by trained research assistants under close supervision by the researcher while the secondary data were collected from different sources such as books, research publications, journals, office reports of zonal and woredas agriculture etc. The collected data was analyzed statistically using SPSS (version 20) for Windows. Both descriptive statistics and general linear model were used to analyzed the data; the values were considered significant at $p < 0.05$.

3. RESULTS AND DISCUSSION

3.1 Local cattle holdings and herd structure

Least square means (±SE) of local cattle holdings and herd structure in different agro-ecology is presented in Table 1, revealed from Table 1, in average 1.723, 1.536, 1.59, 1.589, 1.49 and 5.704 of dairy cows, heifers, calves, oxen, bulls and total cattle, respectively were found in the hand of per interviewed respondents. In similar to the results of present study Dessalegn [13] in Bench-Maji Zone, Southwest Ethiopia reported that in average heads number per house hold for dairy, 1.91 ± 1.2; oxen, 1.54 ± 1.3; heifers, 0.58 ± 0.8; calves, 0.61 ± 0.8 (0.69 ± 0.8 males and 0.53 ± 0.8 females) and for total cattle, 5.64 ± 3.3. In the present study, even though, in all local cattle composition; dairy cows, heifers, calves, oxen, bulls and total cattle numbers were reported higher in lowland than midland areas however, only the numbers of dairy cows, calves and oxen were declare significant variation across agro-ecologies, between midland and low land while heifers and bulls numbers were not significantly vary between midland and low land in the study areas. Kedija et al. [14] from Mieso district Oromia Regional State, Ethiopia who reported that in average 5.69 ± 0.35 heads cattle holds individual households which was in consistent to current finding 5.704 ± 0.266 heads cattle holds in average per household.

However, the result of current study was significantly lower than the result of Tesfaye [15] from Metema district, North west Ethiopia who found that 15.53 ± 0.71 heads cattle holds in average per household. Yoseph et al. [16] in urban and peri-urban dairy production system in the Addis Ababa milk shed who found that proportion of milking cows, 72.2%; heifers, 25.6% and calves 21.6%, to total number of cattle herd however, in the current study it was found as 21.733, 19.374, 20.055, 20.043 and 18.794 % proportionate to dairy cows, heifers, calves, oxen and bulls, respectively (Table 1). In agreement to current result, Yoseph, et al. [16] reported higher % dairy cows than other cattle herd. However, in the current finding, the actual proportion number of dairy cows was lower than Yoseph et al. [16] finding.

The reason for being lower, in the present study is because of all of the cattle are indigenous Zebu in the herd; in turn, these cattle have multiple functions (for threshing, plowing, manure production etc.) for famers besides for milking but a study conducted by Yoseph et al. [16] is relatively market oriented and targeted to produce more milk by keeping more cows since sale of the milk is the major income for the farm. More ever, these farmers keep more exotic dairy genotypes besides local cattle.

3.2 Production and reproduction performances

3.2.1 Production performances

a) Daily milk yield: There were significant variation in daily milk yield performances of these cattle (local) due to agro-ecology effect; the value of average daily milk yield was significantly higher in mid land (2.127 ± 0.094) than low land (1.861 ± 0.084). The overall average daily milk yield per cow reported in the current study area was 1.989 ± 0.063 liters (Table 2) which was in approximate to the results of studies conducted by Dessalegn [13], Brokken and Senaite [17], Mugerwa et al. [18], Azage et al. [16] and Fasil [19] from different part of the country Ethiopia who had reported that 2.06 ± 0.5, 2.0, 2.19, 2 ± 0.07 and1.8 liters, respectively. However, Merha [20] reported the daily average milk yield of Abergele cattle to be 0.75 liters which is lower than the present result of study.

Table 1: Least square means (±SE) of local cattle holdings and herd structure in different agro-ecologies.

Cattle Herds	Agro-ecologies			Percentage
	Mid land N(LSM ± SE)	Low land N(LSM ± SE)	Over all N(LSM ± SE)	
Dairy cows	67 ($1.478^a \pm 0.086$)	74($1.946^b \pm 0.096$)	141 (1.723 ± 0.068)	21.733
Heifers	42(1.333 ± 0.147)	55(1.691 ± 0.113)	97(1.536 ± 0.092)	19.374
Calves	50($1.42^a \pm 0.103$)	67($1.716^b \pm 0.09$)	117(1.59 ± 0.069)	20.055
Oxen	39($1.308^a \pm 0.098$)	51($1.804^b \pm 0.128$)	90(1.589 ± 0.088)	20.043
Bulls	35(1.314 ± 0.121)	63(1.587± 0.095)	98(1.49 ± 0.076)	18.794
Total cattle	76($4.25^a \pm 0.268$)	76($7.158^b \pm 0.395$)	152(5.704 ± 0.266)	100

LSM: Least Square Means, SE: Standard Error, N: Number of Observations, Means within the same row in two agro-ecology, in the same parameters' with different super scripts letter are significantly different ($p < 0.05$).

The value of average daily milk yield obtained from current study was significantly lower than values obtained from Minale and Yilkal [21] in Gomgofa zone, Southern Ethiopia who reported that consecutively 2.4, 3.0, 2.6 for 1st, 2nd, and 3rd lactations in Chencha and 2.07, 2.6, 2.6 for 1st, 2nd and 3rd lactations in average milk yield per day per cow in liters for local breeds in kucha areas. The reason for variation of daily milk production in different study might be due to the effect of feed, health, management aspect and genotype. In general, the cattle productivity, in the study areas is low in terms of milk yield which is in similar to other parts of the country Ethiopia under smallholder management conditions due to lack of proper supplementary feeding for the dairy cattle, poor nutritive value of pastures and forages offered to the animals besides, the farmers keep local animals that are generally low producers of milk [22,23].

b) Lactation length: As presented in Table 2, in the present study, the overall average lactation length was reported as 324.06 ± 15.09 days or 10.802 ± 0.503 months and it was found significantly varied between agro-ecologies. Thus, in midland areas (in the study), the local dairy cattle can be milking for longer period than lowland areas. This might be related with, in midland areas by nature there is exceptional longer rainfall period besides, normal rainfall as result there would be forage availability for milking cows for some extent. Belay et al. [24] from East Shwa, Shiferaw [25] from Jimma Town [24] and Kedja et al. [14] from Measo districts areas of Ethiopia who found that, respectively, 255.75 days of Keriyu cattle breed; 273.9 days of the crossbred dairy cows and 220 days of average location length of indigenous cattle which are significantly lower than the finding of current study (324.06 ± 15.09 days). More ever, Minale and Yilkal [21] found out for local breed cattle in Chencha and kucha districts as 7.42 ± 1.89 and 9.23 ± 2.27 months of average lactation length, respectively; Asaminew and Eyassu [26] also reported an average lactation length for local breed as 9.8 ± 2.26 months; these findings are also significantly lower than the finding of current study (324.06 ± 15.09 days). However, having longer lactation period in the current study probably implies that farmers in the study area milk cows until they are quite dry; it might be also related to feed availability as in most areas of the study, there is long rainfall. nevertheless, such type of milking practice has negative effect on calving interval, next season milk and productive efficiency of the cow.

3.2.2 Reproductive performances

a) Age at first calving: Age at first calving is the indicator of beginning of a cow's reproductive life and is closely related to the generation interval. There is considerable economic importance associated with the factors that control the onset of ovarian activity of cattle.

Table 2: Production and reproduction parameters in different agro-ecologies.

Parameters	Agro-ecologies		
	Midland N (LSM ± SE)	Lowland N (LSM ± SE)	Overall N (LSM ± SE)
Age at first calving (in year)	57(3.818 ± 0.104)	65(4.022 ± 0.142)	122(3.926 ± 0.090)
Average milk yield/day/cow (in litre)	65(2.127^{b} ± 0.094)	70(1.861^{a} ± 0.084)	135(1.989 ± 0.063)
Lactation Length (in month)	64(11.875^{b} ± 0.748)	70(9.821^{a} ± 0.661)	134(10.802 ± 0.503)
Calving interval (in month)	60(20.167 ± 1.177)	60(19.658 ± 1.186)	120(19.913 ± 0.832)

LSM: Least Square Means, SE: Standard Error, N: Number of Observations, Means within the same row in two agro-ecology, in the same parameters' with different super scripts letter are significantly different ($p < 0.05$).

In the present study, the overall average for age at first calving cows was 3.926 ± 0.090 years or 47.112 ± 1.08 months however, there was no significant different between agro-ecologies (Table 2). Dessalegn et al. [27] in North-West Zone of Tigray who reported 21.6 months for Arado cattle breed which is significantly lower than current result of age at first calving cows. However, Dessalegn [27] from Benchi-Maji zone, Southern Ethiopia who found 50.87 ± 7.0 months which is higher than the result of current finding of age at first calving cows (47.112 ± 1.08 months). Minale and Yilkal [21] from Gomgofa zone, Southern Ethiopia indicated that 4.6 ± 0.8 and 4.4 ± 0.6 years of age at first calving for local breed Chencha and kucha districts study sites, respectively; Dejene [9] also from lowland and mid-highland agro-ecologies of Borana zone, Southern Ethiopia reported the mean age of first calving of cows are 4.9 and 4.8 years in lowland and mid-highland, respectively, which are also higher than that of the result obtained from present study (3.926 ± 0.090 years). Different factors could cause to advance or delay age at firs calving such as environmental factors like nutrition, health, routine management, climate; these are pre-determine for prepubertal growth rates, reproductive organ development, and onset of puberty and subsequent fertility.

b) Calving interval: The average calving interval is present in Table 2, reveal from Table 2, overall average value of the calving interval was 19.913 ± 0.832 months. There was no significant difference between agro-ecologies (midland and lowland areas). The result obtained in the present study is in comparable with Dessalegn [9] from Benchi-Maji zone, Southern Ethiopia who indicated 21.18 months and Shiferaw [25] reported the mean calving interval for kariyu cattle in the East Shwa zone of Oromiya to be 18 months of calving interval. However, the value of average calving interval present finding was shorter than from Merha [20] resulted from Abergele breed who found 24 months. Dejene [9] from his study in lowland and mid-highland agro-ecologies of Borana zone, Southern Ethiopia found that the calving interval of cows are 16.8 and 13.8 months in the lowland and mid-highland area, respectively which is significantly lower than finding of current study (19.913 ± 0.832 months). The causes of variation for average value of calving interval in different study may be due to the effect of genotype (breed), feed availability year-round, management (nutrition, health, housing conditions etc.) and climate type where the animal live. The supposed reasons for the shorter calving interval might be for the availability of ample feed resources from crop residues, natural pasture and less prone of an area to recurrent drought and suitability of climate. On the other hand, however, the longer calving interval reduce yearly production cycle and the amount of milk a cow likely to produce in given period of time, which might be associated with environmental factors, poor nutrition, poor housing, lack of sufficient bull and AI services and poor health and reproductive management [28]. Therefore, the calving interval should be shortened for improved reproductive and productive performances [29-31].

4. CONCLUSION

The study showed that there were significant variation in daily milk yield and lactation length performances of these cattle (local) due to agro-ecology effect in the study area; the values were significantly higher in mid land than low land. However, values of age at first calving and calving interval cows were not found significant different between agro-ecologies (midland and lowland areas).

ACKNOWLEDGEMENTS

The authors would like to thank to Agriculture office of Humbo and bolesoserea woredas of Wolaita Ethiopia which support and collaborate in giving vital information and secondary data for the study. The authors also would like to thank the respondents and enumerators who involved in this study for their kind cooperation.

REFERENCES

1. MoARD (2007) Livestock development master plan study. Phase 1 report- Data collection and analysis. 1- Dairy. Addis Ababa, Ethiopia: GRM International BV
2. FAOSTAT (2009) Production statistics. Live animals FAO Statistics Division. Rome, Italy: Food and Agriculture Organization of the United Nations (FAO).
3. Assegid W (2000) Constraints to livestock and its products in Ethiopia:Policy implications. DVM thesis, Faculty of Veterinary Medicine, Addis Ababa University, Addis Ababa, Ethiopia, p:52.
4. International Dairy Federation (1996) The dairy world: Twenty-five years of change. 1985-2010. Bulletin of the International Dairy Federation 316.
5. CSA (2008) Statistical Abstract 2007. Central Statistical Authority, Addis Ababa, Ethiopia.
6. Lemma A, Kebede S (2011) The effect of mating system and herd size on reproductive performance of dairy cows in market oriented urban dairy farms in and around Addis Ababa. Revue de Medicine Veterinaries 162: 526-530.
7. MoA (Ministry of Agriculture) (1998) The role of village dairy co-operatives in dairy development: Prospects for improving dairy in Ethiopia. SDDP (Smallholder Dairy Development Project), Addis Ababa, Ethiopia.
8. Le Blanc S (2007) Economics of improving reproductive performance in dairy herds. Advanced Dairy Technology 19: 201-214.
9. Dejene TG (2014) Assessment of Dairy Cattle Husbandry and Breeding Management Practices of Lowland and Mid-Highland Agro-Ecologies of Borana Zone. Animal and Veterinary Sciences 2: 62-69.

10. Saeed AM, Ward PN, Light D, Durkin JW, Wilson RT (1987) Characterization of Kenana cattle at Um Banein, Sudan. International Livestock Center For Africa, Addis Ababa, Ethiopia.
11. Cavestany D, Galin CS (2001) Evaluation of an artificial insemination programme in a seasonal breeding dairy system through milk progesterone. Reproduction Domestic Animals 36: 79-84.
12. Pursley JR, Kosorok MR, Wiltbank MC (1997) Reproductive management of lactating dairy cows using synchronization of ovulation. Journal of Dairy Science 80:301-306.
13. Weldegebriel DG (2015) Assessment of Production and Reproductive Performances of Cattle and Husbandry Practices in Bench-Maji Zone, Southwest Ethiopia. Global Journal of Animal Scientific Research 3: 441-452.
14. Kedija H, Azage T, Mohammed-Yousuf K, Berhanu G (2008) Traditional cow and camel milk production and marketing in agro pastoral and mixed crop–livestock systems: The case of Mieso District, Oromia Regional State, Ethiopia. IPMS (Improving Productivity and Market Success) of Ethiopian Farmers Project. International Livestock Research Institute, Nairobi, Kenya. p: 56
15. Tesfaye MD (2006) Characterization of cattle milk and meat production, processing and marketing system in metema district, Ethiopia. The annual conference of the Ethiopian Society of Animal Production (ESAP) held in Addis Ababa, Ethiopia, part 2 Technical papers. pp: 109-114.
16. Yoseph M, Azage T, Alemu Y, Umunna NN (2003) Evaluation of the General Farm Characteristics and Dairy Herd Structure in Urban and Peri-Urban Dairy Pro duction System in the Addis Ababa Milk Shed.CGSpace A Repository of Agricultural Research Output.
17. Brokken RF, Senaite S (1992) Dairy marketing in sub-Saharan Africa. In: Proceeding of a symposium held at ILCA. Addis Ababa, Ethiopia. International Livestock Centre for Africa Pp: 123-130.
18. Mukasa-Mugerwa E, Ephrem B, Tadesse T (1983) Reproductive performance of indigenous cattle in the Ada district of central Ethiopian highlands. Mimeograph. International Livestock Centre for Africa (ILCA), Addis Ababa, Ethiopia. Pp: 99.
19. Kebede FG (2006) On-Farm Phenotypic Characterization Of Cattle Genetic Resources And Their Production Systems In Awi, East And West Gojjam Zones Of Amhara Region, Ethiopia M.Sc. Thesis, Haremaya University
20. Merha Z (2006) Genetic Diversity in Northern Ethiopian Cattle Breed: An on-Farm and Molecular Characterization, Ph. D Dissertation, Agricultural University, Wageningen, The Netherlands. Morti More, M., 1991. Livestock economics division. A review of mixed farming systems in the semi-arid zone of sub-Saharan Africa. International Livestock Centre for Africa (ILCA), Addis Ababa, Ethiopia: 181: 29.
21. Minale G, Yilkal T (2015) Dairy Production, Processing and Marketing in Chencha and Kutcha Districts, Southern Ethiopia. Journal of Marketing and Consumer Research Journal 9: 6-16
22. Zelalem Y, Ledin I (1999) Milk production systems, processing techniques and the role of milk and milk products on the small holder farms in central high lands of Ethiopia.
23. Derese T (2008) Present situation of urban and peri-urban milk production and quality of raw milkproduced in west shoa zone, oromia region, Ethiopia
24. Belay D, Yisehak K, Janssens GPJ (2012) Productive and Reproductive Performance of Zebu X Holstein-FriesianCrossbred Dairy Cows in Jimma Town, Oromia, Department of Animal Science, Jimma University, Ethiopia. Global Veterinaria 8: 67-72.
25. Shiferaw G, Hegde BP (2006) IN-Situ Phenotypic characterization of Kereyu cattle type In Fentalle District of Oromia Region, Ethiopia, A M.Sc Thesis, presented to the school of graduate studies of Haramaya University, Ethiopia.
26. Asaminew T, Eyassu S (2009) Small holder Dairy Production System and Emergence of Dairy Cooperatives in Bahir Dar Zuria and Mecha Districts, Northwestern Ethiopia.World Journal of Dairy & Food Sciences 4:185-192.

27. Dessalegn G, Mekonnen H, Kelay B (2012) Morph metric characteristics and livestock keeper perceptions of "Arado" cattle breed in North-West Zone of Tigray, Ethiopia, M.Sc. thesis, School of Veterinary Medicine, Addis Ababa University, Debre-Zeit, Ethiopia.
28. Million T, Thiengtham J, Pinyopummin A, Prasanpanich S (2010) Productive and reproductive performance of Holstein Friesian dairy cows in Ethiopia. Livestock Research for Rural Development 22.
29. Kiwuwa GH, Trail JCM ,Kurtu MY, Getachew W, Anderson FM et al. (1993) Crossbred dairy.
30. Azage T, Tesfaye M, Tesfaye D, Worku T, Eshete D (2009) Transhumance cattle production system in NorthGondar, Amhara Region, Ethiopia: Is it sustainable?. IPMS (Improving Productivity and Market Success) of Ethiopian Farmers Project, ILRI (International Livestock Research Institute), Nairobi, Kenya:1-73.
31. De VA (2006)Economic value of pregnancy in dairy cattle. Journal of Dairy Science 89: 3876-3885.

CHAPTER 5

Public concerns about dairy-cow welfare: How should the industry respond?

D. M. Weary and M. A. G. von Keyserlingk

Animal Welfare Program, Faculty of Land and Food Systems, University of British Columbia, 2357 Main Mall, Vancouver, BC, V6T 1Z4, Canada.

ABSTRACT

Common practices on dairy farms have fallen out of step with public values, such that the dairy industry has now become a target for public criticism. In the present paper, we describe some of the forces that have led to the current situation, and various potential methods to rectify the situation. One approach is to shield industry practices from public scrutiny, for example, by using 'ag-gag' legislation to stem the flow of videos exposing contentious practices. Another is to educate members of the public so that they better understand the nature of these practices and the reasons that they are used on farms. The literature we reviewed indicated that neither of these approaches is likely to be successful. Instead, we suggest that the dairy industry needs to develop methods of meaningful two-way engagement with concerned citizens, including research using social-science methods to document the values of different stakeholders and examine approaches to resolving conflicts. We also reviewed how biological research can help resolve issues, for example, by developing rearing systems that address public concerns around freedom of movement and social contact without putting animals at an increased risk of disease. We end with a discussion of how policy efforts by the dairy industry can be used to ensure compliance with commonly accepted standards, and more ambitiously, develop a common vision of dairying that positions the industry as a leader in animal welfare.

Keywords: animal welfare, assurance, audit, milk, survey, trust.

1. GREATER SCRUTINY AND A BLACK EYE FOR THE INDUSTRY

The positive image of the dairy farm is under threat. People are becoming increasingly interested in how their food is produced (Cembalo et al. 2016), and in some cases, their introduction to the modern dairy farm comes in the form of videos documenting abuse and other contentious practices.

These types of images can have a profound impact. For example, in February 2014, an undercover video taken on a New Zealand owned dairy farm operating in Chile reported that over 6000 calves had been killed using blunt force trauma (Gulliver 2014). This video resulted in public outcry in New Zealand and a change in law protecting farm animals; the New Zealand Animal Welfare Act (New Zealand Ministry of Primary Industries 2016) states that 'it is illegal to kill a calf by blunt force to the head, except in emergency circumstances'. Another undercover video, also released in 2014, showed footage of disturbing animal handling on a dairy farm in British Columbia, Canada (Clark 2014). The largest Canadian milk processor, Saputo Inc., a company that had remained largely silent on the issue of farm animal welfare, revised its animalwelfare policy in response to the consequent public outcry:

Saputo has zero tolerance for any act of animal cruelty. This includes, but is not limited to, willful mistreatment and neglect of animals and acts that maliciously cause pain, injury or suffering. We expect all dairy cattle workers (employers and employees) to adopt and adhere to proper animal care and handling methods at all times (Saputo 2015, p. 1).

In addition to documenting disturbing practices, these videos also show other examples of workers interacting with animals, and allow the viewer to see how animals are housed and cared for. In this way, the video on calves can lead viewers to wonder why the calves are separated from the cow, how they are cared for, and why the male calves are of so little value that they are killed on farms. Similarly, the video on cows (focussed especially on downer and cull cows) raises questions about why many cows are sick and lame, and how these vulnerable animals are cared for.

Video exposés may be especially effective because they resonate with public concerns about the intensification of agriculture. Intensification can mean different things, including increased farm size, automation, confinement and corporate versus family ownership (Thompson 2008). Each of these factors can be a cause for concern. For example, people tend to think that larger farms are worse for animal welfare (even though there is little empirical evidence that animals fare worse on these farms; see review by Robbins et al. 2016a).

In countries such as the United States, few dairy cows have access to pasture; less than 5% of the lactating dairy herd in the USA is able to graze routinely, in direct contrast to what the public wants and expects for dairy cows (Cardoso et al. 2016). Early cow–calf separation provides another example of this divergence between public views and industry practice (Ventura et al. 2013). There is evidence that the public is often unaware of common practices on farms (Ventura et al. 2016a) and failure to address these gaps may affect public trust when the public does become aware (Robbins et al. 2016b).

Willingness to trust is often rooted in shared values (Robbins et al. 2016b), and when it comes to animal welfare, the values of farmers can differ from those of the general public. In contrast to the public, farmers tend to express less concern over animal welfare (Vanhonacker et al. 2008), and to rate different dimensions of animal welfare differently from the lay public (Te Velde et al. 2002; Lassen et al. 2006; Vanhonacker et al. 2008; Bergstra et al. 2015). The public's perspective of good welfare includes that animals live a reasonably natural life (Lassen et al. 2006; Prickett et al. 2010; Verbeke et al. 2010; Cardoso et al. 2016; see review by Clark et al. 2016).

Different stakeholders vary in the amount of trust extended to them by the public. Farmers are often viewed as trustworthy (e.g. Coleman et al. 2015), perhaps because farmers are perceived as being especially knowledgeable about animal production (reviewed by Clark et al. 2016). However, judgments of trustworthiness are also affected by the extent to which individuals or organisations are perceived to be transparent about their practices (Peters et al. 1997; Fisman and Khanna 1999; Maeda and Miyahara 2003; Rawlins 2008). This may explain why animal-protection groups are considered more credible sources of information than are livestock industry groups (McKendree et al. 2014), and why this positive perception increases following animal-abuse scandals (Scudder and Bishop-Mills 2009; Tiplady et al. 2013). These results are consistent with other research, showing that whistle-blowers are viewed favourably by society (Callahan and Dworkin 2000), despite the short-term upheaval caused by exposés (Hersh 2002).

In the following sections of the paper, we review and evaluate various approaches to how the dairy industry might respond to criticism. For example, should dairy practices be concealed behind closed doors in the hope that concerns will fade, should the industry attempt to educate people about common practices in the hope that any concerns are rooted in ignorance, or should the industry attempt to engage with citizens to better understand and address their concerns? We also discuss how research can be used to create solutions that address certain welfare concerns. We end with a discussion on policy interventions that may be required to bring farms into compliance with widely held expectations for good care, and, more broadly, how the dairy industry might develop a vision that allows it to take leadership on animalwelfare issues.

2. HOW TO RESPOND TO THE CRITIQUE?

2.1 Close the barn doors

A natural response to criticism is hiding in the shadows; if people cannot see the practices, then there will be nothing to criticise. This approach may work in the short term; starved of material, including fresh video, the news cycle may pass onto other items. But this approach is less likely to work in the long term. Interest about where food comes from appears to be on the rise (Wheale and Hinton 2007), and this demand for information will likely be filled even if only by stories about secrecy in agriculture and any videos and images that continue to leak out

An example of how this 'closed door' approach has been unsuccessful comes from laboratory animal use. Fear of criticism has largely driven the use of animals in biomedical research 'underground'. Underground figuratively in the sense that little information about research animal use is publically available, including which animals are used, where they are used and when they are used, but also underground literally, in that these laboratories are often in university basements, away from the prying eyes of the critics. This has not made the criticisms of laboratory animal use disappear, and may even encourage the use of misinformation leading people to evaluate conditions to be worse than they really are (Ormandy et al. 2013).

The appearance of secrecy can also accentuate concerns, as closed doors give the outside viewer the sense that there must be something to hide (Broad 2016). Thus, efforts to reduce the risk of reputational harm that results from bad stories getting out can have the perverse effect of eroding public trust in the individuals and institutions that are preventing access to the information. An example of how efforts to reduce transparency can harm the reputation of agriculture comes from the so-called 'ag-gag' laws;

legislative attempts to prevent the publication of undercover videos. One study experimentally assessed the views of participants that were told about these laws (Robbins et al. 2016b), and found that when people were made aware of aggag restrictions they were less likely to trust famers and more likely to support the introduction of new animal-welfare laws (Fig. 1). Knowing about the ag-gag laws also resulted in respondents believing that the welfare of the animals on farms was poorer, and even that environmental conditions were poorer. These results indicated that laws intended to protect livestock industries from public scrutiny lower public trust and support for farmers and farming practices.

Of course, individual farmers may not support ag-gag laws or other industry and government attempts to reduce transparency. But our most recent results suggest that just working in more secretive settings can reduce trust in the individuals working there, such that these workers are viewed as less warm, and having less integrity. It seems that agriculture has much to lose from closing the barn doors.

2.2 Educate the public

If hiding does not solve the problem of public criticism, then perhaps telling others about industry practices will help. Many within the livestock industries believe that the public is ignorant of farming practices, a view consistent with the limited research available on this topic (Harper and Henson 2001; Boogaard et al. 2008; Coleman et al. 2015), so educating the public may be a way of increasing acceptance of animal agriculture (Croney et al. 2012; You et al. 2014; Pieper et al. 2016).

Unfortunately, there are several reasons why efforts to educate the public are unlikely to successfully resolve concerns about practices on farms. One limitation is the ratio of mostly naïve public to industry insiders. In most developed economies, the proportion of the population involved with agriculture is small, such that few citizens feel that they are connected with animal production practices (e.g. Frewer et al. 2005). Can a small minority involved with agriculture really achieve much when trying to 'educate' the mass of others? Perhaps targeted campaigns directed at one or two issues that resonate with scientific evidence and widely held community values may stand a chance, but relying on public education as a way to resolve most concerns is unlikely to be successful.

Second, the public seems to be adopting an increasingly jaded view when it comes to the advertising efforts of specific industries, so the intended beneficiaries of educational efforts may be resistant to industry talking points. This resistance is likely to be increased as a result of the reputational damage already suffered by the livestock industries. Animal-advocacy organisations are seen as more trustworthy sources of information than are animal industries (McKendree et al. 2014), and this difference in credibility is magnified in the wake of animal-abuse exposés when attitudes towards the livestock sector are especially negative (Tiplady et al. 2013). This may explain why negative press has a much greater impact on consumer behaviour than has any positive effect from advertising (Verbeke and Ward 2001).

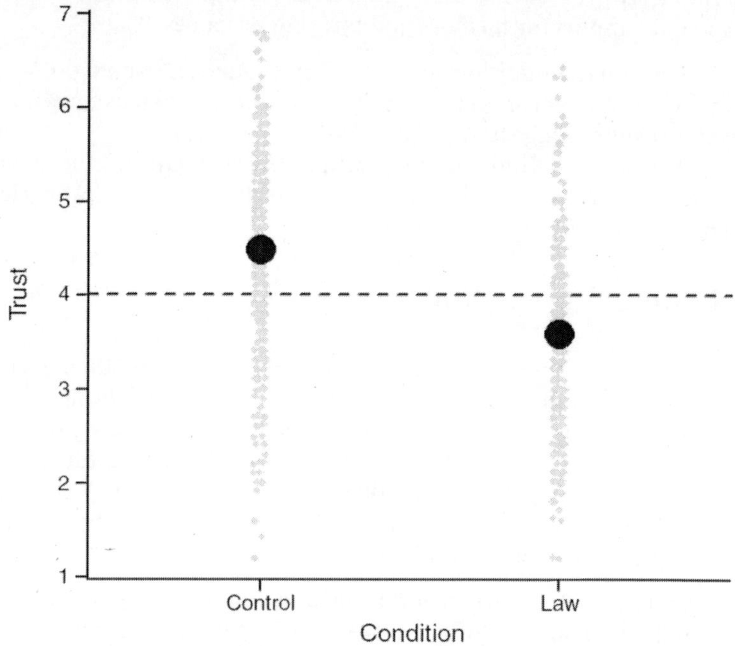

Figure. 1. Trust (measured on a Likert scale from a low of 1 to a high of 7) in farmers decreased after learning about laws that restrict the flow of information coming out of farm facilities (i.e. ag-gag), demonstrating that even the intention to restrict access to information can undermine trust (redrawn from Robbins et al. 2016b). Small grey dots show the responses of individual participants and the large grey dot shows the treatment mean.

Research within our group has found that efforts to educate the public about livestock farming can increase criticism of farming practices. In one study, Ryan et al. (2015) asked Canadian and American respondents about their attitudes to various housing systems for pregnant sows, and found that people became less willing to accept the use of gestation crates after they learnt more about the various options (including viewing photographs, videos and text information).

In another relevant study, Ventura et al. (2016a) examined the views of individuals with little previous exposure to dairy, before and after they toured a working dairy farm. Visiting this farm seemed to address certain concerns, including that cows did not have adequate access to food and water and that they were not handled appropriately. However, when the visitors learned more about the details of the production system, they identified new concerns, including the lack of pasture access and early separation of cow and calf.

This study also tested how knowledgeable participants were about production practices, and found that although participants varied in both knowledge and attitudes there was little association between the two. The lack of relationship makes sense when we consider the broader literature about attitudes and knowledge (Hansen et al. 2003); people's views are highly related to their values, and these values are not easily shifted. Moreover, there is also some evidence suggesting that moral values appear to have a greater impact on negative attitudes towards farm-animal welfare than does factual knowledge (Boogaard et al. 2011), which may explain why criticisms regarding production practices fail to disappear following an educational intervention. As Ventura et al. (2016a) concluded (p. 15), 'livestock industries cannot expect one-way education efforts (even immersive experiences such as a farm tour) to resolve societal concerns about animal welfare'.

2.3 Engage with the public

Two-way conversations with the public, including the consumers who purchase dairy products and the citizens that provide a social licence for the dairy industry to operate, may provide a more productive approach. Much of the earlier work on social attitudes to agriculture focussed onthe consumer,including studies of what people actually buy at the grocery store. In other work, the focus has shifted to the citizen, often assessing attitudes via interviews and surveys. The two types of data are sometimes at odds, as choices at the grocery store may be affected by several factors, including price (Harvey and Hubbard 2013) and labelling (Hoogland et al. 2007).

Those interested in maintaining the status quo have taken some comfort from this disconnect, believing that little will change without direct consumer pressure, but others argue that the consumer's role is overstated. Most people buy their food from retailers, who hire a few buyers to source products that fit with their corporate values and the needs of their customers. As argued by Aerts (2013) 'it is easier [for citizen advocacy groups] to convince five (or fifteen) buying directors than five (or twenty five) million consumers', driving a shift to welfare friendly practices that happens long before the food arrives at the store. Thus, engagement efforts must extend beyond consumers of dairy products, to include citizens that are interested in the issues and who influence corporate and government responses.

Citizen engagement can take many forms, but one of our favourites is the 'open house', in part because this is something that almost any dairy farmer can do. Inviting citizens onto farms allows farmers to explain their practices and to hear the concerns of their visitors. When done right, this type of engagement can feed into a cycle of continuous improvement, starting with a conversation, leading to reflection on the practice, developing and implementing a solution, and then going back for more engagement to ensure that the modified practice resolves the initial concern without creating new issues.

As an example, consider the feedback we get during the public events we host at The University of British Columbia's Dairy Education and Research Centre (where cows are housed in a conventional free-stall barn). During these events, visitors often ask why the cows are not on pasture. This comment reflects what we know to be a widely held concern about the importance of pasture access for cows (Schuppli et al. 2014). As part of our on-going research on the topic, we have experimented with different types of pasture access (such as free choice between the barn and pasture) during the pasture season (Legrand et al. 2009), ways of managing pasture that also result in high levels of milk production (Chapinal et al. 2010), and other types of outdoor (non-pasture) access when the pasture is too wet to use. Although some of these approaches may differ from what the public imagines as ideal (i.e. that cows are on pasture all the time; Cardoso et al. 2016), we have found that people are often receptive to these hybrid solutions when the constraints and context is provided (e.g. that cows often prefer to come inside during the day and escape the hot sun, and that allowing cows to use pasture when the soil is too wet can damage the pasture).

Thus, the good news is that the urban public seems open to hearing about the perspective of the farmer, and thus being able to consider solutions that take this perspective into account. The bad news is that people working within production agriculture seem to be less able to consider the citizen's perspective as legitimate (Benard and de Cock Buning 2013). Poor ability to see the citizen's perspective may be a serious problem for engagement, as simply listening to the concerns will be of little help if these are not taken seriously.

2.4 Engage in social science research

Engaging in the substantive conversations will allow individuals within the dairy industry to better understand public concerns, and to develop practices on their farms that begin to address these concerns. Further progress will also require collecting data in a more rigorous fashion. This means investing in social science research that identifies key areas of public concern and perceived barriers to change on the part of the industry. Only a tiny fraction of dairy research focuses on these questions, but a considerable proportion of our own work now includes at least some social science component. We have summarised highlights from our work in a recent review paper (Weary et al. 2016), so here we briefly summarise just a few of the key methods, findings and conclusions from this work.

One obvious approach to research on human attitudes is to conduct some type of survey, and indeed much research examining attitudes around animal welfare is based on quantitative surveys (for example, asking whether respondents eat meat; Prickett et al. 2010). Unfortunately, this type of quantitative approach is limited when trying to address more complex issues. Animal welfare is a multi-dimensional and value-laden concept, and even experts disagree about which attributes to consider and how these should be weighted in welfare assessments (Fraser et al. 1997). For this reason, we believe that qualitative approaches are more likely to be useful for identifying and understanding different types of welfare concerns.

Qualitative studies typically use interviews or focus groups to allow participants to talk through issues, allowing researchers to better understand their meaning. One useful approach is to begin with qualitative studies to better understand the context, and then proceed to quantitative work to document the prevalence of different views within a population. Another approach is to employ a mixed methodology, asking participants to both state their views and to provide some explanation. For example, in some of our work we have allowed participants to interact online, such that they could read, vote and comment on the reasoning of other participants.

One interesting finding from several studies is that public views often align with those of dairy producers even when discussing contentious practices within the dairy industry. For example, even though tail docking is still common on dairy farms in the USA, both producers and the public tend to agree that the practice is not acceptable (Weary et al. 2011). Similarly, although many calves are still disbudded without any form of pain relief, producers and the public widely agree that pain relief should be provided (Robbins et al. 2015). Finally, even though only a small minority of dairy farms in the USA provide routine access to pasture, producers and the public seem to agree that pasture access is desirable (Schuppli et al. 2014).

In cases where the views of farmers and citizens are in close agreement, the question becomes why are producers continuing to employ the contentious practice? Here, the qualitative responses of producer and other industry participants have been especially informative, as they illustrate some of the perceived barriers to adopting less contentious alternatives. For example, our work has shown that some producers (likely those who continue to dock their cows' tails) believe that docking results in cleaner, healthier cows. These producers would benefit from targeted extension efforts that show why docking is ineffective (and sometimes counterproductive), and that provide examples of more effective strategies for improving cow cleanliness and udder health (Weary et al. 2011). Similarly, some producers stated that they believed that effective methods of pain mitigation for disbudding were either not available, or were prohibitively expensive (Robbins et al. 2015). Here too, targeted producer extension events could help overcome these barriers by showing the efficacy of pain mitigation provided by various agents, and documenting the costs involved. In the case of pasture access, some producers argued that pasture access was impractical, due to various constraints including land access and production requirements (Schuppli et al. 2014). We have argued that these perceived barriers can also be addressed, in this case, through targeted scientific projects designed to develop different approaches to providing outdoor access, including hybrid systems that allow cows free choice between the barn and pasture (Legrand et al. 2009).

For some other contentious issues, there are important disagreements between individuals working within the dairy industry and the general public. A particularly interesting example concerns the early separation of the cow and calf. In this case, the majority of the public sees early separation as a threat to the welfare of both cow and calf, while most working within the industry believe that later separation creates greater problems (Ventura et al. 2013). In the following section, we describe how animal-welfare science can be used to address this and a few other examples. We also discuss some of the limitations of scientific approaches in addressing welfare issues.

3. ANIMAL-WELFARE SCIENCE: APPROACHES AND LIMITS

Scientific research on the welfare of dairy cattle has a well established track record in providing evidence useful for addressing welfare issues (von Keyserlingk et al. 2009). For example, a series of studies have shown that tail-docking has little or no positive effect on cow cleanliness and udder health (reviewed by Sutherland and Tucker 2011), indicating that producers wanting cleaner cows and lower rates of mastitis will need to use other approaches. Similarly, a series of studies have examined the pain associated with various methods of disbudding and dehorning and have documented the efficacy of methods for controlling intra-operative and post-operative pain responses (reviewed by Stafford and Mellor 2011).

These examples are especially clear because these issues are, relative to some other welfare issues, simple and onedimensional. For example, if the concern is that whether cows are dirty, then all that needs to be measured is cow cleanliness. In this case, authors may still disagree (for example, some using a clinical scale and others quantifying the surface area with faecal contamination), but these disagreements tend to be technical as all agree that cleanliness is relevant. Pain is an affective state (or 'feeling'), and, thus, more difficult to study scientifically, but here too, there is wide agreement that pain (however it is measured) is the relevant consideration (Robbins et al. 2015).

Finding scientific solutions to welfare concerns becomes more difficult when practices affect multiple dimensions of animal welfare (Fraser et al. 1997). Understanding the welfare consequences of early cow–calf separation provides a good example. As described by Ventura et al. (2013), early separation is a concern to some because they see this as unnatural, preventing the natural bonding and interactions between cow and calf. These individuals may also believe that early separation is harmful because it robs cows and calves of the opportunity to experience the positive affective states associated with these affiliative relationships. In contrast, those in favour of early separation argue that this benefits the calf by reducing the distress response to separation when it does occur, and by reducing the risk of disease transmission between cows and calves. These differences in values mean that different scientists may study completely different outcomes, and that there is no value-free way of balancing the various outcome measures.

Even within one sphere of concern there may be biases. For example, research on the affective component has focussed almost entirely on the acute distress response at separation, with little research addressing positive emotions associated with maternal–filial interactions (Flower and Weary 2003). Although recognising these biases is important, this does not mean that scientific research on such topics is futile. Indeed, we would suggest that properly recognising biases can inspire new research, for example, examining the positive emotional responses of cows and calves kept together.

The multiple dimensions of animal welfare should not be seen as problematic if we think of these as forming the basis for a check list of issues that need to be addressed, rather than a zero-sum trade-off that requires value-laden assessments and compromise. We are especially interested in the development of systems that work well from the perspective of biological functioning, naturalness and affective state. This includes methods to keep cows and calves together that maintain health and minimise the distress response to separation when it does occur (Johnsen et al. 2015), as well as the development of other types of social housing for young calves that can provide some of the same benefits as does cow–calf rearing (Meagher et al. 2015).

From this perspective of developing systems that resolve multiple types of concern, we have argued elsewhere (Weary et al. 2016) that scientists should be sceptical of welfare 'solutions' that attempt to resolve one type of welfare concern (e.g. disease transmission) at the expense of another (e.g. freedom of movement, a key issue of key public concern; Boogaard et al. 2011; Popescu et al. 2013). Examples abound, but a relevant case for dairy is the calf hutch. Producers and citizens agree that animal health is of fundamental importance, but 'solutions' to health problems that require restrictions of movement (and, in this case, restricted social contact) create new types of concern. In the worst case, the proposed 'solution' introduces new concerns (i.e. restricted movement and social contact), without actually achieving the desired health outcome. The use of individual hutches and pens is standard on dairy farms in the USA, but calf mortality remains high. A recent survey reported that dairy heifer mortality, from 48 h to weaning, is at ~7%, with this number increasing to ~15% if the first 2 days of life are included (USDA/NAHMS 2016), and there is little evidence of increase in disease risk (and many benefits) of keeping calves in pairs or small social groups (Costa et al. 2016).

A more controversial example is the early separation of cow and calf. This is often argued to provide health improvements (e.g. reduce the risk of Johne's disease being transmitted to the calf; Windsor and Whittington 2010). Unfortunately, early separation increases the risk of other ailments, including metritis and mastitis in early lactation (Krohn et al. 1990), and there is little evidence that early separation provides any general benefit to the health of either cow or calf (Flower and Weary 2003). Research on other species has shown that early separation from the young causes neurochemical alterations that have been related to depression in the dam (von Poser Toigo et al. 2012), potentially placing the dam at a greater risk for other diseases.

Recognising that systems need to work well from the perspective of biological functioning, naturalness and affective state can inform the development of new research priorities for the industry. For example, in some regions such as eastern Canada, the north-eastern United States and parts of Europe, many dairy farms still use tie stalls, where cattle are restricted to a single stall for most of their life. Growing concern is expected to lead to a decline in this type of restrictive housing (Spooner et al. 2014).

Building new tie stall barns has been outlawed in Norway, with a complete ban scheduled for 2023 (Simensen et al. 2010). The proportion of farms using tie stalls is likely to decline due to market factors driven by changes in farm size (Barkema et al. 2015), but research efforts that help producers transition to less restrictive housing should become a priority.

Finally, there are some issues for which scientific efforts may be counter-productive. For example, the non-farming public wants assurance that the people who care for dairy cattle, really do care about their animals (Ventura et al. 2016a). People may be tempted to use the argument that farmers care about their animals because good care translates into good productivity. Although the link between attitudes and outcomes is researchable, this approach turns 'good care' from a moral to an economic issue. The implicit contract of animal agriculture is 'I take care of the animals, the animals take care of me' (Rollin 1994), and the moral standing of this enterprise is undermined when this wording becomes 'I provide care to the extent that this benefits me financially'.

4. ROLE FOR POLICY

People expect, at a minimum, humane care, appropriate hygiene and good husbandry (Cardoso et al. 2016), but how can the dairy industry prove that it lives up to these minimum standards? In addition, can the dairy industry position itself so that instead of simply responding to welfare concerns as external threats, it develops a vision that positions itself as a leader on welfare issues? We turn to these two issues below.

4.1 Demonstrating compliance with minimum standards

Different countries have used different vehicles to develop farm animal-welfare standards and ensure compliance (reviewed in part by von Keyserlingk and Hötzel 2015). In countries such as New Zealand, legislation has played a central role (e.g. the 1999 Animal Welfare Act, Parliamentary Counsel Office of New Zealand 1999). Similarly, legislation has been used to control how farm animals are cared for in many European countries. For example, Sweden passed its first dairy cattle welfare law in 1988, effectively ending zero-grazing systems for dairy cattle (Government of Sweden 1988). The European Union, through the European Commission Directive 2001/93/EC, has created provisions regarding housing and practices such as painful procedures. The lack of appropriate enforcement in some countries can undermine confidence in this legislative approach (e.g. Ventura et al. 2016b).

Other regions have relied on a combination of industry and retailer standards. For example, the Dairy Farmers of Canada (DFC) and Canada's National Farm Animal Care Committee (NFACC) worked together to create code of practice for the care and handling of dairy cattle (DFC–NFACC 2009). The development of codes is a multi-step process involving a range of stakeholders, including producers, scientists, government officials, veterinarians, grocery-chain distributors and representatives of the humane movement (NFACC 2014). Before publication, the draft document is opened for public comment to allow for further input.

In the USA, the National Federation of Milk Producers (NMPF) created the FARM program (Farmers Assuring Responsible Management). The guidance document was written by an NMPF-appointed technical writing committee that included representatives from the NMPF board of directors, various milk cooperatives, the National Cattlemen's Beef Association, an audit company as well as scientific and veterinary advisors. External stakeholders were also able to comment on the document (NMPF 2016).

Whether the process used by these dairy industry groups has merit in assuring those that buy their milk, remains to be seen, but standards are likely to be challenged on several fronts. There will be pressure from some to have standards sufficiently slack to allow almost all farms to comply, and pressure from others for stricter standards that keep products from 'bad actors' out of supply chain.

Another challenge will be to promote a culture of selfgovernance within the industry (Fraser 2014). This will be facilitated if farmers have access to critical feedback in a way that does not simultaneously put them at risk for being noncompliant. We see opportunities for rigorous but confidential first-party audits as a method for preparing farmers to meet standards. But these first-party audits will not be sufficient; third-party audits will also be required to provide external stakeholders confidence that standards are being met.

4.2 Developing a common vision

Instead of reacting to welfare issues as they emerge and treating these as external threats, can the dairy industry develop a vision that makes it a leader in animal welfare? The merits of developing a common vision would seem obvious, in part because engaging in big-picture discussions will also allow the industry to avoid problems facing other sectors.

Consider the sobering question 'will my grand-daughter be able to (and want to) take over my farm 30 years from now?'. If current trends persist, the answer to this question is likely to be no (see MacDonald et al. 2016 for a description of trends in the USA), as only a small fraction of farms currently operating are likely to exist in 30 years. A powerful vision may be able to motivate actions that help preserve a meaningful proportion of farms.

First and foremost should be a culture of care. This means more than just providing access to food, water and shelter; it means avoiding easily preventable harms, especially those caused by farmers (e.g. dehorning without pain control) and providing care to the most vulnerable (e.g. the sick cow or calf and the cull cow). For those who work with farm animals, health is likely to be a key issue; no one wants to see calves scour and cows with mastitis. But the question is whether the industry is willing to devote the resources required to address these health issues, even if the payoff is long term (for example, in the case of research), or if there is likely to be little economic return (in the case of animals with low economic value, such as bobby calves and cull cows)?

Even a strong vision built around common values will not be enough if the industry lacks the policy instruments to make the plan reality. Canada has a policy framework that allows for implementing an ambitious vision. It maintains a supply management system based on the cost of production (Skogstad 2008), making it feasible to manage a 'Made in Canada' brand for dairy products, produced according the common values of farmers and citizens. This not only provides an effective way of passing on the costs to consumers associated with improved welfare outcomes (via the cost of production formula), it provides a strong public-good argument in favour of maintaining supply management.

5. CONCLUSIONS

Animal agriculture is coming under greater public scrutiny. Hiding practices, for example, using ag-gag legislation, is unlikely to help and may contribute to a loss in public trust. One-way educational efforts are also unlikely to help, in part because new concerns emerge when people learn more about common practices. Instead, the industry needs to foster sustained engagement among producers, industry specialists, citizens, consumers and the general public. This will require conversations, in which the dairy industry listens to the concerns. The dairy industry will also need to make changes to accommodate public expectations. Developing solutions to animal-welfare issues will also require new biological research, especially studies focussed on developing systems that address multiple types of welfare concern (including biological functioning, naturalness and affective state). The industry requires policy instruments that will allow it to develop standards and put these into practice, including mechanisms to demonstrate that these standards are applied on farms. More ambitiously, we need to develop a common vision for what dairy farming should look like a generation from now, and use this vision as a basis for taking leadership on providing what we can all consider to be a good life for dairy cows.

ACKNOWLEDGEMENTS

We are grateful to our colleagues within UBC's Animal Welfare Program for the many helpful discussions that have inspired this work. Funding for our work is provided by the Industrial Research Chair program of Canada's Natural Sciences and Engineering Research Council, with industry contributions from the Dairy Farmers of Canada (Ottawa, ON, Canada), British Columbia Dairy Association (Burnaby, BC Canada), Westgen Endowment Fund (Milner, BC, Canada), Intervet Canada Corporation (Kirkland, QC, Canada), Novus International Inc. (Oakville, ON, Canada), Zoetis (Kirkland, QC, Canada), BC Cattle Industry Development Fund (Kamloops, BC, Canada), Alberta Milk (Edmonton, AB, Canada), Valacta (St. Anne-de-Bellevue, QC, Canada) and CanWest DHI (Guelph, ON, Canada).

REFERENCES

1. Aerts S (2013) The consumer does not exist: overcoming the citizen/ consumer paradox by shifting focus. In 'The ethics of consumption: the citizen, the market and the law'. (Eds H Rocklinsberg, P Sandin) pp. 172–175. (Wageningen Academic Publishers, Wageningen, The Netherlands)
2. Barkema HW, von Keyserlingk MAG, Kastelic JP, Lam TJGM, Luby C, Roy JP, LeBlanc SJ, Keefe GP, Kelton DF (2015) Invited review: changes in the dairy industry affecting dairy cattle health and welfare. Journal of Dairy Science 98, 7426–7445. doi:10.3168/jds.2015-9377
3. Benard M, de Cock Buning T (2013) Exploring the Potential of Dutch Pig Farmers and Urban-Citizens to Learn Through Frame Reflection. Journal f Agricultural & Environmental Ethics 26, 1015–1036. doi:10.1007/ s10806-013-9438-y
4. Bergstra TJ, Gremmen B, Stassen EN (2015) Moral values and attitudes toward Dutch sow husbandry. Journal of Agricultural & Environmental Ethics 28, 375–401. doi:10.1007/s10806-015-9539-x
5. Boogaard BK, Oosting SJ, Bock BB (2008) Defining sustainability as a sociocultural concept: citizen panels visiting dairy farms in the Netherlands. Livestock Science 117, 24–33. doi:10.1016/j.livsci.2007.11.004
6. Boogaard BK, Bock BB, Oosting SJ, Wiskerke JSC, van der Zijpp AJ (2011) Social acceptance of dairy farming: the ambivalence between the two faces of modernity. Journal of Agricultural & Environmental Ethics 24, 259–282. doi:10.1007/s10806-010-9256-4
7. Broad GM (2016) Animal production, 'ag-gag' laws, and the social production of ignorance: exploring the role of storytelling. Environmental Communication 10, 43–61. doi:10.1080/17524032.2014. 968178

8. Callahan ES, Dworkin TM (2000) The state of state whistleblower protection. American Business Law Journal 38, 99–175. doi:10.1111/ j.1744-1714.2000.tb00286.x
9. Cardoso CS, Hötzel MJ, Weary DM, Robbins JA, von Keyserlingk MAG (2016) Imagining the ideal dairy farm. Journal of Dairy Science 99, 1663–1671. doi:10.3168/jds.2015-9925
10. Cembalo L, Caracciolo F, Lombardi A, Del Giudice T, Grunert KG, Cicia G (2016) Determinants of individual attitudes toward animal welfarefriendly food products. Journal of Agricultural & Environmental Ethics 29, 237–254. doi:10.1007/s10806-015-9598-z
11. Chapinal N, Goldhawk C, de Passillé AM, von Keyserlingk MAG, Weary DM, Rushen J (2010) Overnight access to pasture does not reduce milk production or feed intake in dairy cattle. Livestock Science 129, 104–110. doi:10.1016/j.livsci.2010.01.011
12. Clark M (2014) Chilliwack cattle sales boycott threatened over animal abuse video. Available at http://www.cbc.ca/news/canada/britishcolumbia/chilliwack-cattle-sales-boycott-threatened-over-animal-abusevideo-1.2676340 [Verified 10 February 2017]
13. Clark B, Stewart GB, Panzone LA, Kyriazakis I, Frewer LJ (2016) A systematic review of Public attitudes, perceptions and behaviours towards production diseases associated with farm animal welfare. Journal of Agricultural & Environmental Ethics 29, 455–478. doi:10.1007/ s10806-016-9615-x
14. Coleman G, Rohlf V, Toukhsati S, Blache D (2015) Public attitudes relevant to livestock animal welfare policy. Farm Policy Journal 12, 45–57
15. Costa JHC, von Keyserlingk MAG, Weary DM (2016) Invited review: effects of group housing of dairy calves on behavior, cognition, performance and health. Journal of Dairy Science 99, 2453–2467. doi:10.3168/jds.2015-10144
16. Croney CC, Apley M, Capper JL, Mench JA, Priest S (2012) Bioethics symposium: the ethical food movement: what does it mean for the role of science and scientists in current debates about animal agriculture? Journal of Animal Science 90, 1570–1582. doi:10.2527/jas.2011-4702
17. DFC-NFACC (2009) Code of practice for the care and handling of dairy cattle. Available at http://www.nfacc.ca/pdfs/codes/Dairy%20Code% 20of%20 Practice.pdf [Verified 17 October 2016]
18. European Commission Directive (2001) European Commission Directive 2001/93/EC 2001/93/EC. Available at http://eur-lex.europa.eu/legalcontent/EN/TXT/?uri= celex% 3A52006SC0065 [Verified 17 October 2016]
19. Fisman R, Khanna T (1999) Is trust a historical residue? Information flows and trust levels. Journal of Economic Behavior & Organization 38, 79–92. doi:10.1016/S0167-2681(98)00123-1
20. Flower F, Weary DM (2003) The effects of early separation of the dairy cow and calf. Animal Welfare 12, 339–348.
21. Fraser D (2014) Could animal production become a profession? Livestock Science 169, 155–162. doi:10.1016/j.livsci.2014.09.017
22. Fraser D, Weary DM, Pajor EA, Milligan BN (1997) A scientific conception of animal welfare that reflects ethical concerns. Animal Welfare 6, 187–205.
23. Frewer LJ, Kole A, Van De Kroon SM, De Lauwere C (2005) Consumer attitudes towards the development of animal-friendly husbandry systems. Journal of Agricultural & Environmental Ethics 18, 345–367. doi:10.1007/ s10806-005-1489-2
24. Government of Sweden (1988) The Animal Welfare Act. Unofficial English translation. Available at http://extwprlegs1.fao.org/docs/pdf/swe19544E. pdf [Verified 10 February 2017]
25. Gulliver A (2014) Manuka dairy firm under investigation in Chile. Available at http://www.stuff.co.nz/business/farming/dairy/9634584/Manukadairy-firm-under-investigation-in-Chile [Verified 17 October 2016]
26. Hansen J, Holm L, Frewer L, Robinson P, Sandøe P (2003) Beyond the knowledge deficit: recent research into lay and expert attitudes to food risks. Appetite 41, 111–121. doi:10.1016/S0195-6663(03)00079-5
27. Harper G, Henson S (2001) Consumer concerns about animal welfare and the impact on food choice. EU FAIR CT98-3678 final report. Available at https://ec.europa.eu/

food/sites/food/files/animals/docs/aw_arch_hist_ eu_fair_project_en.pdf [Verified 10 February 2017]

28. Harvey D, Hubbard C (2013) Reconsidering the political economy of farm animal welfare: an anatomy of market failure. Food Policy 38, 105–114. doi:10.1016/ j. foodpol.2012.11.006
29. Hersh MA (2002) Whistleblowers: heroes or traitors?: Individual and collective responsibility for ethical behaviour. Annual Reviews in Control 26, 243–262. doi:10.1016/S1367-5788(02)00025-1
30. Hoogland CT, de Boer J, Boersema JJ (2007) Food and sustainability: do consumers recognize, understand and value on package information on production standards? Appetite 49, 47–57. doi:10.1016/j.appet.2006. 11.009
31. Johnsen JF, Beaver A, Mejdell CM, Rushen J, de Passillé AM, Weary DM (2015) Providing supplementary milk to suckling dairy calves improves performance at separation and weaning. Journal of Dairy Science 98, 4800–4810. doi:10.3168/jds.2014-9128
32. Krohn C, Jonasen B, Munksgaard L (1990) Cow–calf relations. 2: The effect of 0 vs. 5 days suckling on behaviour, milk production and udder health of cows in different stabling. In 'Report no. 678'. (National Institute of Animal Science: Foulum, Denmark)
33. Lassen J, Sandøe P, Forkman B (2006) Happy pigs are dirty! Conflicting perspectives on animal welfare. Livestock Science 103, 221–230. doi:10.1016/j.livsci.2006.05.008
34. Legrand AL, von Keyserlingk MAG, Weary DM (2009) Preference and usage of pasture versus freestall housing by lactating dairy cattle. Journal of Animal Science 92, 3651–3658.
35. MacDonald JM, Cessna J, Mosheim R (2016) Changing structure, financial risks, and government policy for the US dairy industry. Available at https://www.ers. usda. gov/webdocs/publications/err205/56833_err205_ errata.pdf [Verified 10 February 2017]
36. Maeda Y, Miyahara M (2003) Determinants of trust in industry, government, and citizen's groups in Japan. Risk Analysis 23, 303–310. doi:10.1111/ 1539-6924.00310
37. McKendree MGS, Croney CC, Widmar NJO (2014) Effects of demographic factors and information sources on United States consumer perceptions of animal welfare. Journal of Animal Science 92, 3161–3173. doi:10.2527/ jas.2014-6874
38. MeagherRK, DarosRR,Costa JH, Von KeyserlingkMAG, HötzelMJ,Weary DM (2015) Effects of degree and timing of social housing on reversal learning and response to novel objects in dairy calves. PLoS One 10, e0132828. doi:10.1371/journal.pone.0132828
39. New Zealand Ministry of Primary Industries (2016) Animal welfare regulations. Available at http://mpi.govt.nz/law-and-policy/legal-overviews/ animal-welfare/animal-welfare-regulations/ [Verified 10 February 2017]
40. NFACC (2014) NFACC code of practice development process. Available at http://www. nfacc.ca/resources/codes-ofpractice/NFACC_Code_ process_Jan_2014.pdf [Verified 10 February 2017]
41. NMPF (2016) Animal care reference manual. Available at http:// www. nationaldairyfarm.com/sites/default/files/Version-3-Manual.pdf [Verified 10 February 2017]
42. Ormandy EH, Schuppli CA, Weary DM (2013) Public attitudes towards the use of animals in research: effects of invasiveness, genetic modification and regulation. Anthrozoos 26, 165–184. doi:10.2752/ 175303713X13636846944240
43. Parliamentary Counsel Office of New Zealand (1999) Animal Welfare Act 1999. Public Act 1999 no. 142. Available at http://www.legislation.govt. nz/act/public/1999/0142/ latest/DLM49664.html [Verified 17 October 2016]
44. Peters RG, Covello VT, McCallum DB (1997) The determinants of trust and credibility in environmental risk communication. Risk Analysis 17, 43–54. doi:10.1111/j.1539-6924.1997.tb00842.x
45. Pieper L, Doherr MG, Heuwieser W (2016) Consumers' attitudes about milk quality and fertilization methods in dairy cows in Germany. Journal of Dairy Science 99, 3162–3170. doi:10.3168/jds.2015-10169

46. Popescu S, Borda C, Diugan EA, Spinu M, Groza IS, Sandru D (2013) Dairy cows' welfare quality in tie-stall housing system with or without access to exercise. Acta Veterinaria Scandinavica 55, 43. doi:10.1186/ 1751-0147-55-43
47. Prickett RW, Norwood FB, Lusk JL (2010) Consumer preference for farm animal welfare: results from a telephone survey of US households. Animal Welfare 19, 335–347.
48. Rawlins BL (2008) Measuring the relationship between organizational transparency and employee trust. The Public Relations Journal 2, 1–21.
49. Robbins JA, Weary DM, Schuppli CA, von Keyserlingk MAG (2015) Stakeholder views on treating pain due to dehorning dairy calves. Animal Welfare 24, 399–406. doi:10.7120/09627286.24.4.399
50. Robbins JA, von Keyserlingk MAG, Fraser D, Weary DM (2016a) Farm size and animal welfare. Journal of Animal Science 94, 5439–5455. doi:10.2527/jas.2016-0805
51. Robbins JA, Franks B, Weary DM, von Keyserlingk MAG (2016b) Awareness of AG-GAG laws erodes trust in farmers and increases support for animal welfare regulations. Food Policy 61, 121–125. doi:10.1016/j.foodpol.2016.02.008
52. Rollin BE (1994) Animal production and the new social ethic for animals. Journal of Social Philosophy 25, 71–83.
53. Ryan EB, Fraser D, Weary DM (2015) Public attitudes to housing systems for pregnant pigs. PLoS One 10, e0141878. doi:10.1371/journal.pone. 0141878
54. Saputo (2015) Animal welfare policy. Available at http://www.saputo.com/uploadedFiles/Saputo/shared/social/Animal%20welfare%20policy.pdf [Verified 10 February 2017]
55. Schuppli CA, von Keyserlingk MAG, Weary DM (2014) Access to pasture for dairy cows: responses from an on-line engagement. Journal of Animal Science 92, 5185–5192. doi:10.2527/jas.2014-7725
56. Scudder JN, Bishop-Mills C (2009) The credibility of shock advocacy: animal rights attack messages. Public Relations Review 35, 162–164. doi:10.1016/j.pubrev.2008.09.007
57. Simensen E, Østerås O, Bøe KE, Kielland C, Ruud LE, Naess G (2010) Housing system and herd size interactions in Norwegian dairy herds; associations with performance and disease incidence. Acta Veterinaria Scandinavica 52, 14. doi:10.1186/1751-0147-52-14
58. Skogstad G (2008) Canadian agricultural programs and paradigms: the influence of international trade agreements and domestic factors. Canadian Journal of Agricultural Economics 56, 493–507. doi:10.1111/ j.1744-7976.2008.00143.x
59. Spooner JM, Schuppli CA, Fraser D (2014) Attitudes of Canadian citizens toward farm animal welfare: a qualitative study. Livestock Science 163, 150–158. doi: 10.1016/j.livsci.2014.02.011
60. Stafford KJ, Mellor DJ (2011) Addressing the pain associated with disbudding and dehorning in cattle. Applied Animal Behaviour Science 135, 226–231. doi:10.1016/j.applanim.2011.10.018
61. Sutherland MA, Tucker CB (2011) The long and short of it: a review of tail docking in farm animals. Applied Animal Behaviour Science 135, 179–191. doi:10.1016/j. applanim. 2011.10.015
62. Te Velde H, Aarts N, Van Woerkum C (2002) Dealing with ambivalence: farmers' and consumers' perceptions of animal welfare in livestock breeding. Journal of Agricultural & Environmental Ethics 15, 203–219. doi:10.1023/A:1015012403331
63. Thompson PB (2008) 'The ethics of intensification: agricultural development and cultural change.' (Springer: Heidelberg, Germany)
64. Tiplady CM, Walsh DAB, Phillips CJ (2013) Public response to media coverage of animal cruelty. Journal of Agricultural & Environmental Ethics 26, 869–885. doi:10.1007/s10806-012-9412-0
65. USDA/NAHMS (2016) Dairy 2014. Dairy cattle management practices in the United States, 2014. Available at https://www.aphis.usda.gov/ animal_health/nahms/ dairy/downloads/dairy14/Dairy14_dr_PartI.pdf [Verified 1 April 2016]
66. Vanhonacker F, Verbeke W, Van Poucke E, Tuyttens FAM (2008) Do citizens and farmers interpret the concept of farm animal welfare differently? Livestock Science 116, 126–136. doi:10.1016/j.livsci.2007. 09.017

67. Ventura BA, Schuppli C, von Keyserlingk MAG, Weary DM (2013) Views on contentious practices in dairy farming: the case of early cow–calf separation. Journal of Dairy Science 96, 6105–6116. doi:10.3168/ jds.2012-6040
68. Ventura BA, von Keyserlingk MAG, Wittman H, Weary DM (2016a) What difference does a visit make? Changes in animal welfare perceptions after interested citizens tour a dairy farm. PLoS One 11, e0154733. doi:10.1371/journal.pone.0154733
69. Ventura BA, Weary DM, Giovanetti AS, von Keyserlingk MAG (2016b) Veterinary perspectives on cattle welfare challenges and solutions. Livestock Science 193, 95–102. doi:10.1016/j.livsci.2016.10.004
70. Verbeke V, Ward RW (2001) A fresh meat almost ideal demand system incorporating negative TV press and advertising impact. Agricultural Economics 25, 359–374. doi:10.1111/j.1574-0862.2001.tb00215.x
71. Verbeke WF, Pérez-Cueto JA, Barcellos MD, Krystallis A, Grunert KG (2010) European citizen and consumer attitudes and preferences regarding beef and pork. Meat Science 84, 284–292. doi:10.1016/j.meatsci.2009. 05.001
72. von Keyserlingk MAG, Hötzel MJ (2015) The ticking clock: addressing farm animal welfare in emerging countries. Journal of Agricultural & Environmental Ethics 28, 179–195. doi:10.1007/s10806-014-9518-7
73. von Keyserlingk MAG, Rushen J, de Passillé AM, Weary DM (2009) Invited review: the welfare of dairy cattle – key concepts and the role of science. Journal of Dairy Science 92, 4101–4111. doi:10.3168/ jds.2009-2326
74. von Poser Toigo E, Diehl L, Ferreira A, Mackendanz V, Krolow R, Benitz A, Noschang C, Huffell A, Silveira P, Wyse A, Dalmaz C (2012) Maternal depression model: long lasting effects on the mother following separation from pups. Neurochemical Research 37, 126–133. doi:10.1007/s11064- 011-0590-3
75. Weary DM, Schuppli CA, von Keyserlingk MAG (2011) Tail docking dairy cattle: responses from an online engagement. Journal of Animal Science 89, 3831–3837. doi:10.2527/ jas.2011-3858
76. Weary DM, Ventura BA, von Keyserlingk MAG (2016) Societal views and animal welfare science: understanding why the modified cage may fail and other stories.Animal 10, 309–317. doi:10.1017/S1751731115001160
77. Wheale P, Hinton D (2007) Ethical consumers in search of markets. Business Strategy and the Environment 16, 302–315. doi:10.1002/bse.484
78. Windsor PA, Whittington RJ (2010) Evidence for age susceptibility of cattle to Johne's disease. Veterinary Journal (London, England) 184, 37–44. doi:10.1016/j.tvjl.2009.01.007
79. You X, Li Y, Zhang M, Yan H, Zhao R (2014) A survey of Chinese citizens' perceptions on farm animal welfare. PLoS One 9, e109177. doi:10.1371/ journal.pone.0109177

CHAPTER 6

Economic performance of small ruminant production in a protected area: a case study from Tell Ar-Rumman, a Mediterranean ecosystem in Jordan

Khalid M Al-Khalidi[1*] Amani A Alassaf, Mustafa F Al-Shudiefat[1] and Raed J Al-Tabini[1]

1 Community Based Rangeland Rehabilitation Programme, Royal Botanic Garden, P.O. Box 99, 11910-Amman, Jordan

ABSTRACT

This paper discusses the qualitative impact of a grazing/herding management program on the profitability of Small Ruminant (SR) herds near the protected area of the Royal Botanic Garden (RBG) in Tell Ar-Rumman, Jordan. As the ultimate goal of establishing a protected area is to rehabilitate and enhance vegetation cover in a sustainable manner, a unique program was set up to meet the needs of local herders and enable the RBG to pursue its biodiversity conservation goals. Financial data pertaining to five local herders were recorded using the accrual accounting system and analyzed for the years 2009 and 2010. Significant changes occurred in the productivity and profitability of the SR herds, with the herders' net incomes increasing by 6% to 159% as a result of direct and indirect support provided by the RBG in 2010. The grazing/herding management program is not only improving the income of participating pastoralists, but is also contributing to the growth and conservation of vegetation and wildlife in Tell Ar-Rumman.

Keywords: Economic performance, Small ruminant, Grazing-herding management

1. BACKGROUND

Small Ruminant livestock (SR) refers to sheep and goats. In 2010, the Al-Balqa Governorate in Jordan hosted 7.8% of Jordan's total population of SR (DOS, 2011), primarily produced under traditional systems characterized by inadequate feed and water supply, poor veterinarian services and weak management practices (Alassaf, 2012).

The SR population in Jordan dropped from 7.1 million in 2008 to 5.9 million head in 2009 (DOS, 2011), due to limited grazing options and low rainfall. Livestock owners generally have few resources to invest in technologies and other advanced equipment that would maximize herd productivity. So there was a need to record and analyze herd management decisions in order to find ways to raise profitability. A rise in profitability would be a clear indicator of success. The tangible benefits are regular cash income, meat, milk and manure, and the intangible benefit is the role of SR as insurance in case of emergency (Kosgey et al., 2008).

Economic performance has an important impact on the decision to continue farming in marginal areas (Alassaf et al., 2012; Hadjigeorgiou, and Zervas, 2009). In general, livestock owners in Tell Ar-Rumman (Jordan) are small herders and have adopted a low level of production technology. They have suffered from successive weak production seasons, causing a fragile economic status. But there is always room for improving herd economic efficiency and productivity. An economic and financial analysis can assist herders in making profitable decisions, especially in a context of agricultural sustainability in a protected area (Thomson and Nardone, 1999).

Grazing and herding practices have a great impact on the vegetation and biodiversity of protected areas. It has been found that plant cover in ungrazed plots was 20% greater than in grazed plots (Louhaich et al., 2009). However, the absence of grazing may not necessarily improve the variety of plant species available for livestock utilization, as rangeland genetic erosion can be an attribute of traditional production systems (Thornton et al., 1999).

A protected area, called the Royal Botanic Garden (RBG), was set up in Tell Ar-Rumman with the specific goal of enhancing the quantity and diversity of native vegetation cover. Local pastoralists initially voiced their opposition to the RBG, because a fence was erected around an area they had traditionally used for grazing, and they expected that their livelihood would be endangered.

However, the RBG initiated a collaborative grazing and land management program based on a participatory approach with members of the local community. This study tracks the economic results of the first group of farmers who cooperated with the RBG in this program, with particular attention to qualitative changes in herd management attributed to RBG recommendations, and reflected in the herders' revenue and production costs. An efficiently managed grazing system has been established on land in the protected area, with no negative impact on the herders' income.

Since pastoralism constitutes a deep cultural tradition in Jordan, simply removing animals from the range was never considered to be the right solution. It is well-documented in other parts of the world that removing grazing from a landscape which has traditionally been grazed often has a *negative* impact on species diversity (Al-Tabini et al., 2012; Louhaichi et al., 2009; Thornton et al., 1999). Therefore, after initial biomass studies and grazing trials, the RBG introduced a managed grazing system that allows local herds to graze on RBG land at certain times of the year and under specific conditions.

In 2008, the RBG conducted a rapid assessment of how local herders were managing their herds and the current herd status. It was found that the herds suffered from poor feeding management and weak veterinarian services, which resulted in low productivity. Low productivity was forcing herders to intensify grazing, in order to reduce the high costs of feed, and this was detrimental to the RBG's efforts to rehabilitate vegetation cover and restore biodiversity.

Through close cooperation and consulting between the RBG and the pastoralists, the latter have not only seen improvements in herd health, herd management, and increased profits, but have also adopted a conservation mindset and are now actively protecting the RBG site, having recognized the importance and benefits of conserving biodiversity. At the same time, plant surveys on RBG land, where managed grazing is being allowed, have revealed a significant change in plant species composition, with an increase from 436 species in 2006 to 580 species in 2010 (RBG, 2011).

2. METHODS

Tell Ar-Rumman, like most areas in Jordan, suffers from excessive grazing pressure which, combined with drought, leads to the loss of desirable returning grasses and shrubs. The loss of plants and grasses causes increased soil erosion and reduced availability of grazing resources.

The present study was conducted at the RBG located 30 km northwest of Amman, Jordan, near the village of Tell Ar-Rumman (32°10'57" N, 35°49'38"E), on 2 km^2 of land donated by the Ministry of Agriculture for the establishment of the RBG, a new botanical garden and protected area. The RBG is intended to be a center for public education, scientific research, and biodiversity conservation.

Data were gathered in 2009 and 2010 from five herders who used to illegally graze their herds on the site. The RBG conducted training courses in 2009 aimed at introducing skills and practices in SR management, including herd management, veterinarian training on common disease diagnosis and treatment, and herd record keeping. The RBG provided additional veterinarian services: a vaccination program, training for a veterinarian extension worker, and a veterinarian clinic. The RBG also supplied supplementary feed, to compensate the loss of rangeland, to cover the daily feed requirements of the herds, especially in mating and late pregnancy periods.

The RBG staff trained the herders on how to record herd production data using structured records. The herders recorded their production data and related costs on a daily basis under the supervision of the RBG staff. Herd production data were collected and analyzed for each herd separately, and showed the productivity and net revenue per head. The collected data included the following:

- Production features and revenue: herd size and types (sheep, goat), lambs, milk production and prices, sales and prices of livestock production (lambs, sheep, and milk).
- Production costs: the expenses included feed cost, transportation, medical services, water, and labor.

The accounting system was designed to obtain reasonable accuracy for the net income (revenue) over a given period of time. It was assumed that 2009 was the second productive year for each herd. The net revenue calculated in this study included the opportunity costs of herders, in the case of family-based herding. Later the data were recorded and analyzed using MS Excel.

3. RESULTS AND DISCUSSION

3.1 Economic performance of SR production

Small Ruminants are raised for different reasons. Some people in Jordan are motivated to raise SR for additional income, while others depend on SR as a main source of income, for domestic food, and as simple investments they are familiar with.

Herd sizes depend on the purpose for raising the livestock. Small herds are mainly meat producing, as well as fiber and milk producing. More importantly they are considered as an in-hand cash source.

Table 1 represents the herd size, total net revenues and per head revenues for 2009 and 2010.

Table 1 Herd size and revenue in 2009 and 2010

Herder name	Animal type	Herd size			Revenue				
		2009 head	2010 head	% of change	2009 JD®	2010 JD®	Per head (2009) JD	Per head (2010) JD	% of change
HERDER 1	Sheep	250	243	-3	20,290.80	33,547	81.16	138.05	60
HERDER 2	Goat	58	57	-2	2,541.63	3,626.80	43.82	63.63	70
HERDER 3	Sheep	129	133	3	12,423.28**	12,093.40**	58.87	56.25	-3
HERDER 3	Goat	82	82	0					
HERDER 4	Goat	125	146	14	3,445.79**	6,213.47**	27.56	42.55	55
HERDER 5	Sheep	35	34	-3	886.56	489.26	10.19	6.52	-55
HERDER 5	Goat	52	41	-3					

® JD: Local currency, Jordanian Dinar.

**Revenue for sheep and goats.

Source: Income statements of 2009 and 2010.

The percentage of change was calculated to determine changes in herd size and revenues, and thus to represent improvements, if any, in performance due to RBG recommendations on herd management. The results showed a fluctuation in herd size between −3% and 3%, except in the case of Herder 4 who owns a goat herd. In the latter case, a 14% increase was seen in 2010, as goats are known for high twinning rates.

Changes in herd sizes were inconsistent with changes in revenue from 2009 to 2010. The herders who had either sheep or goats (but not both) generated higher revenues in comparison to those who owned both. This indicates that specialization in one type of SR might improve herd management and make it easier to meet expected revenue goals.

In detail, Table 1 shows that Herder 2's goat herd had the highest change in revenue (70%), in spite of having a negative change (−2%) in herd size for the two years 2009 and 2010. This change in revenue may be attributed to improved herd and grazing management practices and a consequent increase in lambs and milk per unit of production. It is important to emphasize that the changes in revenue are not attributed to changes in prices or market availability. The product prices and market outlets for these herders was the same in 2009 and 2010.

On other hand, the change in revenue for Herder 3 was the least (−3%), while Herder 5 had an exceptional reduction in revenue (−55%). Both of their herds were mixed, consisting of both sheep and goats.

The herders who paid most attention to the RBG's recommendations derived greater benefits than those who did not fully implement the recommendations.

4. FACTORS AFFECTING ECONOMIC PERFORMANCE OF SR PRODUCTION

4.1 Production costs

In general, there are two types of costs associated with producing an agricultural product: variable and fixed costs. Variable costs vary according to the size of the herd, whereas fixed costs occur regardless of the level of output. Examples of variable costs include feed, medicine, labor, and supplies. Fixed costs include depreciation on the herd and buildings, land, equipment, and rents. In this group of herders, fixed costs represented a very low portion out of the total costs. The herding style is traditional, where concrete buildings, vehicles and equipment are in limited use, and the herders do not have to pay rent for the land they graze on. The traditional style of herding involves mainly variable costs.

Table 2 presents the percentages of variable cost items out of the total production costs in 2010. The main purpose of this calculation was to compare the relative costs, to reflect different management decisions and their impact on reducing the production cost and thus increasing the net revenue.

Table 2 Percentage of variable cost items out of total costs recorded in 2010

Item	Herder 1 sheep	Herder 2 goat	Herder 3 sheep	Herder 3 goat	Herder 4 goat	Herder 5 sheep	Herder 5 goat
Feed	86%	62%	79%	61%	62%	51%	53%
Labor	11%	34%	18%	36%	33%	42%	44%
Transportation	0%	0%	0%	0%	0%	0%	0%
Water	1%	0%	1%	0%	2%	3%	1%
Maintenance	0%	1%	1%	0%	0%	1%	1%
Medications	1%	3%	1%	1%	2%	2%	0%

Source: Calculated from 2010 records.

The records show that feed and labor accounted for the highest share of the total costs. The feed cost was 51% to 86%, while labor was 11% to 44%. Understanding this contribution to the cost shares gives insights on how herders could better manage costs to improve their profitability. Herders can control feed costs by maximizing the use of pasture, producing their own harvested feeds, buying and storing feed in bulk, and minimizing feed wastage. Even though the cost of veterinarian medications is very low, it accounted for 0% to 3% of the total variable cost, and had a tangible impact on improving herd productivity.

4.2 Lamb sales, breeding and mortality rates

A short lambing season is a result of successful herd management. On the other hand, the number of lambs born per ewe is certainly an economically important attribute in an SR enterprise. Profitability is largely determined by the number of lambs sold per ewe, as selling lambs is the main return from SR farming. Therefore, a great deal of effort should be devoted to taking care of animal health before, during, and after birth. Knowledge of when and how lamb mortality occurs can be helpful in minimizing the mortality rate. The percentages of sales, breeding, and death losses from the total number of lambs give quantitative measures for herd management. For this group of herders, lambs were either sold, died or were replaced into the herd, as the herders decided.

The high animal mortality rate for some herders indicated poor herd management, specifically with respect to veterinarian services. In 2009, the RBG staff conducted trainings on different veterinarian topics, which helped control the mortality rate in almost all the herds. Herder 3, however, had the highest mortality rates due to a mismanagement decision to purchase sick lambs and kids. It is important to know when and how lamb and kid mortality occurs, in order to avoid such losses in the future. This information should be included in the herd's records to handle future losses.

Table 3 shows the lambing rate in 2009 and 2010, and indicates an improvement for most of the herds.

Table 3 Lambing rate in 2009 and 2010, and lamb distribution and sales in 2010

Herder	Type	Total 2009	Total 2010	Lambing rate 2009	Lambing rate 2010	Lamb distribution in 2010			Milk and lamb sales as a % of total sales	
						Sales	Breeding	Mortality	Lamb %	Milk %
HERDER 1	Sheep	250	243	79%	99%	75%	23%	2%	44%	21%
HERDER 2	Goat	58	57	66%	79%	96%	0%	4%	73%	13%
HERDER 3	Sheep	129	133	67%	62%	57%	27%	16%	44%	24%
HERDER 3	Goat	82	82	47%	74%	58%	21%	21%	48%	24%
HERDER 4	Goat	125	146	30%	98%	72%	23%	6%	58%	5%
HERDER 5	Sheep	35	34	96%	112%	74%	11%	8%	58%	10%
HERDER 5	Goat	52	41	82%	70%	38%	58%	5%	26%	21%

In general, the lambing rate was higher in 2010, except for Herders 3 and 5. Herder 3 had a slight decrease from 67% to 62%, while Herder 5 reduced the size of his goat herd from 52 head to 41 head in 2010, which negatively affected the lambing rate for the year.

Lambs and milk sales are the primary source of income for small herds in Jordan, with minor contributions from the sale of manure and wool. Milk productivity depends on a number of factors such as the nutrition and health of the milking flock. Herders need to give more attention to these factors to ensure a productive and constant milking season with high quality milk, and to gain increased revenues.

As demonstrated in Table 3, lamb sales counted for 26% to 73% of the total sales. These figures explore how the revenue from selling lambs could cover the variable costs of production, especially feed costs, leaving the revenue from milk as net revenue for the herders.

4.3 Variable costs as a percentage of revenue

A high level of SR sales does not necessarily lead to high revenue. There is a need to examine the costs associated with these sales. This examination is helpful in determining how the sales could cover the variable costs used in production.

Table 4 shows that sales accounted for a 32% recovery of the variable costs out of total sales for Herder 1, indicating that Herder 1 had the most efficient herd management of all the herders. At the opposite end of the scale, Herder 5 used 93% of his total sales to cover variable costs of sheep production, and 86% of goat sales to cover variable costs. In other words, a major share of his sales revenue was used to recover variable costs, which decreased the herd's profitability.

Table 4 Variable costs as a percentage of revenue in 2010

Name	*Type*	Variable costs as a percentage of revenue (%)
HERDER 1	Sheep	32
HERDER 2	Goat	53
HERDER 3	Sheep	48
HERDER 3	Goat	76
HERDER 4	Goat	47
HERDER 5	Sheep	93
HERDER 5	Goat	86

The RBG introduced a grazing/feeding plan for the herders, to measure the impact of managed grazing and herding on both SR profitability and the biodiversity of the RBG site. The plan was based on direct and indirect feed supplements. RBG support meant a reduction in feed cost that would improve the net revenue. The direct supplement was the provision of barley based on the daily feed requirements for mating and late pregnancy periods, while the indirect supplement referred to periods of managed grazing inside the RBG site. These grazing periods were defined by a specific estimation of the biomass and stocking rate in the RBG site on a yearly basis. The herders were allowed to graze for a specific number of hours per day on pre-determined days, for the equivalent of 50% of the daily requirements per SR head.

The RBG grazing/herding management assisted the herders to decrease their variable costs and improve profits in 2010. Table 5 shows the impact of RBG in-kind support, in the form of direct and indirect supplements, on the revenue of the herders. The results range between 6% and 159%. Since Herder 5 did not follow RBG recommendations and ended up with very low revenue, the RBG support appears extremely high in this table. Supplementary grazing (indirect support) improved the revenue by 4% to 34%, while grazing plus supplemental feed (direct and indirect support) resulted in a 6% to 159% improvement in revenue. Some herders would have made zero net revenue, or suffered a loss, if the RBG grazing/herding management had not been offered. Moreover, without the RBG's program, such a reduction in herd revenue would have been detrimental to the RBG's vegetation recovery rate, because the herders would have cut fences to enter RBG land and graze everywhere possible in an uncontrolled manner.

Table 5 Impact of rbg support on net income of herders

RBG support	Herder 1	Herder 2	Herder 3	Herder 4	Herder 5
Indirect: Pasture (JD®)	1,346	269	673	875	404
Direct: Barley (JD®)	805	74	833	722	610
TOTAL JD	**2,151**	**343**	**1,506**	**1,597**	**1014**
Indirect revenue (%)	4	7	6	14	34
Direct revenue (%)	2	2	9	12	125
OVERALL NET REVENUE %	**6**	**9**	**15**	**26**	**159**

®JD: Local currency, Jordanian Dinar.

5. CONCLUSION

To understand economic performance, this study analyzed the profitability of SR production for five herders at close coordination with the RBG. Profits were driven by cost control and gross production. The study suggests that traditionally measured parameters such as lambing rate, cost per head and death loss, although they do have an effect on net profit, are not enough to determine best practice management in SR flocks.

Herders are strongly advised to use a profit analysis method to determine and monitor the profitability of their flocks, as well as obtain clear insights for future business and management decisions.

In addition, the study suggests that SR herders should maintain historical records allowing them to measure progress over time. Herders have to understand how to increase the productivity of each head rather than just increase the number of head per herd. The role of specialists in animal health and nutrition is very important in helping the livestock owners reach this goal.

This multi-faceted support and managed grazing program for livestock herders, using a participatory approach, has been shown to raise the income of herding families while helping to conserve and augment the vegetation cover and wildlife in Tell Ar-Rumman. Future biodiversity conservation polices and initiatives should therefore consider introducing a similar program to promote better livelihoods for herders and enhance the sustainable growth and conservation of biodiversity.

ACKNOWLEDGEMENTS

We would like to express our gratitude to Her Royal Highness Princess Basma bint Ali, founder of the Royal Botanic Garden (RBG) of Jordan, for her encouragement and continued support, to RBG Executive Director Tariq Abu Taleb, and the Association française de développement and Fonds français pour l'environnement mondial for funding the project. Thanks is also extended to Tell Ar-Rumman's community for their help and Habiba Dingwall for her kind support in editing the paper.

REFERENCES

1. Al-Assaf A: Economic implications of small ruminant diseases in the Northern Area of Jordan. Journal of Food, Agriculture & Environment. 2012, 10(1):323–326.
2. Al-Assaf A, Majadalawi M, Nawash O: Factors affecting farmer's decision to continue farm activity in marginal areas of Jordan. African Journal of Agricultural Research 2012, 6(12):2755–2760.
3. Al-Tabini R, Al Khalidi K, Al-Shudiefat M: Livestock, medicinal plants, and rangeland viability in Jordan's Badia: through the lens of traditional and local knowledge. Pastoralism: Research, Policy and Practice2012,2:4.
4. DOS (Department of Statistics): Annual Statistical Report. Amman Jordan 2011, 1: 17–18.
5. Hadjigeorgiou I, Zervas G: Evaluation of production systems in protected areas: Case studies on the Greek Natura 2000 network. In: Pacheco F. (ed.), Moran d-Fehr P. (ed.). Changes in sheep and goat farming systems at the beginning of the 21st century: research, tools, methods and initiatives in favour of a sustainable development. Zaragoza: CIHEAM /DRAP-Norte /FAO, 2009. 2009, 101–111.
6. Kosgey I, Rowlands G, Van Arendonk J, Baker R: Small ruminant production in smallholder and pastoral/extensive farming systems in Kenya. Small Ruminant Res 2008, 77: 11–24.
7. Louhaichi M, Salkini K, Petersen S: International Journal of Agricultural & Biology. Effect of Small Ruminant Grazing on the Plant Community Characteristics of Semiarid Mediterranean Ecosystems. 2009, 6: 681–689.
8. Garden RB: Yearly plant survey. Amman Jordan: Unpublished Report; 2011.
9. Thomson P, Nardone A: Sustainable livestock production: Methodological and ethical challenges. Livestock Production Science 61st edition. 1999, 111–119.

CHAPTER 7

Status of Beef Cattle Production in Argentina Over the Last Decade and Its Prospects

J.C. Guevara[1, 2] and E.G. Grünwaldt[1, 3]

1 Argentinean Institute for Arid Land Research (IADIZA-CONICET), Argentina

2 Faculty of Agricultural Sciences, National University of Cuyo, Argentina

3 Argentinean Institute of Nivology, Glaciology and Environmental Sciences (IANIGLA-CONICET), Argentina

1. INTRODUCTION

Historically, beef cattle production has been one of the traditional activities and an important support to the economic growth of Argentina. This activity led the country to being inserted in the international market as a beef supplier, and placed it in the past as one of the world's largest beef exporters.

During 2001-2010, Argentina devoted an average of 84% of its beef production to the domestic market, on account of which it was exposed more to within-country changes than to international ones; an opposite situation to that of other South American countries where most beef is allocated to global markets.

The increase in soybean planting in Argentina led to its positioning as the crop with the largest planted area. It expanded from less than 40 thousand hectares at the beginning of the 70's [1] to 18.3 million hectares in the 2009/10 crop season [2]. Because of the steady increase in soybean production, cattle are being displaced from traditional production areas in Argentina's Pampa plains to other regions of the country.

In the course of the year 2006, misleading public policies intensified a process of strong intervention to ensure lower beef prices in the domestic market, which affected exports as well as domestic trade.

Although valuable information has been reported by several sources that emphasized different aspects of Argentina's beef cattle production [1, 3, 4, 5, 6, 7], this chapter is based mainly on local sources and it reviews and analyzes the information available on beef cattle: stock and its composition, relationship between cattle stock and human population, extraction rate, domestic consumption, production systems, territorial distribution, meat exports and health status over the period 2001-2010 and prospects. It also analyzes a particular non-traditional case of a province located in the west of the country.

2. CATTLE STOCK AND ITS COMPOSITION

The number of beef cattle in Argentina experienced a sustained rise over the period 2001-2007. Since 2007, a sharp decline is recorded in cattle stock, which by the year 2010 had decreased by nearly 10 million head (Table 1).

Table 1. Cattle stock evolution over the period 2001-2010 (million head) by category

Category	Year									
	2001	2002	2003	2004	2005	2006	2007	2008	2009	2010
Cows	21.9	20.2	22.9	23.1	23.5	24.2	24.4	23.9	22.2	20.5
Calves	10.8	10.7	14.7	14.6	14.5	15.3	15.4	15.4	14.1	12.6
Heifers	7.0	7.2	8.0	8.2	7.9	8.0	8.3	7.9	7.7	6.9
Steers	7.8	8.9	10.5	11.1	11.3	10.9	10.8	10.7	10.6	9.1
Bulls	1.2	1.2	1.2	1.3	1.3	1.3	1.3	1.3	1.2	1.2
Without identification	0.1	0.3	--	--	--	--	--	--	--	--
Total stock	48.8	48.5	57.3	58.3	58.5	59.7	60.2	59.2	55.8	50.3

Source: Own preparation based on [8] Year 2001; [9] Year 2002; [10] Years 2003-2010

The fluctuation of the cattle population in Argentina observed in Table 1 is not something new, because, to just mention an example, in 1977 there occurred the greatest liquidation of beef-cow herds in the country's cattle history, since cattle stock between that year and 1988 fell from 61.1 to 47.1 million head, which represented a 22.9% stock contraction [11].

From November 2005 on, misleading public policies intensified a process of strong intervention to ensure lower prices of beef in the domestic market, affecting exports as well as domestic trade. Some of the measures adopted by the National Government in terms of foreign trade were [11]:

- November 2005, a rise from 5 to 15% in export duties on beef cuts
- February 2006, creation of the Export Operations Register (ROE)
- March 2006, banning of beef exports for 180 days, except Hilton cuts
- May 2006, restriction of exports in the period June-November 2006 to 40% of the volume exported in the same period in 2005
- November 2006, restriction of monthly exports in the period December 2006-May 2007 to 50% of the average monthly volume exported in the period January-December 2005
- May 2008, restriction of exports to 540,000 tons per year

The public policies imposed were effective in the short term in keeping beef prices low in the domestic market, although in the medium term they favoured the process of beef cow liquidation. However, these policies had no influence on the high agricultural profitability, and did not reverse the existing difference between the last one and that from cattle production. As consequence of the government's intervention policy, the live weight price dropped, which reduced the profitability of cattle rearing, causing a strong sell-off of breeding cows, factors that explain the cattle stock decrease.

On the other hand, during the period 2007-2010, the national cattle herd drop was exacerbated by the worst drought in 70 years over 2008-2009 that affected about one third of the farm belt forcing some ranchers to sell off cattle [12].

3. RELATIONSHIP BETWEEN CATTLE STOCK AND HUMAN POPULATION

The human population in the country was 36.3 million and 40.1 million in 2001 and 2010, respectively, which means an increase close to 11%. In turn, the cattle stock increased by only approximately 3% over the same period.

The beef cattle/inhabitant relationship was respectively 1.34, 1.53 and 1.25 for the years 2001, 2007 and 2010 [13, 14, 15]. The number of cattle head per inhabitant over the whole period analyzed here (2001-2010) was notoriously lower than the one the country had in 1952, which was 2.56 [3].

During 2010, Argentina's per capita cattle stock was higher than those in Brazil, Australia, United States of America (USA) and India, and lower than those in Uruguay and Paraguay, some of them competitors of Argentina in the world beef market (Table 2).

Table 2. Cattle population per capita in some of the main beef exporting countries in 2010

Country*	Cattle population (million head)**	Human population (million inhabitants)***	Cattle per capita
Uruguay	11.80	3.36	3.51
Paraguay	12.31	6.5	1.91
Argentina	48.95	40.41	1.21ᶠ
Australia	26.73	22.30	1.20
Brazil	209.54	194.95	1.07
USA	93.88	309.35	0.30
India	210.20	1,224.62	0.17

ᶠdiffers from the 1.25 value previously cited because of variation in the data source.
Source: Own preparation based on *[16]; **[17]; ***[18]

4. CATTLE EXTRACTION RATE

The number of animals slaughtered and the extraction rate in the country over the study period are shown in Table 3. The extraction rate (slaughter/beef cattle stock) was obtained from the stock cited in Table 1. In the year 2010 the extraction rate in the USA (37.6%) and Australia (31.1%) was higher than that in Argentina, whereas that Uruguay (18.6%), Brazil (14.0%), Paraguay (12.2%) and India (5.0%) had lower extraction rate [17, 19].

5. BEEF DOMESTIC CONSUMPTION

For many years, Argentina was the country with the highest per capita meat consumption worldwide. During the period considered, the year 2007 recorded the highest meat consumption, 68.3 kg per capita, and 2010 the lowest, 56.3 kg per capita (Table 4). These figures contrast with the historical maximum recorded in 1956, with 100.8 kg per capita [21].

Table 3. Argentina's cattle extraction rate over the period 2001-2010

Year	Slaughter (million head*)	Extraction rate (%)
2001	11.6	23.8
2002	11.5	23.7
2003	12.5	21.8
2004	14.3	24.5
2005	14.4	24.6
2006	13.4	22.4
2007	15.0	24.9
2008	14.7	24.8
2009	16.1	28.9
2010	11.9	23.7

Source: Own preparation based on *[20]

Resurgence of the demand for beef since 2002 led to a rise in price that allowed a quick recovery of the cattle production profitability, influencing on the price of land in the Pampas cattle-rearing region, which quintupled in value between 2002 and 2008, with US$ values of 377 and 1,950 per hectare, respectively [22].

For a long time, an undisputed paradigm of the beef market in Argentina was the inelasticity of the demand. Because beef is deeply rooted in the diet of the Argentines, a rise in price did not affect the amount of meat demanded, which continued to be strong. The decreasing per capita meat consumption has resulted in Argentina losing the first place in the ranking of countries that most consume meat. This structural change could be explained by an alteration of the factors determining the price-elasticity of beef demand, mainly thanks to availability of substitutes at competitive prices and to a new appraisal regarding the participation of the different products composing the typical diet of the consumers [21].

In relation to availability of substitutes, the great competitor for beef over the last years has been poultry meat. Consumption of poultry meat increased by 34% between 2001 and 2010, with consumption values being respectively 25.7 and 34.5 kg per capita [23, 24]. One kilogram of beef was equivalent to 2.1 and 2.5 kg of chicken in 2001 [25] and 2010 [24], respectively.

Despite meat consumption in Argentina decreased by some 11% in the cited period (Table 4), in 2010 the country continued to be, along with Uruguay with 55.5 kg per capita, the leading beef consumers in the world, compared for instance to the USA with 38.5, Brazil with 37.3, Australia with 35.3 and India with 1.8 kg per capita [26].

Table 4. Per capita beef meat apparent consumption in Argentina over the period 2001-2010

Year	Total apparent consumption (tons carcass weight equivalent*)	Human population**	Per capita consumption (kg)
2001	2,347,819	37,156,195	63.2
2002	2,181,066	37,515,632	58.1
2003	2,280,345	37,869,730	60.2
2004	2,395,806	38,226,051	62.7
2005	2,379,375	38,592,150	61.7
2006	2,475,541	38,970,611	63.5
2007	2,687,746	39,356,383	68.3
2008	2,705,482	39,745,613	68.1
2009	2,715,874	40,134,425	67.7
2010	2,305,917	40,518,951	56.9

Source: Own preparation based on *[27]; ** [14]

6. CATTLE PRODUCTION SYSTEMS

6.1. Classification

Feedlots are excluded in the methodology used in [27] to classify production systems but it takes into account:

a) Cow-calf: ranchers with cows and without steers+yearling steers (17% of total national beef cattle stock)

b) Ranchers with cows, steers and yearling steers. The variable selected to subdivide this stratum was the steer+yearling steer/total cows ratio

b1. Predominantly cow-calf: ratio lower than 0.2; cow-calf and finishing of part of own production (28% of total national beef cattle stock)

b2. Complete cycle: ratio between 0.2 and 0.4; cow-calf and finishing of total or great part of own production (15% of total national beef cattle stock)

b3. Finishing+cow-calf: ratio between 0.4 and 0.8; cow-calf and finishing of own and purchased production (17% of total national beef cattle stock)

b4. Predominantly finishing: ratio higher than 0.8; cow-calf and finishing of own production and purchased production higher than b.3. (19% of total national beef cattle stock)

b5. Finishing: ranchers with steer+yearling steer and without cows (4% of total national beef cattle stock)

Although the classification in [27] does not include feedlots, it must be highlighted that for some time now they have been making an important contribution to cattle production, bringing a change to the traditional cattle system in Argentina, which had been eminently pastoral for years. The feedlot activity was not immune to government intervention, which resulted in fluctuations in use of the available infrastructure. At the beginning of the contribution of state subsidies in 2007, feedlots contributed 14% to total slaughter [1]. From that time on, the number of animals from feedlot destined for slaughter increased (Table 5). During 2010, cattle feedlot occupancy amounted to 56%, a lower figure than the average for 2006-2010, which was 67% [28]. Although feedlots ceased to be subsidized in 2010, they continue to make a relevant contribution to the number of annually slaughtered animals in the country.

Table 5. Participation of feedlots in Argentina beef cattle slaughter

Year	National stock (head)*	Annual slaughter (head)**	Number of feedlot facilities	Slaughter of feedlot cattle	Slaughter from feedlots/annual slaughter (%)
2001	48,851,400	11.586.732		527,700***	4.6
2008	59,261,268	14,660,284	1,653&	3,436,125&	23.4
2009	55,803,147	16,053,031	2,213#	4,991,227#	31.1
2010	50,268,465	11,882,706	2,147#	3,714,557#	31.3

Source: Own preparation based on *[10]; **[20]; ***[8]; #[29]; &[30]

6.2. Description and 2001-2010 cost evolution of some production systems

6.2.1. Pastoral finishing

The technical scheme contemplates 80% of perennial pastures based on alfalfa, 20% of annual winter pastures and alfalfa hay, a stocking rate of 2 heads ha^{-1}, live weight at entry and exit of 180 and 440 kg, respectively, and a fattening cycle of 17.3 months. Costs in the period 2001-2010 were (mean and SD) US$ 451.5 ± 130.4 ha^{-1} equivalent to 521.3 ± 26.8 kg steer^{-1}. Despite in January 2011 the cost increased to 1,046 US$ ha^{-1}, the cost in terms of kg steer-1 showed no substantial variation as consequence of the fact that the price of steers rose as well [31].

6.2.2. Pastoral finishing with supplementation

This system is based on 70% of perennial pastures based on alfalfa, 30% of annual winter pastures and a supplementation (alfalfa hay, maize grain and protein nucleus) and a stocking rate of 3.5 heads ha^{-1}, live weight at entry and exit of 180 and 410 kg, respectively, and a fattening cycle of 13.1 months. The costs in the period 2001-2010 were (mean and SD) US$ 1,050.0 ± 34.7 ha^{-1} equivalent to 1,190.3 ± 39.9 kg $steer^{-1}$. Despite in January 2011 the cost increased to US$ 2,493 ha^{-1}, the cost in terms of kg $steer^{-1}$ increased only to US$ 1,234 as consequence of the fact steer prices were also raised [31].

Both systems are located in the Pampas region (West of Buenos Aires Province and South of Córdoba Province) [31].

6.2.3. Cow-calf production

This system is carried out in rangeland areas of the Pampas region (Southeast of Buenos Aires Province). Cows are fed some alfalfa hay as supplementary food. Calf crop is 80% and stocking rate is 0.5 cow ha^{-1}. Costs in the period 2001-2010 were (mean and SD) US$ 52.8 ± 13.4 ha^{-1} equivalent to 58.1 ± 11.5 kg $calf^{-1}$ [31].

In the three production systems there was a significant linear increase in the price per hectare in the studied period: R^2 Adj.= 0.51, p=0.01 for pastoral finishing, R^2 Adj.= 0.57, p=0.007 for pastoral finishing with supplementation, and R^2 Adj.= 0.35, p=0.04 for cow-calf production.

6.2.4. Gross margin and meat production for some of the production systems

Based on the information in [27] was estimated the gross margin per kg of sold meat for all alternatives of cattle finishing and feedlot shown in Table 6. The mean values were US$ 0.71 and 0.54 for pastoral finishing and feedlot, respectively.

6.2.5. Calf/steer price ratio 2001-2010

Regarding cattle production systems, the purchase-to-sale ratio is a factor that should not be excluded for the analysis because of its effect on their viability. Historic prices (1985-2005) indicate that the calf price has been 10% higher than that of the steer [32]. The average price ratio of Aberdeen Angus calves and steers (Table 7) over the period 2001-2009 concurs with the cited historic value, even though in the years 2001, 2005 and 2006 there were values above that average, but always favoring the calf over the steer. The increased calf-steer ratio in 2010 begins to be a hint of the price ratio that followed later, until exceeding 40%. As of 2010, the highest rise in price for the calf compared to the steer can be explained by a lower supply of the former.

Table 6 .Gross margin (US$) and meat production for some cattle production systems in June 2010

	Systems									
	Cow-calf[1]	Cattle finishing[2]		Complete cycle[3]		Feedlot[4]				
		Without suppl.	With suppl.	Area 1	Area 2	Hotel		Own		
						Cattle purchase		Cattle purchase		
						March	July	March	July	
Gross margin ha^{-1}	88.0	211.2	375.6	150.1	65.9					
Gross margin head^{-1}						37.7	68.7	76.3	113.7	
Meat production ha^{-1} year^{-1}	72.2	278.0	571.0	165.0	55.8					

Source: Own preparation based on [27]
[1]Cow-calf typical area (Cuenca del Salado, Buenos Aires Province)
[2]Western area of Buenos Aires Province
[3]Area 1: Central Southern Córdoba Province; Area 2: Semiarid La Pampa and San Luis Provinces
[4]Hotel: Rent structure and "know-how" offering cattle fattening services; Own: freelance entrepreneur; March and July: calf supply is higher in March than in July
suppl. = supplementation

Table 7. Cattle prices (US$ kg-1) and calf/steer ratio over the period 2001-2010

Year	Aberdeen Angus calf	Steer	Calf/steer ratio
2001	0.96	0.77	1.25
2002	0.50	0.47	1.06
2003	0.66	0.65	1.02
2004	0.73	0.71	1.03
2005	0.99	0.77	1.29
2006	0.92	0.79	1.16
2007	0.94	0.92	1.02
2008	1.08	1.08	1.00
2009	0.98	0.92	1.07
2010	2.18	1.71	1.27

Source: [33]

7. TERRITORIAL DISTRIBUTION OF CATTLE POPULATION

To analyze the territorial distribution of the beef cattle population, the provinces accounting for approximately 95% of the cattle stock were grouped into two zones. The Central-Eastern (CE) zone comprises the provinces historically producing beef cattle (Buenos Aires, Santa Fe, Córdoba, La Pampa and Entre Ríos) which keep the greatest number of cows. The other area, which was called North Eastern (NE) – North Western (NW) zone, to which cattle production displaced over the study period (Table 8) as result of agriculture intensification in the Central Eastern area.

During the period 2003-2010, the CE zone reduced its cattle inventories from 76.6 to 69.5%, whilst the NE-NW zone increased its cattle herd from 18.4 to 25.3%. In the CE zone, the Provinces with higher stock losses were La Pampa, Córdoba and Buenos Aires with 38.4, 28.1 and 20.5%, respectively. Moreover, in the NE-NW zone, Misiones and Salta provinces increased their cattle stock by 69.8 and 86.8%, respectively.

Table 8. Cattle displacement by zones over the period 2002-2010 (million head)

Zone and Province	Year								
	2002	2003	2004	2005	2006	2007	2008	2009	2010
Central Eastern (CE)									
Buenos Aires	16.61	21.07	21.79	21.56	21.58	21.50	20.39	18.87	16.74
Córdoba	6.10	7.02	6.83	6.59	6.60	6.42	5.93	5.69	5.05
Santa Fe	6.15	6.96	7.14	7.25	7.45	7.70	7.59	7.07	6.44
Entre Ríos	3.81	4.61	4.67	4.64	4.87	4.78	4.78	4.63	4.14
La Pampa	3.69	4.19	4.00	4.04	4.06	4.02	3.85	3.18	2.58
Subtotal	36.36	43.86	44.42	44.09	44.55	44.42	42.53	39.44	34.95
CE/total (%)	74.9	76.6	76.3	75.4	74.6	73.8	71.8	70.7	69.5
North Eastern (NE) and North Western (NW)									
Corrientes	3.61	4.42	4.58	4.68	5.07	5.20	5.61	5.35	5.07
Chaco	1.98	2.20	2.45	2.37	2.49	2.56	2.76	2.66	2.42
Formosa	1.34	1.40	1.46	1.67	1.70	1.68	1.80	1.79	1.75
Santiago del Estero	1.04	1.16	1.23	1.22	1.25	1.37	1.51	1.53	1.40
Salta	0.49	0.56	0.62	0.69	0.76	0.88	0.97	1.02	1.05
Misiones	0.35	0.22	0.27	0.31	0.31	0.34	0.37	0.38	0.38
La Rioja	0.25	0.22	0.20	0.19	0.19	0.18	0.20	0.21	0.21
Catamarca	0.23	0.19	0.20	0.22	0.21	0.24	0.24	0.24	0.22
Tucumán	0.10	0.12	0.10	0.12	0.14	0.15	0.14	0.15	0.15
Jujuy	0.09	0.07	0.07	0.07	0.07	0.06	0.04	0.06	0.05
Subtotal	9.49	10.56	11.17	11.54	12.19	12.65	13.65	13.39	12.70
(NE) and (NW)/total (%)	19.6	18.4	19.2	19.8	20.4	21.0	23.0	24.0	25.3
Total country	48.54	57.25	58.24	58.44	59.72	60.17	59.26	55.80	50.27

Source: Own preparation based on [8, 10]

In spite of the fact that the NE-NW zone has increased its cattle herd, this increase does not compensate for the loss suffered by the CE zone and cannot be explained by territorial space because their land areas are comparable in size, 828.1 and 849.6 thousand km^2 respectively for the CE and NE-NW zones. The reason is that the conditions of production are not equal in different aspects that influence on production efficiency such as infrastructure, health, food, among others. Evidence to this is the calf-cow ratio during 2010, 65.8 and 51.8%, respectively, for CE and NE-NW zones (Table 9). This implies 14 less calves every 100 cows that have been displaced from the CE to the NE-NW zone if it is assumed that the calf-cow ratio is a variable that approximates the weaning index.

Table 9. Calf-cow ratio by zones and Provinces in 2003 and 2010

Zone and Province	Cows		Calves		Calf/cow (%)	
	2003	2010	2003	2010	2003	2010
Central Eastern (CE)						
Buenos Aires	8,459,775	7,004,706	6,362,850	5,199,002	75.2	74.2
Córdoba	2,576,079	1,958,439	1,574,957	1,151,538	61.1	58.8
Santa Fe	2,566,576	2,407,153	1,507,229	1,384,517	58.7	57.5
Entre Ríos	1,813,823	1,641,825	1,124,019	935,557	62.0	57.0
La Pampa	1,478,333	896,702	970,841	478,737	65.7	53.4
Subtotal	16,894,586	13,908,825	11,539,896	9,149,351	68.3	65.8
CE/total (%)	73.9	67.9	78.4	72.6		
North Eastern (NE) and North Western (NW)						
Corrientes	2,076,142	2,256,967	955,591	1,054,629	46.0	46.7
Chaco	926,542	1,019,506	508,366	555,558	54.9	54.5
Formosa	610,463	737,928	336,740	385,175	55.2	52.2
Santiago del Estero	471,551	544,312	277,480	318,355	58.8	58.5
Salta	224,425	390,423	117,139	236,767	52.2	60.6
Misiones	103,762	153,492	36,462	76,307	35.1	49.7
La Rioja	103,323	110,155	59,501	59,929	57.6	54.4
Catamarca	80,986	93,199	39,962	56,965	49.3	61.1
Tucumán	48,890	58,085	25,967	34,916	53.1	60.1
Jujuy	28,640	20,391	16,024	11,360	55.9	55.7
Subtotal	4,674,724	5,384,458	2,373,232	2,789,961	50.8	51.8
NE and NW/total (%)	20.4	26.3	16.1	22.2		
Total country	22,864,159	20.469.240	14,719,545	12,595,096	64.4	61.5

Source: Own preparation based on [10]

Fluctuation in the stock also involves the slaughter of female cattle, which varied among 42, 46, 42, 49 and a little more than 43% for the years 2001, 2004, 2006, 2009 and 2010, respectively considering that the limit value for maintaining the stock is about 43% [1]. Between 2009 and 2010 the number of cows and heifers continued to fall, which makes future restocking difficult. Table 10 illustrates the evolution of cows stock over the study period and its territorial distribution by zone.

Thus, the aforementioned is reinforced by the cow loss in the country that occurred between 2007 and 2010, which amounted to 3,883,266 cows, with the provinces losing the highest number of cows being Buenos Aires and La Pampa, both making up 63.4% of the total loss (2,460,086 cows).

8. CATTLE MEAT EXPORTS

In analyzing Argentina's beef exports over the decade, it is observed that both ends of the decade, years 2001 and 2010, the lowest export values in tons carcass weight equivalent were recorded (Table 11). Thus, exports were of 150,025 and 308,663 tons respectively for 2001 and 2010. The export average of the decade was of 468,439 tons with two peak export values being recorded in 2005 and 2009. Exports in 2005 and 2001 were the highest and lowest export values recorded since 1934 [34] with 771,942 tons for 2005. The drop in exports in 2001 occurred in a context of foreign markets closed by an outbreak of Foot and Mouth Disease (FMD). The increase in foreign sales, mostly between 2002 and 2005, was due to favorable conditions as result of increased international prices and lower worldwide supply because of animal health problems in some of the major beef exporting countries.

The export value, as percentage of cattle meat annual production, during 2010 was lower than those of Australia, Canada, India and Brazil that allocated 65.5, 41.1, 32.3 and 17.1%, respectively, but higher than those of USA, Mexico and European Union (EU), with 8.7, 5.9 and 4.2%, respectively [35].

Cattle meat exports by product category and their value over the period 2001-2010 and Argentina beef exports, and their value and destination countries during 2010 are presented in Table 12 and Table 13, respectively. Except for 2001, the greatest export volumes correspond to chilled and frozen meat (Table 12). The decline in sales during 2010 is indicative of the loss of presence of Argentina cattle meat in the international beef market. In 2010, the primary destination of non-Hilton chilled cuts and frozen meat was Russia, whereas the major purchaser of Hilton cuts was Germany (Table 13).

The Hilton Quota is an export quota of high-quality high-value boneless beef cuts that the EU grants to beef producing and exporting countries. Argentina is the country having the highest percentage of this quota, with 28,000 tons year-1 in 2010. Other supplying countries are Brazil, Uruguay, Paraguay, USA, Canada, Australia and New Zealand. Beef cuts included in the quota are rump and loin, strip loin, rump, tender loin, silver side, top side and knuckle. The 25,639 tons of Hilton cuts exported in 2010 did not fulfill the quota allotted for that year.

Table 10. Beef cow stock by zones over the period 2002-2010 (million head)

Zone and Province	Year								
	2002	2003	2004	2005	2006	2007	2008	2009	2010
Central Eastern (CE)									
Buenos Aires	7.08	8.46	8.75	8.79	8.94	8.91	8.47	7.75	7.00
Córdoba	2.22	2.58	2.49	2.44	2.46	2.40	2.27	2.15	1.96
Santa Fe	2.33	2.57	2.59	2.65	2.75	2.84	2.82	2.54	2.41
Entre Ríos	1.62	1.81	1.81	1.84	1.92	1.90	1.85	1.76	1.64
La Pampa	1.33	1.48	1.40	1.45	1.47	1.45	1.41	1.14	0.90
Subtotal	14.58	16,89	17.04	17.16	17.54	17.50	16.81	15.34	13.91
CE/total (%)	72.2	73.9	73.8	73.1	72.6	71.9	70.4	69.0	67.9
North Eastern (NE) and North Western (NW)									
Corrientes	1.80	2.08	2.11	2.15	2.28	2.35	2.46	2.36	2.26
Chaco	0.87	0.93	0.99	0.97	1.03	1.07	1.14	1.11	1.02
Formosa	0.59	0.61	0.63	0.71	0.72	0.73	0.76	0.75	0.74
Santiago del Estero	0.42	0.47	0.50	0.50	0.51	0.54	0.59	0.59	0.54
Salta	0.20	0.22	0.24	0.26	0.29	0.33	0.35	0.36	0.39
Misiones	0.13	0.10	0.11	0.14	0.14	0.14	0.15	0.15	0.15
La Rioja	0.12	0.10	0.10	0.10	0.08	0.09	0.08	0.09	0.11
Catamarca	0.09	0.08	0.08	0.09	0.09	0.10	0.10	0.10	0.09
Tucumán	0.04	0.05	0.04	0.05	0.06	0.06	0.06	0.06	0.06
Jujuy	0.03	0.03	0.03	0.03	0.03	0.03	0.02	0.03	0.02
Subtotal	4.29	4.67	4.83	5.00	5.23	5.45	5.70	5.60	5.38
NE-NW/total (%)	21.3	20.4	20.9	21.3	21.6	22.4	23.9	25.2	26.1
Total country	20.18	22.86	23.08	23.47	24.16	24.35	23.88	22.23	20.47

Source: Own preparation based on [8, 10]

Table 11. Production in tons of carcass weight equivalent, apparent consumption and exports of beef cattle meat over the period 2001-2010

Year	Production	Consumption (%)	Exports (%)
2001	2,500,418	94.0	6.0
2002	2,532,207	86.1	13.9
2003	2,672,328	85.3	14.7
2004	3,026,836	79.2	20.8
2005	3,150,784	75.5	24.5
2006	3,040,598	81.4	18.6
2007	3,226,757	83.3	16.7
2008	3,134,482	86.3	13.7
2009	3,377,252	80.4	19.6
2010	2,615,791	88.2	11.8
Mean 2001-2010	2,927,745.3	84.0	16.0

Source: [27]

Table 12. Cattle meat exports by product category and their value over the period 2001-2010

Year	Chilled and frozen meat (tons)	Processed meat (tons)	Total meat exports (tons)	FOB value (thousand US$)
2001	38,414	39,028	77,442	215,733
2002	158,321	46,497	204,818	452,735
2003	184,118	46,919	231,037	577,206
2004	322,713	59,885	382,598	972,522
2005	432,653	50,494	483,147	1,294,966
2006	316,504	37,471	353,975	1,199,889
2007	296,592	38,927	335,519	1,281,042
2008	229,991	34,920	264,911	1,486,335
2009	383,501	35,836	419,337	1,652,731
2010	166,265	25,494	191,759	1,187,454

Source: Own preparation based on [27]

Table 13. Destination countries of Argentina's cattle meat exports and their value in the year 2010

Country	Exports (tons)	(Value (thousand US$)	Value (US$ per ton)
	Non-Hilton chilled cuts and frozen meat		
Russia	35,678	119,785	3,357
Israel	26,558	132,343	4,983
Chile	18,007	89,222	4,955
Venezuela	11,762	54,796	4,659
Germany	10,325	110,558	10,708
Others	38,296	236,024	6,163
Total	140,626	742,728	5,282
	Hilton chilled cuts		
Germany	14,776	194,145	13,139
Netherlands	5,625	71,180	12,654
Italy	4,255	55,300	12,996
Spain	495	6,280	12,687
Others	488	6,293	12,895
Total	25,639	333,198	12,996
	Processed meat		
Total	25,494	111,528	4,375
Total Argentina	191,759	1,187,454	6,192

Source: Own preparation based on [27]

9. CATTLE HEALTH STATE

Foot and Mouth Disease, the last outbreak of which occurred in February 2006, is the disease of greatest economic importance because it hampers exports to FMD–free circuits. The Agri-food Health and Quality National Service (SENASA) has implemented health control programs against different pathologies and has taken actions towards preventing the entry of exotic diseases. Most important among them to the international market for animals and their byproducts are Bovine Spongiform Encephalopathy (BSE) and Scrapie. According to World Organization for Animal Health (OIE), Argentina is a member recognized as having a negligible BSE risk and as being an FMD-free zone where vaccination is not practiced to the south of Parallel 42° and an FMD-free zone with vaccination in the rest of the country [36]. Also SENASA has implemented health control programs against other diseases which are not restrictive on beef exports but do limit herd productivity, such as Brucellosis and Tuberculosis.

10. A NON-TRADITIONAL CATTLE PRODUCTION CASE

Mendoza lies in the West Central of the country with 148,827 km^2. An important portion of Mendoza falls within the Central Eastern part of the Monte Phytogeographic Province, the most arid rangeland of the country. Of the total surface of Mendoza, around 9 million hectares could be devoted to cow-calf production systems and about 450,000 ha are under irrigation, of which 75,000 ha are uncultivated at present. Cow–calf operations under rangeland conditions are the dominant production system [37]. The steer+yearling steer/cows ratio ranged from 0.08 to 0.16 in 2002 and 2010, respectively.

Mendoza has beef cattle but not in enough numbers to be considered a high cattle-producing province. Proof of this is the beef cattle stock/human population ratio, 0.25 and 0.31 for 2002 and 2010 respectively, values far below the 1.21 ratio for the country during 2010. The 2002-2010 evolution of bovine stock and human population was 404,710 to 533,488 heads [38] and 1,595,448 to 1,738,929 people, respectively [39].

The displacement of beef cattle production as consequence of the advance of agriculture is a process of no return. This situation led to the Argentinean Institute for Arid Land Research (IADIZA) to analyze the possibility of developing a non-traditional cattle activity in Mendoza, a province where competition with agricultural activities is high. At present, 10% of the animals consumed in Mendoza are finished locally, and hence the aim arises to enhance the production of steers for increasing the local supply. This framework promoted several investigations for analyzing the profitability of different production systems on cultivated pastures such as [40] beef cattle post-weaning, feedlot [41] and early weaning of calves combined with post-weaning production [37].

11. PROSPECTS

Based on the analysis of the information developed here and on the opinion of leading specialists [1, 4, 5, 42] in the topic addressed, the need is highlighted for the growth of Argentina's beef cattle production to fulfill the needs of both domestic consumption and exports. Evolution of the stock through retention of females is slow, somewhat faster by keeping cow culling rates low, although a better result is obtained by improving weaning. For this reason it is indispensable to improve this index as soon as possible.

Different hypothetical projections can be made in relation to cattle stock and beef production. One scenario that would allow achieving a production of 3.31 million tons of beef by 2020, with higher export surpluses and possibilities of expanding domestic consumption, should consider national weaning indices of 65%, a 25% of regional production rate, a 25% of cattle extraction rate, a 77% of retention of females, and a carcass weight of 225 kg. This would provide an export surplus of 800,000 tons, similar to that exported by Argentina in 2005.

Increased beef demand from developing countries has significantly impacted on the rise of international prices. Upon the basis of a scenario of low supply, a domestic demand that has validated current price levels and an international demand that has approve prices that were unimaginable a few years back (US$ 2,680 and 6,192 per ton carcass weight equivalent in 2005 and 2010 respectively) it is possible to project sustained cattle prices for the next 3 to 4 years and, therefore, cattle production systems would continue to be profitable.

The input/output ratio is going through a propitious time, with values well above those of previous years. Thus, for instance, in December 2001, purchasing a tractor of 100 HP required 50,393 kg of steer and 44,724 kg of calf, whereas in July 2010 these numbers had dropped to 23,937 and 19,552 respectively. On the other hand, the amounts needed to buy a pickup truck on the mentioned dates were 25,352 and 14,545 kg of steer and 22,500 and 11,881 kg of calf [43]. Therefore, with a scenario of good cattle prices ahead for the coming years, it is advantageous to now make investments and adopt technology so that, from the production standpoint, cattle farms are provided with a more solid platform for their growth.

A high-impact factor for economic results in production models is the relationship between steer sale price and calf purchase price. With a buying/selling ratio exceeding 25-30%, as has occurred in 2010 (29.5%), it is convenient to add more kilograms to the animals at finishing stages, which would diminish the impact of the above ratio. However, the limitation to this higher weight per head is given by a lower sale price for heavier animals, which is directly related to the purchasing power of exporters.

Because pastoral finishing is more profitable that feedlot, in terms of gross margin per kg of meat sold, is expected that the number of pasture-finished cattle destined for slaughter increase in the future.

International prices are excellent and there is an unfulfilled demand. Notwithstanding, export meat processing plants have a limited purchasing power due to the export-restricting policies. These policies have been wrong because the countries with high purchasing power consume the highest-value cuts. If it were possible to export these cuts at high prices, the lower value cuts could be destined for domestic consumption.

Argentina has gradually lost its place in the world markets. However, the country still has possibilities of recovery its position because the world will run short of meat due to deceleration of the production processes in Europe and a growing demand from countries like China and India. Russia needs increasing amounts of imported beef. The only reservoirs for production of red meat are in Latin America. This turned Brazil, despite not having high quality meat, into the major exporter worldwide, whilst Uruguay earned a place by exporting 63% of its production. The strategy is focused on the EU potential demand, on the important volumes imported into Russia and on the possibility of entering NAFTA, but it must be considered that the strategy for the next decade is toward Japan and Korea.

At the XVIII Meat World Congress held in Buenos Aires in September 2010, it was stated that food production will increase strongly over the next years and that the Argentina beef-producing sector has the necessary technology to improve efficiency and increase cattle stock. Nonetheless, some of the challenges facing the Argentina meat chain are the expansion of agriculture at the expense of grazing land devoted to cattle farming and the need to set up a sustainable system including all facets: economic, social, health and animal welfare. The future of the meat market of Mercosur has the challenge to coordinate the combat of FMD disease and remove the barriers for enhancing production [27], highlighting the possibility and strength it represents for Argentina to be close to being an FMD-free country without vaccination.

At global scale, raising awareness, compromise, joint action, cooperation, a long term vision, predictability and equity are some of the concepts that will determine the magnitude of the change and its success. Argentina will have to take the challenge, to not miss opportunities, to be part of the change and become established as a major player in supplying high quality meat to the world [27].

While all links in the chain should be adjusted to increase meat production, it is evident that production, and rearing within it, is the limitation to be solved. Increased calf production will have to come from improved production efficiency of the already existing herd. It is likely that the cattle area continues to decrease, but this should not be viewed as an obstacle to the sector's growth.

In brief, the displacement of cattle population to marginal lands and reduction of stock numbers are some of the changes in beef cattle production occurred over the last decade. Specific policies are needed to increase the cattle production in view of this new frame of situation.

ACKNOWLEDGEMENT

The authors wish to express their thanks to Nélida Horak for assisting with the English version and to Guillermo F. Grünwaldt for designing tables and formatting the manuscript.

REFERENCES

1. Charvay P (2011) Los Cambios en la Producción Ganadera en la Posconvertibilidad. La Expansión Sojera y su Impacto sobre la Ganadería. Available: http://www.vocesenelfenix.com/sites/default/files/pdf/03_3.pdf.
2. MAGYP (2011) Informe Semanal al 03/02/11. Avance de la Siembra de Soja. Available: http://www.siia.gov.ar/estimaciones_agricolas/01 semanal/_archivo//110000_2011//000200_Febrero//110203%20Informe%20semanal %2003 -Feb-11.pdf. Accessed 2012 May 8.
3. Melo O, Soetto C, Gómez Demmel A (2008) Análisis de la Ganadería Bovina de Carne Argentina. Available: http://www.produccion-animal.com.ar/informacion_tecnica/origenes_evolucion_y_ estadisticas_de_la_ganaderia/50-analisis.pdf.
4. Pacífico C (2010) La Recomposición de la Ganadería Vacuna Argentina. Una Visión por Regiones. Available: http://www.publitec.com/contenido/objetos/ Larecompo-sicin-delaganaderavacunaargent ina.pdf.
5. Rearte D. (2010) Situación actual y prospectivas de la producción de carne vacuna. Available: http://inta.gob.ar/documentos/situacion-actual-y-prospectiva-de-la-produ-ccion-de-carne-vacuna/. Accessed 2011 Dic 7.
6. Arelovich H M, Bravo R D, Martínez M F (2011) Development, Characteristics, and Trends for Beef Cattle Production in Argentina. Available: http://animalfrontiers.org/content/1/2/37.full.pdf+html.

7. Milano R (2011) El Nuevo Escenario de la Ganadería en Argentina. Available: http://www.bcr.com.ar/Secretara%20de%20Cultura/Revista%20Institucional/2011/Ag ost o/GANADERIA.pdf.
8. INDEC (2001) Encuesta Nacional Agropecuaria 2001. Available: http://www.indec.gov.ar/nuevaweb/cuadros/11/ena_02_02.pdf. Accessed 2011 Nov 25.
9. INDEC (2002) Censo Nacional Agropecuario 2002. Available: http://www.indec.gov.ar/agropecuario/cna_principal.asp.
10. SENASA (2012) Datos de las Campañas de Vacunación Antiaftosa. Available: http://www.senasa.gov.ar/contenido.php?to=n&in=673&io=7320.
11. IERAL (2011) Una Argentina Competitiva, Productiva y Federal. Cadena de la Carne Bovina. Available: http://www.ieral.org/images_db/noticias_archivos/1861.pdf.
12. Beef Production and Market Situation in Argentina (2010). Available: http://www.argentine-embassyuk. org/docs/economia_comercio/files/networking/erc.pdf.
13. INDEC (2001) Censo Nacional de Población, Hogares y Viviendas en la Argentina. Available: http://www.indec.gov.ar/censo2001S2/ampliada_index.asp?mode=01. Accessed 2012 May 8. Status of Beef Cattle Production in Argentina Over the Last Decade and Its Prospects 133
14. INDEC (2004) Estimaciones y Proyecciones de Población. Total del País 1950 – 2015. Available: http://www.indec.gov.ar/nuevaweb/cuadros/2/proyecyestimaciones_1950-2015.pdf.
15. INDEC (2010) Censo Nacional de Población, Hogares y Viviendas en la Argentina 2010. Available: http://www.indec.gov.ar/.
16. Kearney A T (2010) Diagnóstico del Sector en el Mundo y Punto de Partida y Diagnóstico del Sector en Colombia. Sector: Carne Bovina. Available: http://www.minagricultura.gov.co/archivos/Plan_carne_bovina.pdf.
17. Organización de las Naciones Unidas para la Alimentación y la Agricultura (2012) FAOSTAT. Available: http://faostat.fao.org/site/573/default.aspx#ancor.
18. El Banco Mundial (2011) Población Total. Available: http://datos.bancomundial.org/indicador/SP.POP.TOTL.
19. Organización de las Naciones Unidas para la Alimentación y la Agricultura (2012) FAOSTAT. Available: http://faostat.fao.org/site/569/default.aspx#ancor.
20. ONCCA (2011) Serie Histórica de Faena Bovina 1998-2011. Available: http://www.oncca.gov.ar/principal.php?nvx_ver=6605&m=463.
21. Bolsa de Comercio de Rosario (2011) Cambios en el Comportamiento del Consumidor Plantean un Nuevo Escenario para la Ganadería Bovina. Available: http://www.bcr.com.ar/Publicaciones/Informativo%20semanal/bcr2011_08_19.pdf.
22. Márgenes Agropecuarios (2012) El Valor de la Tierra en la Pradera Pampeana 321: 39.
23. MAGYP (2010) Indicadores de la Actividad Avícola – Series Históricas. Available: http://64.76.123.202/site/ganaderia/aves/01-Estad%C3%ADsticas/_archivos/000002_Indicadores/000002_indicadores%20(hist%C3% B3ricos).pdf.
24. MAGYP (2012) Indicadores de la Actividad Avícola. Available: http://64.76.123.202/site/ganaderia/aves/01- Estad%C3%ADsticas/_archivos/000002_Indicadores/000001_indicadores%20(actuales).pdf.
25. Iriarte I (2008) Comercialización de Ganados y Carnes. Available: http://www.cacg.org.ar/comercio25/html/18067capitulo11.pdf.
26. USDA (2011) Livestock and Poultry: World Markets and Trade. Available: http://www.fas.usda.gov/dlp/circular/2010/livestock_poultryfull101510.pdf.
27. MAGYP (2010) Ganados y Carnes Anuario 2010. Available: http://64.76.123.202/site/ganaderia/anuario/index.php.
28. Cámara Argentina de Feedlot (2011) Informe Mensual del Sector Feedlot. Available: http://www.feedlot.com.ar/sitio/wp-content/uploads/Inf-CAF-Diciembre-2011.pdf. Accessed 2012 Mar 6.
29. SENASA (2011) Movimientos de Ganado Bovino Año 2010. Available: http://www.senasa.gov.ar/Archivos/File/File5146-inf-esta-18.pdf.

30. SENASA (2010) Movimiento de Ganado Bovino Año 2009. Available: http://www.senasa.gov.ar/Archivos/File/File3851-File3851-mov-ganado-bovino- 2009.pdf.
31. Arbolave F (2011) Costos Ganaderos 2011. Márgenes Agropecuarios. Suplemento Ganadero: 16-19.
32. Halle A G (2009) Ganadería; Análisis de Coyuntura Junio 2009. Available: http://www.econoagro.com/downloads/act_gan_06_09.pdf.
33. Márgenes Agropecuarios (2011) Relaciones Producto/Producto. Precio de la Hacienda. Relaciones Ganaderas. Suplemento Ganadero: 28.
34. IPCVA (2011) Argentina. Exportaciones de Carne Vacuna. Available: http://www.ipcva.com.ar/documentos/1035_informemensualdeexportacionesdiciembre 2011.pdf.
35. USDA (2011) Livestock and Poultry: World Markets and Trade. Available: http://www.fas.usda.gov/psdonline/circulars/livestock_poultry.pdf.
36. Guevara J C, Bertiller M B, Estevez O R, Grünwaldt E G, Allegretti, L I (2006) Pastizales y Producción Animal en las Zonas Áridas de Argentina. Sécheresse. 17 (1-2): 242-256. Available: http://www.jle.com/en/revues/agro_biotech/sec/e-docs/00/04/1F/18/article.phtml
37. Guevara J C, Grünwaldt E G (2012) The Desert Environment of Mendoza, Argentina: Status and Prospects for Sustainable Beef Cattle Production. In: Guevara J C, Grünwaldt E G, Sivaperuman C, editors. Desert: Fauna, Flora and Environment. New York: Nova Science Publishers, Inc. pp. 115-127. Available: https://www.novapublishers.com/catalog/product_info.php?products_id=26023
38. MAGYP (2011) Sistema Integrado de Información Agropecuaria. Available: http://www.siia.gov.ar/index.php/series-por-tema/ganaderia.
39. Dirección de Estadísticas e Investigaciones Económicas Mendoza (2011). Available: http://www.deie.mendoza.gov.ar/tematicas/menu_tematicas.asp?filtro=Censo%20Nacional%20de%20Poblaci%F3n.
40. Guevara J C, Grünwaldt E G, Bifaretti A. (2010) Determinación de la rentabilidad de la recría de bovinos de carne en la provincia de Mendoza, Argentina. Revista de la Facultad de Ciencias Agrarias, Universidad Nacional de Cuyo. 42 (2): 23-37. Available: http://revista.fca.uncu.edu.ar/images/stories/pdfs/2010-02/T42_2_03_Guevara.pdf
41. Grünwaldt E G, Guevara J C (2011) Rentabilidad del engorde a corral de bovinos de carne en la provincia de Mendoza, Argentina. Revista de la Facultad de Ciencias Agrarias, Universidad Nacional de Cuyo. 43 (2): 21-34. Available: http://bdigital.uncu.edu.ar/objetos_digitales/4310/t43-2-02-grunwaldt-guevara.pdf
42. Tonelli V (2011) La Ganadería de los Próximos 4 Años. Márgenes Agropecuarios. Suplemento Ganadero: 20-22.
43. Márgenes Agropecuarios (2011) Relaciones Insumo/Producto. Suplemento Ganadero: 50.

CHAPTER 8

Sesame oil in diets for lambari: Effects on growth parameters, corporal chemical composition and physiological alterations

Mariene Miyoko Natori[1], Rachel Cristina Prehl Alves[1], Ricardo Henrique Franco de Oliveira[2], Julio Guerra Segura[1] Elisabete Maria Macedo Viegas[1*]

[1]*Departamento de Zootecnia, Faculdade de Zootecnia e Engenharia de Alimentos, Universidade de São Paulo (USP), 13635-900, Pirassununga, SP, Brasil.*

[2]*Departamento de Ciências Básicas, Faculdade de Zootecnia e Engenharia de Alimentos, Universidade de São Paulo (USP), Pirassununga, SP, Brasil.*

ABSTRACT

The inclusion of sesame oil associated with soybean, linseed and freshwater fish residue oil in the diets fed to Lambaris ***Astyanax altiparanae*** was evaluated by the growth performance parameters, body composition and possible physiological changes (GARUTTI & BRISTSKI, 2000). The experiment was a completely randomized design in two factorial parameters tested: three oil types (soy oil (SO), linseed oil (LO) and freshwater fish residue oil (FRO)), combined or not with sesame oil (SEO), totalizing six treatments and four replications 24 cages, capacity of 0.70m^3 (density of 251 fish m^{-3}). The fish (mean weight 2.35g±0.62g and mean length 5.25cm±0.68cm) were fed with the experimental diets twice a day. After 75 days, the following parameters were determined: body chemical composition and fatty acid profile, glycemia, liver (LG) and muscle glycogen (MG) levels, and lipid peroxidation (TBARS). Fish fed with diets containing SEO had higher desaturation index values of LNA while those fed with SEO combined with LO displayed reduced hepatic lipid oxidation. Inclusion of SEO improved the fatty acid profile and stability, without causing problems related to fish performance and health.

Keywords: Astyanax altiparanae; fatty acids; nutrition; sesamin.

1. INTRODUCTION

The interest for vegetal oils as an alternative lipid source has increased (NAYLOR et al., 2000) because the inclusion of fishmeal and fish oil in the diet fed to fish led to increasing ration prices. However, it has been reported that in some marine fish species, the substitution of these ingredients can interfere with growth parameters, cause problems related to fish metabolism and consequently, the oxidative homeostasis, promoting the appearance of reactive oxygen species (ROS) (OLSVIK et al., 2011).

In tropical fish species such as Nile tilapia ***Oreochromis niloticus*** , the linolenic acid (LNA) included in the diets is converted to high unsaturated fatty acid (HUFA) such as the docosahexanoic acid (DHA). The LNA addition decreased the linoleic acid (LA) desaturation to HUFA n-6 due to the high affinity of the enzymes involved in desaturation and elongation processes with the n-3 fatty acids (CHEN et al., 2013). Likewise, the lambari ***Astyanax altiparanae*** (Garutti & Bristski, 2000) is also able to convert LNA into eicosapentaenoic acid (EPA) and DHA, and LA into arachidonic acid (AA) (GONÇALVES et al., 2012; CAMPELO et al., 2014).

The therapeutic properties of sesame oil (SEO) have been verified in experiments with rats. The SEO-containing diets reduced glycemia, blood cholesterol and triglyceride levels and prevented pathological problems like formation of hepatic steatosis, and renal sepsis caused by lipopolysaccharide (LPS) (HSU et al., 2005; GUIMARÃES & MACEDO, 2013; PERIASAMY et al., 2014). Moreover, the lignans contained in SEO such as sesamin, episamin and sesamolin, may promote the formation of fatty acids as the DHA, from the desaturation and elongation of LNA (TRATTNER et al., 2008a; KÖSE & YILDIZ, 2013).

Therefore, this study evaluated how the inclusion of sesame oil associated with soy oil (SO), linseed oil (LO) or FRO (freshwater fish residue oil) in diets fed to lambari ***Astyanax altiparanae*** (Garutti & Bristski, 2000) affected the growth performance parameters, body chemical composition and determined possible physiological alterations.

2. MATERIALS AND METHODS

Table 1 shows the formulation (COTAN et al., 2006) and composition of the extruded experimental diets (A.O.A.C, 1990), and lipid profiles (A.O.A.C, 2005) are shown in table 2. During the experimental period of 75 days, 4,224 fish (mean weight 0.95g±0.46g and mean length 4.21cm±2.77cm) were distributed in 24 cages (capacity 0.70m^3, density 251 fishes m^{-3}). Fish were fed *ad libitum* twice a day (at 9:00 and 16:00). Water quality parameters such as temperature, dissolved oxygen rate, and pH were measured using the Multi-Parameter Water Quality Monitoring Meter (Horiba U-10), on alternate days. Ammonia and nitrite levels were monitored using kits and meters from Hanna Instruments, United States.

Table 1 Formulation and composition of experimental diets.

Ingredients (%)	SO*	LO*	FRO*	SO/SEO*	LO/SEO*	FRO/SEO*
Corn meal	26.71	26.71	26.71	26.71	26.71	26.71
Viscera meal	12.00	12.00	12.00	12.00	12.00	12.00
Soy meal	13.73	13.73	13.73	13.73	13.73	13.73
Wheat meal	22.00	22.00	22.00	22.00	22.00	22.00
Sugar cane yeast	0.50	0.50	0.50	0.50	0.50	0.50
Rice meal	8.00	8.00	8.00	8.00	8.00	8.00
Meat meal	7.00	7.00	7.00	7.00	7.00	7.00
Fish meal	5.50	5.50	5.50	5.50	5.50	5.50
Dicalcium phosphate	0.41	0.41	0.41	0.41	0.41	0.41
Soy oil (SO)	3.00	-	-	1.50	-	-
Linseed oil (LO)	-	3.00	-	-	1.50	-
Fleshwater fish residue oil (FRO)	-		3.00	-	-	1.50
Sesame oil (SEO)	-	-	-	1.50	1.50	1.50
Choline chloride	0.10	0.10	0.10	0.10	0.10	0.10
L-Lysine	0.11	0.11	0.11	0.11	0.11	0.11
DL-Methionine	0.13	0.13	0.13	0.13	0.13	0.13
Antioxidant	0.01	0.01	0.01	0.01	0.01	0.01
Antifungal	0.10	0.10	0.10	0.10	0.10	0.10
Vitamin and mineral supplement[1]	0.50	0.50	0.50	0.50	0.50	0.50
Common salt	0.20	0.20	0.20	0.20	0.20	0.20
Chemical Composition (%)						
Dry matter	94.19	94.35	95.70	95.54	94.78	94.06
Crude protein	29.39	29.65	29.09	29.92	29.29	29.28
Crude energy (kcal kg^{-1})	4375	4365	4395	4365	4385	4414
Ether extract	9.17	9.24	9.13	9.20	9.29	9.24
Mineral matter	10.18	10.35	10.15	10.02	10.00	10.18

*Treatments: Soy oil (SO 3%), linseed oil (LO 3%), freshwater fish residue oil (FRO 3%), soy oil and sesame oil (SO/SEO, 1.5%/1.5%), linseed oil and sesame oil (LO/SEO, 1.5%/1.5%), freshwater fish residue oil and sesame oil (FRO/SEO, 1.5%/1.5%). [1]Vitamin and mineral supplement (0.50g/100g): vitamin A, 12,000UI; vitamin D3, 3,000UI; Vitamin E, 150mg; Vitamin K3, 15mg; Vitamin B2, 20mg; Vitamin B6, 17.50mg; Vitamin B12, 40mcg; Vitamin C, 300mg; Nicotinic acid, 100mg; Calcium pantotenate, 50mg; Biotin, 1.00mg; Folic acid, 6mg; Sulphate Copper, 17.50mg; Iron sulphate, 100mg; Manganese sulfate, 50mg; Zinc sulfate, 120mg; Calcium iodide, 0,80mg; Sodium Sulfate, 0.50mg; Cobalt sulfate, 0,40mg; Inositol 125mg; Choline Chloride, 500mg; vehicle.

Fish were weighed on days 0, 35 and 75, anesthetized with clove oil (50mg L^{-1} of water) (PEREIRA-DA-SILVA et al., 2009) cinco grupos de 30 alevinos (0,6/ufffd0,1g and euthanized via cranial drilling.

Table 2 Lipid profile of experimental diets, expressed in g/100g of total fatty acids.

Fatty acids	SO*	LO*	FRO*	SO/SEO*	LO/SEO*	FRO/SEO*
C12:0	0.00	0.06	0.01	0.00	0.00	0.00
C14:0	0.67	0.50	1.29	0.56	0.52	0.97
C14:1	0.05	0.04	0.13	0.07	0.06	0.27
C15:0	0.00	0.09	0.14	0.11	0.13	0.14
C16:0	13.97	10.89	19.78	13.68	11.62	16.07
C16:1	1.42	1.19	5.70	1.15	1.06	3.69
C17:0	0.18	0.27	0.21	0.14	0.14	0.16
C18:0	3.51	4.81	5.65	4.77	4.7	5.21
C18:1-n9 (cis+trans)	24.40	21.30	33.41	29.90	26.91	33.50
C18:2n-6	49.20	22.69	26.37	45.06	31.64	35.15
C18:3n-6	0.00	0.03	0.51	0.00	0.00	0.28
C18:3n-3	5.39	36.79	1.83	3.15	21.89	1.55
C20:0	0.16	0.15	0.16	0.28	0.22	0.29
C20:2n-6	0.00	0.05	0.72	0.00	0.00	0.35
C20:3n-6	0.00	0.04	0.68	0.00	0.00	0.32
C20:4n-6	0.26	0.23	0.98	0.22	0.21	0.59
C20:5n-3	0.19	0.12	0.15	0.15	0.13	0.20
C22:0	0.04	0.09	0.09	0.16	0.10	0.12
C22:1n-9	0.00	0.01	0.11	0.00	0.00	0.00
C22:2n-6	0.00	0.00	0.05	0.00	0.00	0.00
C24:0	0.00	0.05	0.00	0.00	0.00	0.00
C22:6n-3	0.20	0.20	0.64	0.25	0.25	0.48
ΣSFA[1]	18.53	16.91	27.33	19.70	17.43	22.96
ΣMUFA[1]	25.87	22.54	39.35	31.12	28.03	37.46
ΣPUFA[1]	55.24	60.15	31.93	48.83	54.12	38.92
$PUFA^1_{n-6}$	49.46	23.04	29.31	45.28	31.85	36.69
$PUFA^1_{n-3}$	5.78	37.11	2.62	3.55	22.27	2.23
$PUFA^1_{n-6/n-3}$	8.56	0.62	11.19	12.75	1.43	16.45

*Treatments: Soy oil (SO 3%), linseed oil (LO 3%), freshwater fish residue oil (FRO 3%), soy oil and sesame oil (SO/SEO, 1.5%/1.5%), linseed oil and sesame oil (LO/SEO, 1.5%/1.5%), freshwater fish residue oil and sesame oil (FRO/SEO, 1,5%/1.5%). [1]SFA: saturated fatty acids. MUFA: monounsaturated fatty acids. PUFA: polyunsaturated fatty acids.

The sampled fish were collected from those that fasted for 24 hours prior to sampling. The following variables were determined: diet consumption (DC), apparent feed conversion (AFC), weight gain (WG), specific growth rate (SGR) and protein efficiency rate (PER). The body chemical composition of whole fish was performed following the A.O.A.C. (1990) methodology to determine dry matter (DM), crude protein (CP), mineral matter (MM), ethereal extract (EE), and crude energy (CE). A total of 25 fish were sampled per experimental unit, eviscerated and used to determine the fatty acid profile following the A.O.A.C. (2005) methodology. Lipid was extracted according to the BLIGH & DYER (1959) methodology.

Fish glycemia levels (n=10 per treatment) were measured by a portable glucose meter (Roche, Brazil) in a 5-μL blood aliquot that was collected in the caudal vein region using heparinized insulin type syringes. The glycogen was quantified in 10mg of liver samples (LG) (n=14 per treatment) and 100mg of muscles (MG) (n=4 per treatment) following DUBOIS et al. (1956) modified by BIDINOTTO et al. (1997).

The thiobarbituric acid reactive substances (TBARS) were determined in hepatic tissues (n=6 per treatment) following the VYNCKE (1970) methodology adapted to microplate reader by modifying the proportions described by this author. The data were submitted to ANOVA (GLM procedure), using the SAS (2002) software. The F-test results showed significant differences among treatments. The LG and TBARS data were previously treated by the SAS/LAB command.

3. RESULTS AND DISCUSSION

Water quality parameters such as temperature, dissolved oxygen and pH were kept between 23.42 and 27.07ºC; 3.62 and 6.45mg L^{-1}; and, 6.63 and 7.23, respectively. Ammonia and nitrite levels did not exceed 0.34 and 0.41mg L^{-1}, respectively. The growth parameters and body chemical composition ($P<0.05$) were not significantly different for fish fed with diets containing SO, LO and FRO, combined or not with SEO. However, the addition of SEO to the diets affected significantly glycemia and lipid peroxidation of hepatic tissue (TBARS) (Table 3). In addition, independent of the treatment, the glycemic levels above 49mg DL^{-1} (PEREIRA-DA-SILVA et al., 2014) indicated that the fish underwent handling and capture stress during sampling (BARTON, 2002). Addition of SEO to the LO diet increased the blood glucose levels in fish. For the lambari, the lipid sources did not change the hepatic and muscular glycogen levels significantly. However, some fatty acids may have interfered with glycogen formation, such as the oleic acid, which modifies the activity of the enzymes involved in the glucose metabolism and hepatic reserves of rainbow trout (LIBRÁN-PÉREZ et al., 2013).

The high lipid peroxidation observed in the fish fed with LO diet can be explained by the probable high quantity of PUFA n-3 in the hepatic tissues. Lower quantity of PUFA n-3 in the diets reduces fish susceptibility to these oxidative reactions. This fact has been demonstrated in a seabass ***Lateolabrax japonicus*** study, in which the fish were fed with diets that contained palm oil associated with fish oil and presented lower malondialdehyde levels (MDA) (GAO et al., 2012). The ingested PUFAs are stocked in the liver and are susceptible to peroxidation due to the activation of the peroxisome proliferator-activated receptor (PPARα) that produces hydrogen peroxide and causes tissue damages. Thus, liver is the first organ susceptible to stress oxidative reactions and, consequently, lipid peroxidation (TAKAHASHI et al., 2002). In this research, SEO had a positive effect only on/when combined with LO.

Table 3 Mean values of growth performance variables, body chemical composition and physiologic state.

Growth performance	Treatments*							
	SO	SO/SEO	LO	LO/SEO	FRO	FRO/SEO	P- value	RSD[1]
DC[2] (g)	6.89	7.02	6.71	6.98	7.46	7.40	0.51[ns]	0.62
WG[2] (g)	3.89	4.40	4.25	4.11	5.20	4.52	0.24[ns]	0.74
AFC[2]	1.61	1.54	1.57	1.61	1.46	1.53	0.40[ns]	0.11
PER[2]	1.93	2.00	2.13	1.99	2.40	2.09	0.39[ns]	0.31
SGR[2] (% / Day)	1.29	1.4	1.37	1.33	1.55	1.42	0.28[ns]	0.15
Chemical composition (%)	SO	SO/SEO	LO	LO/SEO	FRO	FRO/SEO	P- value	RSD[1]
DM[2]	34.78	35.46	35.41	35.51	35.64	35.41	0.74[ns]	0.82
CP[2]	16.69	17.05	16.74	16.89	16.47	16.69	0.73[ns]	0.53
MM[2]	3.78	3.71	3.89	3.88	3.9	3.83	0.56[ns]	0.17
EE[2]	13.6	15.09	14.24	14.8	14.39	14.29	0.30[ns]	0.70
CE[2] (kcal $100g^{-1}$)	226.3	232.05	229.47	230.75	231.58	228.14	0.87[ns]	7.26
Physiologic state	SO	SO/SEO	LO	LO/SEO	FRO	FRO/SEO	P- value	RSD[1]
MG[2] (mg glicosyl g^{-1})	6.99	6.18	5.46	6.29	7.56	7.51	0.59[ns]	1.77
LG[2] (mg glicosyl g^{-1})	157.06	103.41	122.05	128.21	111.03	106.71	0.31[ns]	61
Interaction Effect Oil types* SEO								
Physiologic state	SO	SO/SEO	LO	LO/SEO	FRO	FRO/SEO	P- value	RSD[1]
Glycaemia (mg DL^{-1})	104^{A}	98^{A}	90^{B}	106^{A}	100^{A}	100.20^{A}	0.03	13
TBARS[2] (µmol g^{-1})	7.53^{A}	14.79^{A}	30.57^{A}	10.93^{B}	6.62^{A}	5.9^{A}	0.035	12

*Treatments: Soy oil (SO 3%), linseed oil (LO 3%), freshwater fish residue oil (FRO 3%), soy oil and sesame oil (SO/SEO, 1.5%/1,5%), linseed oil and sesame oil (LO/SEO, 1.5%/1.5%), freshwater fish residue oil and sesame oil (FRO/SEO, 1,5%/1.5%). [1]Residual Standard Deviation (RSD). [ns]not significant ($P<0.05$). **Means with different letter in the same line represent significant differences ($p<0.05$). [2]Diet consumption (DC), apparent feed conversion (AFC), weight gain (WG), specific growth rate (SGR), protein efficiency rate (PER), dry matter (DM), crude protein (CP), mineral matter (MM), ethereal extract (EE), crude energy (CE), muscular glycogen (MG), liver glycogen (LG), thiobarbituric acid reactive substances (TBARS).

Lipid profile of oils as shown in the lipid profile of the diets, changed significantly the saturated (SFA), monounsaturated (MUFA) and polyunsaturated (PUFA) fatty acids profiles of lambari, especially, the C14:0, C16:0, C16:1, oleic acid (C18:1n-9), linoleic acid (LA), linolenic acid (LNA), arachidonic acid (AA), eicosapentaenoic acid (EPA) and docosahexaenoic acid (DHA) (Table 4). The addition of SEO increased the MUFA levels in fish fed the SO and LO diets. This oil is considered a source of MUFA since it contains about 46% in the lipid profile (KOCHHAR, 2002). However, the same tendency was not observed for the diets FRO combined or not with SEO due to the similar MUFA quantities in these diets composition (Table 1). The higher desaturation rates of LNA to HUFA n-3 were observed in fish that were fed with SEO in the diets. Also, it was verified that the SEO efficiently promotes DHA formation in fish fed with diets with reduced LNA levels (Table 4).

Table 4 Profiles of eviscerated fishes, describing main fatty acids, expressed in g/100g of total fatty acids.

Fatty acids	Oil type effect			P- value	SEO effect		P- value	RSD[1]
	SO	LO	FRO		Without	With		
C12:0	0.08	0.05	0.06	0.53^{ns}	0.07	0.06	0.72^{ns}	0.06
C16:0	20.03^{B}	19.45^{B}	21.12^{A}	0.001	20.38	20.01	0.11^{ns}	0.34
C18:0	8.15	7.89	7.99	0.23^{ns}	7.94	8.09	0.23^{ns}	0.19
C18:1-n9 (trans)	0.06	0.09	0.11	0.53^{ns}	0.1	0.08	0.51^{ns}	0.05
C18:3n-6	0.25^{AB}	0.19^{B}	0.31^{A}	0.023	0.26	0.25	0.80^{ns}	0.04
C20:2n-6	0.19	0.22	0.28	0.06^{ns}	0.23	0.23	0.94^{ns}	0.04
C20:3n-3	0.00^{B}	0.12^{A}	0.00^{B}	0.002	0.05	0.03	0.35^{ns}	0.03
C22:6n-3 (DHA[2])	0.69^{B}	1.00^{A}	0.73^{B}	0.02	0.85	0.77	0.25^{ns}	0.12
SFA[2]	29.95^{B}	28.94^{C}	30.88^{A}	0.003	30.07	29.78	0.31^{ns}	0.46
PUFA[2]n-6/PUFA[2] n-3	9.55^{B}	2.62^{A}	9.26^{B}	<0.0001	6.15^{B}	8.14^{A}	0.003	0.71
HUFA[2] n-3/LNA	0.75^{B}	0.32^{C}	1.07^{A}	<0.0001	0.64^{B}	0.80^{A}	0.03	0.09

Interaction effect oil type* SEO

Fatty acids	SO	SO/SEO	LO	LO/SEO	FRO	FRO/SEO	P- value	RSD[1]
C14:0	1.15^{A}	0.97^{B}	1.02^{A}	1.07^{A}	1.24^{A}	1.15^{A}	0.048	0.051
C18:1-n9 (cis)	44.19^{B}	46.91^{A}	43.93^{A}	45.74^{A}	46.76^{A}	46.80^{A}	0.007	0.392
C18:2n-6 (LA[2])	17.84^{A}	16.61^{B}	14.31^{B}	15.50^{A}	13.48^{B}	15.02^{A}	<0.0001	0.189
C18:3n-3 (LNA[2])	1.45^{A}	0.90^{B}	7.06^{A}	3.35^{B}	0.88^{A}	0.79^{A}	<0.0001	0.152
C20:3n-6	0.55^{A}	0.55^{A}	0.31^{B}	0.48^{A}	0.60^{A}	0.56^{A}	0.003	0.027
C20:4n-6 (AA[2])	0.77^{A}	0.65^{B}	0.37^{B}	0.51^{A}	0.85^{A}	0.81^{A}	0.004	0.033
C20:5n-3 (EPA[2])	0.20^{A}	0.16^{A}	0.46^{A}	0.32^{B}	0.14^{A}	0.16^{B}	0.03	0.03
MUFA[2]	47.79^{B}	50.16^{A}	47.11^{A}	49.18^{A}	51.32^{A}	50.90^{A}	0.003	0.37
PUFA[2] total	22.05^{A}	19.92^{B}	24.12^{A}	21.68^{B}	17.32^{A}	18.67^{A}	0.003	0.33
PUFA[2] n -6	19.40^{A}	18.09^{B}	15.21^{B}	16.68^{A}	15.27^{A}	16.69^{B}	<0.0001	0.25
PUFA[2] n-3	2.46^{A}	1.63^{A}	8.70^{A}	4.75^{B}	1.75^{A}	1.71^{A}	<0.0001	0.20
HUFA[2] n-6/LA[2]	0.08^{A}	0.09^{A}	0.06^{A}	0.07^{A}	0.13^{A}	0.11^{B}	0.01	0.09

*Treatments: Soy oil (SO 3%), linseed oil (LO 3%), freshwater fish residue oil (FRO 3%), soy oil and sesame oil (SO/SEO, 1.5%/1.5%), linseed oil and sesame oil (LO/SEO, 1.5%/1.5%), freshwater fish residue oil and sesame oil (FRO/SEO, 1,5%/1,5%). [1]Residual Standard Deviation (RSD). [ns]not significant (P<0.05). **Means with different letter in the same line represent significant differences (P<0.05). [2]Linoleic acid (LA), linolenic acid (LNA), araquidonic acid (AA), eicosapentaenoic acid (EPA), docosahexaenoic acid (DHA), saturated fatty acids (SFA), polyunsaturated fatty acids (PUFA), high unsaturated fatty acids (HUFA).

The probable higher desaturation of LNA to HUFA-3 and the formation of DHA could be related to the presence of the bioactive compounds in SEO, such as sesamin. The addition of sesamin in diets containing linseed and sunflower oil modified significantly lipid metabolism and increased the DHA quantities in the white muscle of rainbow trouts ***Oncorhynchus mykiss*** and hepatocytes of Atlantic salmon ***Salmo salar*** L. (TRATTNER et al., 2008a, 2008b) hepatocytes were incubated without or with a mixture of sesamin and episesamin in order to test for possible effects on lipid metabolism. Sesamin/episesamin exposure (0.05 mM, final concentration. This compound also increased the activity of enzymes related to β mitochondrial and peroxisomal oxidation, by activating PPARα in rats, and fish as the rainbow trout ***Oncorhynchus mykiss*** and ***Lates calcarifer*** (ASHAKUMARY et al., 1999; TRATTNER et al., 2008a; ALHAZZAA et al., 2012).

4. CONCLUSION

The inclusion of sesame oil increased the DHA levels in fish tissues verified by the higher rates of desaturation of LNA to HUFAn-3, improving the lipid quality and promoting the oxidative stability of hepatic tissue, without altering the growth performance and fish health.

ACKNOWLEDGMENTS

The authors thanks the Fundação de Amparo à Pesquisa do Estado de São Paulo (FAPESP) for the financial support (Process nº 2012/11101-0) and Conselho Nacional de Desenvolvimento Científico e Tecnológico (CNPq), by the grant for the first author.

REFERENCES

1. A.O.A.C. (ASSOCIATION OF OFFICIAL ANALYTICAL CHEMISTS). Official methods of the analysis. 15.ed. Arlington, Virginia: Association of Official Analytical chemists, 1990, v.1, 1117p.
2. A.O.A.C. (ASSOCIATION OF OFFICIAL ANALYTICAL CHEMISTS). Official methods of analysis of the Associaton of Official Analytical Chemists. 18.ed., Gaitherburg, Mayland: AOAC, 2005; method 996.06, p20-25.
3. ALHAZZAA, R. et al. Sesamin modulation of lipid class and fatty acid profile in early juvenile teleost, Lates calcarifer, fed different dietary oils. Food Chemistry, v.134, p.2057-2065, 2012. Available from: <Available from: http://dx.doi.org/10.1016/j.foodchem. 2012.04.004 >. Accessed: Sept. 22, 2014. doi: 10.1016/j.foodchem.2012.04.004.
4. ASHAKUMARY, L. et al. Sesamin, a sesame lignan, is a potent inducer of hepatic fatty acid oxidation in the rat. Metabolism: Clinical and Experimental, v.48, n.10, p.1303-1313, 1999. Available from: <Available from: http://www.metabolismjournal.com/ article/ S0026-0495(99)90272-X/abstract >. Accessed: Aug. 24, 2014. doi: 10.1016/S0026-0495(99)90272-X.
5. BARTON, B.A. Stress in fishes: a diversity of responses with particular reference to changes in circulating corticosteroids. Integrative and Comparative Biology, v.42, p.517-525, 2002. Available from: <Available from: http://10.1093/icb/42.3.517 >. Accessed: Dec. 21, 2013. doi: 10.1093/icb/42.3.517>.
6. BIDINOTTO, M. et al. Hepatic glycogen and glucose in eight tropical fresh-water teleost cist for field determination of micro-samples. Boletim Tecnológico do Centro de Pesquisa e Treinamento em Aquicultura (CEPTA), v.10, p.53-70, 1997. Available from: <Available from: http://www.icmbio.gov.br/cepta/images/stories/producao_cientifica/ hepatic_ 1997_01.pdf >. Accessed: Nov. 24, 2012.
7. BLIGH, E.G.; DYER, W.J. A rapid method of total lipid extraction and purification. Canadian Journal of Biochemistry and Physiology, v.37, p.911-917, 1959. Available from: <Available from: http://10.1139/o59-099>. Accessed: Apr. 21, 2014. doi: 10.1139/o59-099.
8. CAMPELO, D.A.V. et al. Conjugated linoleic acid in diets for lambari (Astyanax altiparanae) (Garutti & Britski, 2000). Aquaculture Nutrition, p. n/a-n/a, 2014. Available from:

<Available from: http://doi.wiley.com/10.1111/anu.12203 >. Accessed: May 25, 2015. doi: 10.1111/anu.12203.

9. COTAN, J.L.V. et al. Níveis de energia digestível e proteína bruta em rações para alevinos de lambari tambiú. Revista Brasileira de Zootecnia, v.35, n.3, p.634-640, 2006. Available from: <Available from: http://www.scielo.br/pdf/rbz/v35n3/30051.pdf >. Accessed: Jun. 21, 2015.

10. CHEN, C. et al. N-3 essential fatty acids in Nile tilapia, Oreochromis niloticus: Quantification of optimum requirement of dietary linolenic acid in juvenile fish. Aquaculture, v.416-417, p.99-104, 2013. Available from: <Available from: http://10.1016/j.aquaculture.2013.09.003 >. Accessed: Sept. 23, 2014. doi: 10.1016/j.aquaculture.2013.09.003.

11. GAO, J. et al. Effects of dietary palm oil supplements with oxidized and non-oxidized fish oil on growth performances and fatty acid compositions of juvenile Japanese sea bass, Lateolabrax japonicus., Aquaculture v.324-325, p.97-103, 2012. Available from: <Available from: http://www.sciencedirect.com/science/article/pii/S004484861 1008 477 >. Accessed: Sept. 24, 2014. doi: 10.1016/j.aquaculture.2011.10.031.

12. GONÇALVES, L.U. et al. Effect of the inclusion of fish residue oils in diets on the fatty acid profile of muscles of males and females lambari (Astyanax altiparanae)., Revista Brasileira de Zootecnia v.41, p.1967-1974, 2012. Available from: <Available from: http://10.1590/S1516-35982012000900001 >. Accessed: Feb. 23, 2014. doi: 10.1590/S1516-35982012000900001.

13. GUIMARÃES, R.; MACEDO, M. Sesame and flaxseed oil: nutritional quality and effects on serum lipids and glucose in rats. Food Science and techonology, v.33, n.1, p.209-217, 2013. Available from: <Available from: http://dx.doi.org/10.1590/S0101-2061201 3005000029 >. Accessed: Jul. 21, 2014. doi: 10.1590/S0101-20612013005000029.

14. HSU, D.-Z. et al. Effect of sesame oil on oxidative-stress-associated renal injury in endotoxemic rats: involvement of nitric oxide and proinflammatory cytokines. Shock (Augusta, Ga.), v.24, n.3, p.276-280, 2005. Available from: <Available from: http://10.1097/01.shk.0000172366.73881.c7 >. Accessed: Aug. 22, 2013. doi: 10.1097/01.shk.0000172366.73881.c7.

15. IDE, T. et al. Interaction of dietary fat types and sesamin on hepatic fatty acid oxidation in rats. Biochimica et Biophysica Acta - Molecular and Cell Biology of Lipids, v.1682, p.80-91, 2004. Available from: <Available from: http://10.1016/j.bbalip.2004.02.001 >. Accessed: Nov. 24, 2012. doi: 10.1016/j.bbalip.2004.02.001.

16. KOCHHAR, S.P. Sesame, rice-bran and flaxseed oils. In: GUNSTONE, F.D. (Ed.). Vegetable oils in food technology: composition, propreties and uses. Florida: Blackwell, 2002. p.297-326.

17. KÖSE, I.; YILDIZ, M. Effect of diets containing sesame oil on growth and fatty acid composition of rainbow trout (Oncorhynchus mykiss). Journal of Applied Ichthyology, v.29, p.1318-1324, 2013. Available from: <Available from: http://10.1111/jai.12184 >. Accessed: Mar. 23, 2015. doi: 10.1111/jai.12184.

18. LIBRÁN-PÉREZ, M. et al. Response of hepatic lipid and glucose metabolism to a mixture or single fatty acids: possible presence of fatty acid-sensing mechanisms. Comparative Biochemistry and Physiology - A Molecular and Integrative Physiology, v.164, n.1, p.241248, 2013. Available from: <Available from: http://10.1371/journal.pone.0059507 >. Accessed: Dec. 21, 2014. doi: 10.1371/journal.pone.0059507.

19. NAYLOR, R.L. et al. Effect of aquaculture on world fish supplies. Nature, v.405, p.1017-1024, 2000. Available from: <Available from: http://10.1038/35016500 >. Accessed: Nov. 21, 2009. doi: 10.1038/35016500.

20. OLSVIK, P.A. et al. Hepatic oxidative stress in Atlantic salmon (Salmo salar L.) transferred from a diet based on marine feed ingredients to a diet based on plant ingredients., Aquaculture Nutrition v.17, p.424-436, 2011. Available from: <Available from: http://10.1111/j.1365-2095.2010.00778.x >. Accessed: Feb. 12, 2012. doi: 10.1111/j.1365-2095.2010.00778.x.

21. PEREIRA-DA-SILVA, E.M. et al. Menthol as anaesthetic for lambari Astyanax altiparanae (Garutti & Britski 2000): attenuation of stress responses. Aquaculture Research, p. n/a-

n/a, 2014. Available from: <Available from: http://10.1111/are.12599 >. Accessed: Jun. 23, 2014. doi: 10.1111/are.12599.
22. PEREIRA-SILVA, E.M. et al. Efeito anestésico do óleo de cravo em alevinos de lambari. Ciência Rural, v.39, p.1851-1856, 2009. Available from: <Available from: http://10.1590/S0103-84782009005000127 >. Accessed: Mar. 23, 2012. doi: 10.1590/S0103-84782009005000127.
23. PERIASAMY, S. et al. Sesame oil mitigates nutritional steatohepatitis via attenuation of oxidative stress and inflammation: a tale of two-hit hypothesis. Journal of Nutritional Biochemistry, v.25, n.2, p.232-240, 2014. Available from: <Available from: http://10.1016/j.jnutbio.2013.10.013 >. Accessed: Mar. 22, 2015. doi: 10.1016/j.jnutbio.2013.10.013.
24. SAS INSTITUTE. SAS/STAT user's guide. Cary, NC, 2002. Version 9.2.
25. TAKAHASHI, M. et al. Fish oil feeding alters liver gene expressions to defend against PPARalpha activation and ROS production. American Journal of Physiology. Gastrointestinal and liver physiology, v.282, p.G338-G348, 2002. Available from: <Available from: http://10.1152/ajpgi.00376.2001 >. Accessed: Nov. 23, 2014. doi: 10.1152/ajpgi.00376.2001.
26. TRATTNER, S. et al. Sesamin increases alpha-linolenic acid conversion to docosahexaenoic acid in atlantic salmon (Salmo salar L.) hepatocytes: role of altered gene expression. Lipids, v.43, p.999-1008, 2008a. Available from: <Available from: http://10.1007/s11745-008-3229-7 >. Accessed: May 21, 2012. doi: 10.1007/s11745-008-3229-7.
27. TRATTNER, S. et al. Sesamin supplementation increases white muscle docosahexaenoic acid (DHA) levels in rainbow trout (Oncorhynchus mykiss) fed high alpha-linolenic acid (ALA) containing vegetable oil: metabolic actions., Lipids v.43, p.989-997, 2008b. Available from: <Available from: http://10.1007/s11745-008-3228-8>. Accessed: Apr. 22, 2012. doi: 10.1007/s11745-008-3228-8.
28. VYNCKE, W. Direct determination of the thiobarbituric acid value in trichloracetic acid extracts of fish as a measure of oxidative rancidity. Fatte Seifen Anstrichmittel, v.72, p.1084-1087, 1970. Available from: <Available from: http://10.1002/lipi.19700721218 >. Accessed: Nov. 21, 2012. doi: 10.1002/lipi.19700721218.

CHAPTER 9

Artificial insemination of Holstein heifers with sex-sorted semen during the hot season in a subtropical region

Lian-Ben Chang[1], Chih-Jen Chou[2], Jia-Shian Shiu[1,3], Po-An Tu[1,4], Shi-Xuan Gao[3], Shao-Yu Peng[5], Shinn-Chih Wu[1,2]

1 Department of Animal Science and Technology, National Taiwan University, No. 50, Lane 155, Sec. 3, Keelung Road, Taipei 10672, Taiwan, Republic of China

2 Institute of Biotechnology, National Taiwan University, No. 81, Chang-Xing St, Taipei 10672, Taiwan, Republic of China

3 Agriculture Experiment Station, College of Bioresources and Agriculture, National Taiwan University, No. 50, Lane 155, Sec. 3, Keelung Road, Taipei 10672, Taiwan, Republic of China

4 Hsinchu Branch, Livestock Research Institute, Council of Agriculture, Executive Yuan, No. 207-5, Bi-tou-mian, Wu-hoo Village, Si-hoo Township, Miao-li County 36848, Taiwan, Republic of China

5 Department of Animal Science, National Pingtung University of Science and Technology, 1, Shuefu Road, Neipu, Pingtung 91201, Taiwan, Republic of China

ABSTRACT

Our aim was to investigate insemination techniques in order to improve pregnancy rates of artificial insemination (AI) using sex-sorted semen (sexed AI) in cattle in tropical and subtropical (T/ST) regions. In T/ST regions, the pregnancy rates by sexed AI are reportedly the lowest in the hottest months of the year, with less than 15% in cows and 35–40% in heifers (PMID 24048822). We compared sexed AI by depositing the semen into the uterine body (UB-AI, n = 12) versus the unilateral uterine horn (UUH-AI, n = 14) of pre-ovulation heifers. The ovary and follicle were assessed by rectal ultrasound before AI. After insemination, pregnancy was determined by ultrasound at approximately 40 days and approximately 70 days. In the present study, we demonstrated that high pregnancy rates (>70%) by sexed AI in the hottest season in a subtropical region such as Taiwan can be achieved when heifers with pre-ovulation follicles are used. The overall pregnancy rates were 54% higher in the UUH-AI (71%) group than in the UB-AI (42%) group (P = 0.06), examined on approximately 40 days post-sexed AI. Surprisingly, however, the pregnancy outcome appeared to be higher in the hot season (62%) than in the cool season (46%) although this difference was not statistically significant. Based on the present study, we recommend that cattle breeders perform UUH-AI using sex-sorted semen for heifers with pre-ovulation follicles in order to achieve satisfactory pregnancy outcome in the hot seasons in T/ST regions.

Keywords: Holstein heifers, Sex-sorted semen, Insemination, deposition, Pre-ovulation, follicle, Subtropical regions

1. INTRODUCTION

Sex-sorted semen is now widely used in cattle breeding to enhance industrial competitiveness (Rath and Johnson 2008) and achieve high female fertility (Seidel 2003a, b). X chromosome sex-sorted semen with artificial insemination (AI) can effectively control the sex ratio of the offspring of dairy cows. In the process of flow cytometry sorting, a number of insults including sperm dilution, centrifugation, dyeing, pressure, laser, discharge, and other processes often lead to damage to sperm structure, decreasing its viability. Consequently, AI using sex-sorted semen (sexed AI) has a pregnancy rate lower than that using conventional unsorted semen (Carvalho et al. 2010; DeJarnette et al. 2011). In addition, the process of flow cytometry sorting is very slow, at a rate of 10–18 million/h. A very low number (~2.0 million) of sperm is packaged in a sex-sorted semen straw only 10% that in conventional unsorted semen (Seidel 2011). The pregnancy rate of sexed AI is approximately 70–75% of that of sex-unsorted semen (DeJarnette et al. 2009). Despite the relatively low pregnancy outcome, the dairy industry is witnessing a trend of switching to sexed AI; this is in part due to higher than 90% female rates in calves (Healy et al. 2013).

Previous studies indicate that high temperature has a negative effect on the reproduction process of dairy cows owing to undetected estrous cycle, decreased estrous period intensity (De Rensis and Scaramuzzi 2003; De Rensis et al. 2015), effects on the physical activity (García-Ispierto et al. 2006; García-Ispierto et al. 2007), and low fertility (López-Gatius 2012; López-Gatius 2013). High environmental temperature has a detrimental effect on embryo survival (García-Ispierto et al. 2006) as well as on the survival and motility of sex-sorted sperm (García-Ispierto et al. 2007; Mellado et al. 2013). Cerchiaro et al. (2007) reported that the pregnancy rate of sexed AI in heifers was lowest in summer (44.2%) and significantly lower than that in spring (53.9%), autumn (50.8%), and winter (50.7%). Similarly, Donovan et al. (2003) reported that the pregnancy rate of sexed AI in heifers in summer (33%) was only approximately half of that in winter (64%). The pregnancy rates in dairy cows are generally lower than those in heifers, but show the same trend in that summer is the worst season: 16% in summer and 21% in winter in a recent study (Mellado et al. 2014).

Taiwan is located in a subtropical region. In 2007–2016, the average monthly temperature and relative humidity in summer reached 27.7 °C and 72.8%, respectively. Such an environment with continuous high temperatures (Table 1) and high humidity has posed a serious challenge to the reproductive performance of dairy cattle. The low amount of sex-sorted semen per straw and the compromised sperm quality conditions present a further burden.

Table 1 Comparison pregnancy per AI (P/AI) of females inseminated with sex-sorted semen in hot regions

Area	Climate	Animal category	Pregnancy per AI based on sex-sorted semen used in AI % (*n*/*n*)	References
Florida	Subtropical region	Heifers	Winter: 64 Summer: 33	Donovan et al. 2003
Northern Italy	Subtropical region	Heifers	Spring: 53.9 (82/152) Summer: 44.2 (19/43) Autumn: 50.8 (62/122) Winter: 50.7 (111/219)	Cerchiaro et al. 2007
Northern Mexico	Desert climate	Heifer cows	Overall: 41.6 Winter: 21 Summer: 16 Overall: 17.3	Mellado et al. 2014

Previous studies have shown that the deposition sites and estrus status can greatly affect the outcome of sexed AI. Kurykin et al. (2007) reported that pregnancy rates are higher than 100% when the animals are in strong estrus (45.9%) than when they are in weak estrus (20.8%).

To date, few studies have sought to improve sexed AI in dairy cattle in the subtropical regions. The aim of the present study was to assess the effects of animal status, deposition site, and season of sexed AI in Taiwan. We hypothesized that heifers with obvious estrous cycle condition should have pre-ovulation follicles; hence, by monitoring pre-ovulation follicles, the pregnancy rate using sex-sorted semen can be improved. We further compared the effects of different deposition sites of pregnancy rates.

2. MATERIALS AND METHODS

2.1 Handling and management of experimental animals

We selected 26 heifers with pre-ovulation follicles after spontaneous estrus for 8–12 h. Heifers 12 to 14 months old with body weight between 355 and 410 kg and normal estrous cycle were also chosen. They were housed at the ranch of the National Taiwan University (Agriculture Experiment Station, College of Bioresources and Agriculture). During the experiment, heifers were fed twice a day (05:00 and 16:00), and estrus was observed four times a day (05:00, 11:00, 16:00, and 22:00) for 20 to 30 min daily. Heifers with spontaneous estrus after 8–12 h were inseminated with sex-sorted semen. We used sex-sorted semen from a single sire (Frontier Agriculture Systems, Inc., Taipei, Taiwan). There were approximately 2.3 million sperm per straw of sex-sorted semen. AI of heifers was performed by a professional. Before insemination, the ovary and follicle were assessed by rectal palpation and rectal ultrasound (portable ultrasound scanner equipped with a 5.0-MHz linear array transducer; SonoSite, Inc., USA). After diagnosis, we were able to confirm deposition in the pre-ovulatory follicle. After thawing the semen for approximately 30 s at 36 °C, a small amount of semen was examined to confirm if it was still viable for AI. Twenty-six heifers were randomly assigned into two groups: (1) sexed AI to uterus body pre-ovulation (UB-AI, n = 12) and (2) sexed AI to unilateral uterine horn pre-ovulation (UUH-AI, n = 14). Pregnancy was determined by ultrasound at approximately 40 days and approximately 70 days after insemination. We recorded the pregnancy and embryonic loss rates during the pregnancy period and the live birth of calves and female rates after parturition.

2.2 Statistical analysis

SAS (Statistical Analysis Systems Institute Inc., Cary, NC, USA) was used in the analysis. Chi-squared analysis (χ^2) was used for comparing the results of different depositions of semen and insemination in the cool season or the hot season in relation to pregnancy rate. Differences were considered statistically significant when $P < 0.05$.

3. RESULTS

3.1 Diameter of pre-ovulation follicles

An ultrasound scan can be used to observe the structure of spontaneous ovarian cycle, with the follicle including the corpus luteum and ovarian stroma. Ovarian stroma imaged via ultrasound showed no follicles (Fig. 1a). Dominant follicles measure 18.2 mm in diameter (Fig. 1b). A small follicle (7.6 mm) in the ovary is also detected via ultrasound (Fig. 1c). The pre-ovulation follicles of all heifers were between 12 and 18.2 mm in our experiment. Follicle diameter of heifers in the cool and hot seasons is shown in Table 2. The average diameter of pre-ovulation follicles was 13.8 mm.

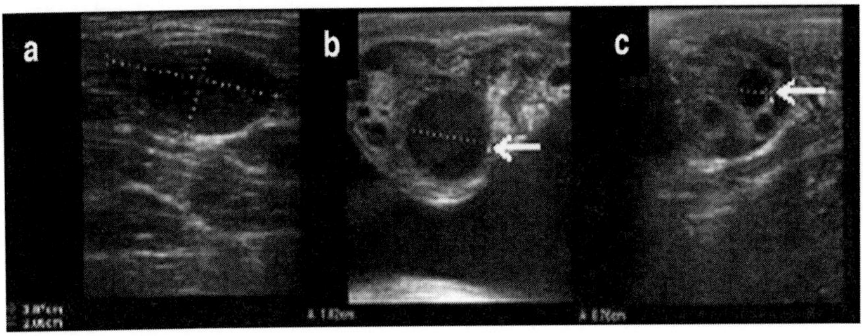

Figure. 1 There is no follicle in the ovary via ultrasound detection (**A**), there is a dominant follicle (18.2 mm, *arrow*) in the ovary via ultrasound detection (**B**), and there is a small follicle (7.6 mm, *arrow*) in the ovary via ultrasound detection (**C**)

Table 2 Follicle diameter of heifers in cool and hot season

Follicle diameter (mm)	Follicle diameter of all heifers	Follicle diameter of pregnant heifers	Follicle diameter of not pregnant heifers
In hot season	13.3 ± 1.4	13.1 ± 1.5	13.6 ± 1.5
In cool season	14.2 ± 1.9	14.5 ± 2.3	14 ± 1.5

3.2 Pregnancy rate, embryonic loss rate, live-born calf rate, and live-born heifer rate

Sex-sorted semen was inseminated in the UB-AI and UUH-AI groups after approximately 40 and 70 days. The pregnancy and embryonic loss rates are shown in Table 3. The pregnancy rates of sex-sorted semen after heifer insemination at approximately 40 days in the UB-AI and UUH-AI groups were 42 and 71%, respectively ($P = 0.06$) (Fig. 2).

Table 3 Pregnancy rate、embryonic loss rate、live offspring rate, and female rate of heifers on approximately 40 and 70 days after frozen-thawed sex-sorted semen deposition at UB and UUH with pre-ovulatory follicle

Trop Anim Health Prod

Site of deposition[a]	UB-AI, *n* (%)	UUH-AI, *n* (%)	In total, *n* (%)
Number of AI	12	14	26
Number of pregnant heifers at approximately 40 days (% per AI)	5 (42)	10 (71)	15 (58)
Number of pregnant heifers at approximately 70 days (% per AI)	5 (42)	9 (64)	14 (54)
Number of embryonic loss rate (% per AI)	0 (0)	1 (10)	1 (7)
Number of live offspring rate (% per born)	5 (100)	9 (100)	14 (100)
Number of female (% offspring rate)	5 (100)	8 (89)	13 (93)

UB semen deposition in the uterine body, *UUH* semen deposition in the unilateral uterine horn

[a] Deposition sites within the uterus

The pregnancy rates after heifer insemination at approximately 70 days of the UB-AI and UUH-AI groups were 42 and 64%, respectively ($P = 0.32$) (Fig. 3). The embryonic loss rate was 7% in our experiment. The number of live-born calves and live-born heifers inseminated with sex-sorted semen are also shown in Table 3. The female ratio of live calves after confirmation was 93% in our experiment.

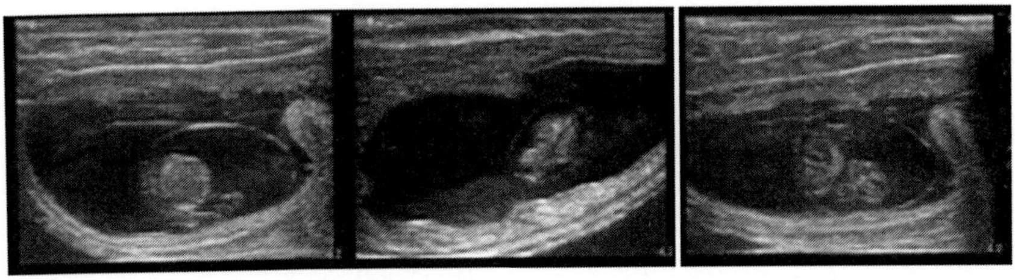

Figure. 2 Pregnancy of heifer on approximately 40 days after inseminated with sex-sorted semen

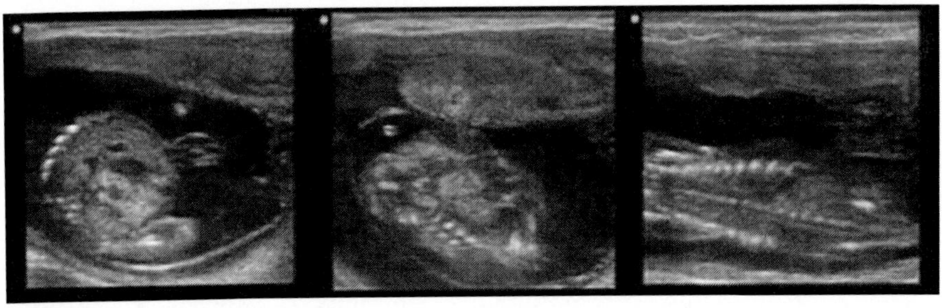

Figure. 3 Pregnancy of heifer on approximately 70 days after inseminated with sex-sorted semen

3.3 Pregnancy rates of heifers inseminated in the cool season and the hot season

Temperature and humidity index during the cool and the hot seasons in Taipei, Taiwan from 2007 to 2016 were measured. The average temperature and humidity index in the hot season were 27.7 ± 2.3 °C and 72.8 ± 3.5%, respectively. The maximum temperature and humidity index were 36.0 ± 1.9 °C and 95.0 ± 3.5%, respectively. Sex-sorted semen was UB-AI and UUH-AI in the cool season and the hot season. The resulting pregnancy rates were compared. The pregnancy rates of heifers inseminated in the cool season and the hot season were 46 and 62%, respectively (P = 0.43). There was no significant effect in the interaction between different sites of semen deposition in the cool season and hot season and pregnancy rate (P = 0.58).

4. DISCUSSION

In the early study, the three critical parameters for measurement of follicular growth by trans-rectal ultrasound are follicle appearance (4 mm), follicle growth (9 mm), and pre-ovulatory follicles (from 10 to 20 mm) (Lucy 2007). In our experiment, if the follicle diameter was less than 10 mm via ultrasound image detection, the AI procedure was not implemented. Sex-sorted semen was used to inseminate heifers with pre-ovulation follicles after spontaneous estrus for 8–12 h. Seidel and Schenk (2008) reported that 1 million sperm per straw of sex-sorted semen was inseminated in the UB-AI and UUH-AI groups by estrus synchronization. The pregnancy rates of heifers after 28 to 31 days and 56 to 58 days in the UB-AI and UUH-AI groups were 46, 32, 39, and 28%, respectively. In their study, 3 million sperm per straw of sex-sorted semen was in the UB-AI and UUH-AI groups by additional estrus synchronization. After 28 to 31 days and 56 to 58 days, the pregnancy rates of heifers in the UB-AI and UUH-AI groups were 47, 51, 44, and 43%, respectively. Using different amounts of sex-sorted sperm in their experiment, it was determined whether 1 million sperm per straw in the UUH-AI had significant declined pregnancy rate than other groups after inseminated 56 to 58 days.

Bodmer et al. (2005) found that 2 million sex-sorted sperm are inseminated in the UB of heifers and cows by non-estrus synchronization. The pregnancy rates of heifer insemination after 30 to 40 days and 70 to 90 days were 33.3 and 29.6%, respectively, whereas the pregnancy rates of cows were 27.6 and 23.8%. In their experiment, the pregnancy rate following heifer insemination with sex-sorted semen was slightly higher than that for cows, but there was no significant difference. Further, in their experiment, the pregnancy rates following heifer insemination after 30 to 40 days and 70 to 90 days were slightly lower than our 2.3 million sperm per straw of sexed AI in the UB-AI group (42 and 42%). A possible reason is that the amount of sperm in their experiment was 0.3 million sperm per straw lower than that in our experiment. Therefore, one of the main factors influencing a higher pregnancy rate is a higher sperm count in a single straw (Seidel and Schenk 2008).

In a study by Kurykin et al. (2007), estrus synchronization in heifer insemination was conducted using approximately 2.2 million sperm per straw of sex-sorted semen in the UB, UUH, and utero-tubal junction. In their experiment, heifers exhibiting strong signs of estrus reached 45.9%, but weak signs of estrus only reached 20.8% of pregnancy rate. Thus, strong or weak signs of estrus affect pregnancy rate. In most heifers with weak signs of estrus, it is not possible to evaluate whether ovulation time is appropriate for AI; thus, when using fewer sex-sorted sperm, the key lies in the timing of fertilization (Kurykin et al. 2003).

In Taiwan, the hot season mostly is with high temperature and high humidity. The pregnancy rate had no significant effect on heifers inseminated with sex-sorted semen in the cool and hot seasons. In the annual research of Mellado et al. (2014), the pregnancy rate of heifers inseminated using sex-sorted semen was significantly higher than that of cows; there was no significant effect of insemination using sex-sorted semen on the pregnancy rate of heifers in the cool season and the hot season. This result is similar to that of our experiment; there was no significant effect of insemination using sex-sorted semen on the pregnancy rate of heifers in the cool season or the hot season in subtropical regions. The possible reason was that the heifers had obvious signs of estrus and had pre-ovulation follicles under hot stress. We used sexed AI in heifers with pre-ovulation follicles after spontaneous estrus for 8–12 h, and no reduction in pregnancy rate was observed in the hot season. The possible reason was that the heifers were not under lactation pressure or uterine without involution.

Most studies did not mention rates of live-born calves using sexed AI. There were equally live calves born using AI on lactating cows by sex-unsorted semen using the estrus synchronized in the research of Stewart et al. (2011). It was clear that AI using sex-sorted semen did not affect the performance of live-born calves. The proportion of live female calves after confirmation was 93% in this experiment. This result was similar to the proportion of females (93.2%) following the use of sex-sorted semen in the estrus synchronization program in the experiment of Kurykin et al. (2007). Other studies showed that the proportion of females following the use of sexed AI in lactating cows can reach 90% (DeJarnette et al. 2009; Norman et al. 2010).

In conclusion, heifer breeders should perform UUH-AI using sex-sorted semen on heifers that have pre-ovulation follicles to achieve satisfactory pregnancy outcomes in subtropical regions. Large-scale follow-up studies are needed to confirm these findings.

ACKNOWLEDGEMENTS

We particularly thank the staff of the College of Bioresources and Agriculture dairy farm (Taiwan, Republic of China) for providing ultrasound equipment and heifer husbandry in the Agriculture Experiment Station, College of Bioresources and Agriculture, National Taiwan University. This work was supported by Council of Agriculture grant 105AS-2.1.4-AD-U1.

COMPLIANCE WITH ETHICAL STANDARDS

All experimental procedures with cows were approved by the Institutional Animal Care and Use Committee of National Taiwan University (NTU-103-EL-34). The methods were carried out in accordance with the approved guidelines.

REFERENCES

1. Bodmer, M., Jannet, F., Hassig, M., Den Daas, N., Reichert, P. and Thun R., 2005. Fertility in heifers and cows after low dose insemination with sex-sorted and non-sorted sperm under field condition. Theriogenology, 64,1647–1655
2. Carvalho, J.O., Sartori, R., Machado, G.M., Mourao, G.B. and Dode M., 2010. Quality assessment of bovine cryopreserved sperm after sexing by flow cytometry and their use in in vitro embryonic production. Theriogenology, 74, 1521–1530
3. Cerchiaro, I., Cassandro, M., Dal Zotto, R., Carnier, P. and Gallo L., 2007. A field study on fertility and purity of sex-sorted cattle sperm. Journal of Dairy Science, 90, 2538–2542
4. De Rensis, F. and Scaramuzzi, J.R., 2003. Heat stress and seasonal effects on reproduction in the dairy cow: a review. Theriogenology, 60, 1139–1151
5. De Rensis, F., Garcia-Ispierto, I. and López-Gatius, F., 2015. Seasonal heat stress: clinical implications and hormone treatments for the fertility of dairy cows. Theriogenology, 84, 659–666
6. DeJarnette, J.M., Nebel, R.L. and Marshall, C.E., 2009. Evaluating the success of sex-sorted semen in US dairy herds from on farm records. Theriogenology, 71, 49–58
7. DeJarnette, J.M., Leach, M.A., Nebel, R.L., Marshall, C.E., McCleary, C.R. and Moreno, J.F., 2011. Effects of sex-sorting and sperm dosage on pregnancy rates of Holstein heifers: is comparable fertility of sex-sorted and conventional semen plausible? Journal of Dairy Science, 94, 3477–3483
8. Donovan, G.A., Bennett, F.L. and Springer, F.S., 2003. Factors associated with first service conception in artificially inseminated nulliparous Holstein heifers. Theriogenology, 60, 67–75
9. García-Ispierto, I., López-Gatius, F., Santolaria, P., Yániz, J.L., Nogareda, C. and López-Béjar, M., 2006. Relationship between heat stress during the peri-implantation period and early fetal loss in dairy cattle. Theriogenology, 65, 799–807
10. García-Ispierto, I., López-Gatius, F., Bech-Sabat, G., Santolaria, P., Yániz, J.L., Nogareda, C., Rensis, F. De. and López-Béjar, M., 2007. Climate factors affecting pregnancy rate of high-producing dairy cows in northeastern Spain. Theriogenology, 67, 1379–1385

11. Healy, A.A., House, J.K. and Thomson, P.C., 2013. Artificial insemination field data on the use of sexed and conventional semen in nulliparous Holstein heifers. Journal of Dairy Science, 96, 1905–1914
12. Kurykin, J., Jaakma, U., Majas, L., Jalakas, M., Aidnik, M. and Waldmann, A., 2003. Fixed time deep intracornual insemination of heifers at synchronized estrus. Theriogenology, 60, 1261–1268
13. Kurykin, J., Jaakma, U., Jalakas, M., Aidnik, M., Waldmann, A. and Majas, L., 2007. Pregnancy percentage following deposition of sex-sorted sperm at different sites within the uterus in estrus-synchronized heifers. Theriogenology, 67, 754–759
14. López-Gatius, F., 2012. Factors of a noninfectious nature affecting fertility after artificial insemination in lactating dairy cows. A review. Theriogenology, 77, 1029–1041
15. López-Gatius, F., 2013. Approaches to increase reproductive efficiency in artificially inseminated dairy cows. Animal Reproduction, 10, 143–147
16. Lucy, M.C., 2007. Fertility in high-producing dairy cows: reasons for decline and corrective strategies for sustainable improvement. Society of Reproduction and Fertility supplement, 64, 237–254
17. Mellado, M., Sepulveda, E., Meza-Herrera, C., Veliz, FG., Arevalo, J.R. and Mellado, J., 2013. Effects of heat stress on reproductive efficiency in high milk-yielding Holstein cows in a hot arid environment. Revista Colombiana de Ciecias Pecuarias, 26, 193–200
18. Mellado, M., Sepulveda, E., Macias-Cruz, U., Avendaño, L., Garcia, J.E. and Francisco, G., 2014. Effects of month of insemination on reproductive efficiency of Holstein cows and heifers inseminated with sex-sorted or conventional semen in a hot environment. Tropical Animal Health and Production, 46, 265–269
19. Norman, H.D., Hutchison, J.L. and Miller, R.H., 2010. Use of sexed semen and its effect on conception rate, calf sex, dystocia, and stillbirth of Holsteins in the United States. Journal of Dairy Science, 93, 3880–3890
20. Rath, D. and Johnson, L.A., 2008. Application and commercialization of flow cytometrically sex-sorted semen. Reproduction in Domestic Animals, 43, 338–346
21. Seidel, G.E. Jr., 2003a. Economics of selecting for sex: the most important genetic trait. Theriogenology, 59, 585–598
22. Seidel, G.E. Jr., 2003b. Sexing mammalian sperm—intertwining of commerce, technology, and biology. Animal Reproduction Science 79, 145–156
23. Seidel, G.E., Jr. 2011. Profitable uses of sex-sorted semen. In: Proc. Applied Reproductive Strategies in Beef Cattle, Joplin, MO. University of Missouri, p. 349–353
24. Seidel, G.E. Jr. and J. L. Schenk. 2008. Pregnancy rates in cattle with cryopreserved sexed sperm: effects of sperm numbers per inseminate and site of sperm deposition. Animal Reproduction Science, 105, 129–138
25. Stewart, B.M., Block, J., Morelli, P., Navarette, A.E., Amstalden, M. and Bonilla, L., 2011. Efficacy of embryo transfer in lactating dairy cows during summer using fresh or vitrified embryos produced in vitro with sex-sorted semen. Journal of Dairy Science, 94, 3437–3445

CHAPTER 10

Challenges of feeding dairy cows in Australia and New Zealand

W. J. Wales [A C] and E. S. Kolver [B]

[A] *Department of Economic Development, Jobs, Transport and Resources, Ellinbank, Vic. 3821, Australia.*

[B] *DairyNZ, Private Bag 3221, Hamilton 3240, New Zealand*

ABSTRACT

There is a continuing evolution of feeding systems in both Australian and New Zealand dairy industries and this presents challenges for the future. Since the turn of the century, the two countries have diverged in industry growth characteristics, with Australian dairying having contracted, with 10% less milk being produced because of 20% fewer cows producing 15% more per cow, whereas New Zealand dairying has expanded, producing 83% more milk driven by a 54% increase in cow numbers and a 31% increase in milk production per cow. Solutions to optimise feed efficiency included the common themes of (1) growing more forage on farm, (2) increasing its utilisation and (3) more efficient use of supplements resulting in increases in DM intake, and they remain relevant. In New Zealand, many of the recent research activities have aimed at improving feed supply while limiting environmental impacts driven by increasing societal concern surrounding the environmental footprint of a growing and intensifying agricultural sector. In Australia, many of the recent research activities have aimed at improving feed efficiency, with a focus on understanding situations where partial mixed ration feeding systems (Australian Farm

Systems 3 and 4) are sustainable. Simply growing more feed on farm can no longer be a sole objective; farms must be operated with a view to reduce the environmental footprint, with New Zealand dairy farmers increasingly needing to farm within nitrogen limits. The present review revisits and reinforces many of the concepts developed in previous reviews, but also examines the evolution of feeding systems in both countries and opportunities to improve feed efficiency and profit, while satisfying public expectations around environmental stewardship. We also identify some of the gaps in the current knowledge that warrant further research

Keywords: feeding systems, grazing, intake, milk response.

1. INTRODUCTION

There is a continuing evolution of feeding systems in both Australia and New Zealand dairy industries and this presents challenges for the future. Several reviews have explored these challenges. For example, Thorrold and Doyle (2007) presented a snapshot of the current and future state of dairy farming in New Zealand and Australia. This was followed up by Clark et al. (2013), who focussed on drivers of dairy production in both countries. The present review revisits and reinforces many of the concepts developed in the previous work, but also looks at the evolution in feeding systems in both countries and opportunities to improve feed efficiency and profit, while satisfying public expectations around environmental stewardship. We also identify some of the gaps in the current knowledge that warrant further research.

2. EVOLUTION OF FARM AND FEEDING SYSTEMS SINCE THE 1980S

2.1 National

Dairying in Australia and New Zealand has undergone a significant structural change over the past three decades, summarised by declines in the number of farms and an increase in total milk produced (Table 1). However, since the turn of the century, the two countries have diverged in industry growth characteristics. Australian dairying has contracted, with 10% less milk being produced because of 20% fewer cows producing 15% more per cow (L/cow; Dairy Australia 2015a). Over a similar period, New

Zealand dairying has dramatically expanded, producing 83% more milk, driven by a 54% increase in cow numbers and a 31% increase in milk production per cow (kg fat+protein)/cow; DairyNZ 2015). Both countries have seen the development of fewer, larger herds, although the effect has arguably been greater in Australia where there were fewer than half the number of dairy farms operating in 2014–2015 than there were 15 years earlier. In New Zealand, the expansion of dairying into Canterbury and Southland has accelerated the increase in cow numbers and herd size.

Table 1. Australian and New Zealand (NZ) dairy-industry statistics for 1979–1980, 1999–2000 and 2014–2015 (Dairy Australia 2015a; DairyNZ 2015) F+P, milk fat + protein

Parameter	1979–1980	1999–2000	2014–2015
	Australia		
Farm number	21 994	12 896	6128
Cow number (million)	1.88	2.17	1.74
Total milk production (million L)	5432	10 847	9731
Milk production per cow (L)	2848	4996	5731
Milk production per cow (kg, F+P)[A]	–	–	430
	New Zealand		
Herd number (NZ)	16 506	13 861	11 970
Cow numbers (million, NZ)	2.05	3.27	5.02
Total milk production (million L)	5997	11 630	21 253
Milk production per cow (L)	3000[A]	3600[A]	4235
Milk production per cow (kg, F+P)	260	288	377

[A]Calculated.

During the past 35 years, farming systems in both countries have intensified in terms of stocking rate and the proportion of imported supplementary feed used on-farm, and a range of farming systems now exists in Australia and New Zealand. Both countries have attempted to describe the characteristics of these farm systems on the basis of differences in key feed inputs, although it is noted that these categories are somewhat arbitrary as there is a continuum of feed combinations in practice (Dairy Australia 2015b; DairyNZ 2016a). The five farm-system categories that have been developed in each country define the timing, purpose and amount of grazing and supplementary feed use, ranging from grazing only and little use of supplementary feed to zero grazing and a high reliance on purchased feeds. The range in systems in Australia is broader than that in New Zealand, with New Zealand categories restricted to approximately the first three systems defined for Australia. The five Australian feeding systems are as follows:

- System 1: low bail system, including grazed pasture + other forages + up to 1.0 t/cow.year of concentrates fed in the dairy shed during milking;
- System 2: moderate-high bail system, including grazed pasture + other forages + more than 1 t/cow.year concentrates fed in the dairy shed during milking;
- System 3: partial mixed ration (PMR) system, including pasture grazed for all year + PMR on feed pad ± concentrates fed in the dairy shed;
- System 4: hybrid system, including pasture grazed for a period of less than 9 months per year + PMR on feed pad ± concentrates fed in the dairy shed;
- System 5: total mixed ration (TMR) system, i.e. zero grazing, where cows are housed and fed a total mixed ration.
- The five New Zealand feeding systems are as follows:
- System 1: all grass self-contained, all stock on the dairy platform;
- System 2: ~4–14% of total feed is imported, either supplement or grazing off, and fed to dry cows;
- System 3: ~10–20% of total feed is imported to extend lactation (typically autumn feed) and for dry cows;
- System 4: ~20–30% of total feed is imported and used at both ends of lactation and for dry cows;
- System 5: at least 30% of total feed is imported and used all year, throughout lactation and for dry cows.

In the most recent survey conducted during the 2014–2015 season, Dairy Australia (2015b) reported that 23% of farms across Australia employed feeding systems where less than 1 t grain/cow per year was fed (Australian System 1), representing a 5% unit decline from the 2010–2011 survey (Dairy Australia 2011). In the current survey, 64% of dairy farmers fed moderate to high amounts of concentrate in the dairy (>1 t/cow per year; Australian System 2); an increase of 10% units compared with the results from the 2010–2011 survey. Farms where grazed pasture was supplemented with PMR offered on a feed pad (Australian Systems 3 and 4) comprised 12% of dairy farms and 1% of farms employed systems characterised as Australian System 5 (TMR).

Farms identified as being in Australian Systems 1 and 2 have increased from 82% to 87% nationally in the 4 years up until 2014–2015, with a concomitant reduction from 18% to 13% in farms employing Systems 3–5. This reflects improvements in home-grown forage supply due to better seasonal conditions, meaning that costs associated with feeding supplements with a mixer wagon could be reduced or avoided.

At a regional level in Australia, ~10% of dairy farms in Tasmania and Gippsland in Victoria do not feed any grain, which contrasts with the remaining regions where this number is less than 4%. Approximately 20% of dairy farms in each of the regions of Queensland, New South Wales, Western Australia and the Murray region of northern Victoria–southern New South Wales use PMR and hybrid systems compared with the other regions, where this percentage is less than 10% (Dairy Australia 2015b). Much of the recent nutrition research has focussed on these regions that use more PMR.

In New Zealand, over the same period (2010–2011 to 2014–2015), the proportion of herds in low-input systems (Systems 1 and 2) reduced from 42% to 32%, was maintained at 39% in medium-input systems (System 3), and increased from 19% to 29% in high-input systems (Systems 4 and 5; DairyNZ 2016a). New Zealand farm systems are now broadly represented across low-, medium- and high-input classifications. Farmers have intensified their systems through increased supplement use and higher stocking rates in response to high milk and land prices, dry seasonal conditions, increased feeding infrastructure on farms, and because of dairy expansion in the South Island, which requires off-farm winter grazing. Conversely, many farmers during the 2015–2016 and 2016–2017 season reviewed their farm systems and adjusted to the low milk prices by reducing stocking rates, keeping more young stock and wintered cows on the milking platform as well as reducing expenditure on harvested and imported supplementary feeds. It is unclear at this stage, whether these adjustments will be sustained in the future or whether this is a short-term adjustment. Also driving the growth of higher-input systems has been the expansion of dairying in the South Island, which requires more off-farm winter grazing.

At a national level, this intensification has resulted in broadly similar or improved profitability. The average financial performance of these low-, medium- and high-input systems over the past 10 years (2005–2006 to 2014–2015) shows that profitability for low, medium, high systems has been $NZ1823/ha (range $NZ871–2905/ha), $NZ2004/ha ($NZ814–

3363/ha) and $NZ2202/ha ($NZ228–3797/ha) respectively; Return on assets has been 6.6% (–4.9–14.8%), 6.6% (–7.0–16.2%), and 7.4% (–7.8–18.7%) respectively; equity growth has been 6.4% (–12.9–17.7%), 6.9% (–14.4–19.3%) and 8.5% (–15.0–25.2%) per annum; and closing term liabilities (as a measure of risk) have been $NZ19.4/kg milk fat + protein (F+P; $NZ12.9–23.2/kg F+P), $NZ18.8/kg F+P ($NZ13.5–21.5/kg F+P), and $NZ18.8/kg F+P ($NZ14.5–21.3/kg F+P) (from DairyNZ 2008, 2010, 2012, 2014, 2016a; note return on assets and equity growth not reported for 2005–2006). At a national level, this implies that a profit response to the additional feed eaten has resulted from the on-farm changes that have occurred during the past 10 years; however, there is a wide range around these average 10-year returns. In particular, returns from the more intensive systems have been more volatile than those produced from lower-input systems. The milk price environment within which these returns were achieved was representative of the range of expected milk prices, averaging $NZ6.09/kg F+P ($NZ4.14–7.69/kg F+P)

3. FEEDS

The evolution in feeding systems from farms dominated by homegrown forage only, to those relying on more concentrates and imported feeds, has been greatest in Australian systems, which now average 1.6 t DM/cow annually, having increased from 0.7 t DM/cow 25 years ago (Dharma et al. 2012). These changes have occurred due to the relatively low cost of cereal grain compared with other energy-based supplements and motivation to increase production per cow and per hectare (Jacobs 2014).

In New Zealand, despite the increased use of supplementary feed, pasture continues to be the dominant feed offered to cows (82% of the diet). The total amount of feed eaten by dairy cows at a national level has increased 161% over the past 25 years; a growth rate of 3.8% (compound annual growth rate, CAGR; DairyNZ 2016b; Table 2). This increase has occurred primarily (80%) through more cows (5.02 vs 2.4M cows), from both new dairy land and increased stocking rates. On a per cow basis, feed eaten has increased 1 t (0.8% CAGR) to 4.93 t DM/cow.

Imported supplements, including palm-kernel extract (*Elaeis guineensis*), increased 7.7% CAGR, harvested supplement including maize (*Zea mays*) silage and barley (*Hordeum vulgare*) increased 6.2% CAGR, while grown crops including fodder beet (*Beta vulgaris*), kale (*Brassica oleracea*) and swedes (*Brassica napobrassica*) increased 4.9% CAGR between 1990–1991 and 2014–2015.

Table 2. Feeds eaten in New Zealand in 1990–1991 compared with 2014–2015 (DairyNZ 2016b)

Time period	Total feed eaten	Pasture eaten	Crop eaten	Harvested supplement	Imported supplement
		t DM/cow			
1990–1991	3.94	3.77	0.03	0.08	0.05
2014–2015	4.93	4.04	0.18	0.26	0.44
		t DM/ha			
1990–1991	9.24	8.86	0.08	0.19	0.11
2014–2015	14.16	11.61	0.52	0.76	1.27

Non-pasture feeds consumed by dairy cows increased from 0.16 t DM/cow in 1990–1991 to 0.89 t DM/cow in 2014–2015. As a proportion of a cow's total diet, non-pasture feeds increased from 4.2% in 1990–1991 to 18.0% in 2014–2015, being made up of harvested crop, imported supplements and grazed crop.

The amount of pasture eaten per cow increased in the 1990s, but has eased during the 2000s. However, pasture eaten per hectare has increased at a rate of 1.7% per year to 2006–2007, but growth has slowed to a rate of 0.7% since.

Feed eaten per hectare has increased 53% (+1.7% CAGR) or 4.9 t DM/ha from 1990–1991 to 2014–2015. This is a faster increase than a per cow basis as it includes an increase in stocking rate. Pasture eaten accounted for just over half the increase in feed eaten per hectare, while imported supplements (+1.2 t DM/ha), harvested supplement (+0.6 t DM/ha) and grown crops (+0.4 t DM/ha) have all increased over the past 25 years.

Compared with Australia, greater use has been made in New Zealand of palm-kernel extract, maize silage and homegrown crops as grain and concentrate supplements are comparatively more expensive.

The increased use of maize silage since the 1990s has driven the growth in non-pasture feed, as has palm-kernel extract since the drought of 2008 (Table 3). More recently, the rapid increase in fodder beet use in winter, and increasingly in summer, has displaced kale and other brassica crops.

Table 3. Non-pasture feed (t DM) eaten in New Zealand, and the proportional composition of non-pasture feed in 1990–1991, 2002–2003 and 2014–2015 (DairyNZ 2016b) Other brassicas include rape, turnips, swedes and other brassicas. Other supplements include brewer's grain, tapioca, soybean meal, cottonseed, molasses, proliq, imported grains, fishmeal, vegetable by-products, horticultural wastes and any other feed type fed to dairy cows

Parameter	1990–1991	2002–2003	2014–2015
Non-pasture feed (t DM)	395 830	1 211 207	4 454 136
Palm-kernel extract (%)	0.0	1.8	32.6
Maize silage (%)	11.8	28.4	23.0
Grains (%)	27.8	13.4	5.7
Cereal whole crop silage (%)	10.2	9.7	0.9
Fodder beet (%)	0.0	0.1	15.7
Kale (%)	11.5	5.4	2.7
Other brassica crops (%)	9.6	6.1	2.0
Other supplements (%)	29.1	35.1	17.3

3.1 Feed conversion efficiency

The combined effects of cow genetic gain over the past 25 years, improved farm management, longer lactations, improved pastures and increased use of supplementary feed have increased the feed conversion efficiency of the New Zealand national herd. The national herd in 2014–2015 produced more milk F+P as a percentage of liveweight (82% vs 52% of liveweight), produced more milk fat+protein from each tonne of DM (76 vs. 62 kg MS/t DM eaten) and required less feed to produce a kilogram of milk fat+protein (13 vs 16 kg DM/kg MS) than it did in 1990–1991 (DairyNZ 2016b).

The differences between Australian and New Zealand dairying systems reflect slightly different economic drivers. As an export nation subject to large year-to-year commodity-price fluctuations, New Zealand has evolved low to moderate production cost systems where pasture makes up the majority (82%) of the diet. Although subject to similar commodity volatility, tempered somewhat by a larger domestic dairy market, Australia has evolved a wider range of farm systems that include higher feed-input systems, which have more of a focus on per cow production and efficiency. This has been a result of the domestic production of low-cost cereal grains along with volatility of climate and availability of water for forage production. These differences have meant that, in recent years, the focus of feeding research in New Zealand has largely been on growing and utilising homegrown forage within anticipated future environmental constraints, while Australia has focussed on improving the efficiency of responses to feeds imported into the farm system.

Profitability in dairy production systems in both countries has in the past, and is still, dependent on efficient use of the farm resources that include cows, land and water, labour, capital and knowledge (Beever and Doyle 2007). As milk production has increased through improvements in genetic merit, cows require more digestible nutrients to avoid excessive body condition loss and ensure appropriate reproductive performance and longevity. This can be achieved through the provision of a well formulated TMR (Kolver and Muller 1998; Bargo et al. 2002); however, the challenge remains of how to increase nutrient intake in systems where pasture makes up a large part of the diet. It is well established that grazing cows consume less than their potential intake, due to factors listed under sward, animal and management factors (Wales et al. 2005; Beever and Doyle 2007). Overcoming the seasonality of pasture growth and of nutritive characteristics also continues to be a challenge for both countries (Chapman et al. 2009).

3.2 Limitations within current feeding systems

In 2005, Garcia and Fulkerson reviewed the opportunities for improvement in the Australian dairy industry, following 10 years of only modest increases in total factor productivity. Their analysis identified below-optimal feed efficiency as a key issue affecting profitability of dairy farmers into the future. The solutions to optimise feed efficiency included the common themes of (1) growing more forage on farm, (2) increasing its utilisation and (3) more efficient use of supplements. These solutions result in increases in DM intake and remain relevant to both Australia and New Zealand.

Dairy cows approach their genetic potential for DM intake and milk production on TMR feeding systems. For example, Kolver and Muller (1998) reported that cows in early lactation consuming TMRs produce 44 kg milk/day, compared with 30 kg milk/day by similar cows grazing pasture without supplements. This increase in milk production was a result of nutrient intake approaching a maximum through a combination of greater DM intake and a ration formulated to provide a more optimal mix of nutrients to meet requirements. This raises the question as to what factors are limiting milk production in our grazing systems and what can be done to remove these limitations?.

Milk production of cows relying solely on grazed herbage is constrained to less than 30 kg/day (~2.3 kg milk fat+protein/day) by seasonal variations in pasture growth and nutritive characteristics (Doyle et al. 2001). A key limitation is the quantity (both under and over supply) of nutrients supplied from grazed pasture, particularly in the more variable Australian environment. Although temperate pasture can provide a high-quality, balanced feed for moderate-yielding cows (Hills et al. 2015), supplementary feeds are used in grazing systems as a grazing management tool, when the supply of pasture is insufficient for herd demand, or when pasture does not supply adequate nutrition to support high production, as is the case on many Australian farms.

A key factor influencing the nutrient supply is amount of pasture available for grazing in each season. The seasonal growth or net pasture accumulation rate of pasture varies considerably in all dairy regions in both countries, with no 2 years following the same pattern of DM supply (Chapman et al. 2009). Figure 1, adapted from Cullen et al. (2008), highlights this seasonal variation and is a result of variations in local climatic conditions, nutrient supply, pasture species, access to irrigation water and management practices. Overcoming this seasonal variation in nutrient supply from pasture remains an important challenge. In addition to variability in DM supply, the nutritive characteristics of the component of the sward consumed by grazing cows is also affected by climatic, soil and plant factors due to such things as adequacy of fertiliser and water, and species and growth stage of the plant. Typically, the metabolisable energy (ME) and crude protein concentrations of the sward decline through spring, reaches nadir during summer and increases in autumn. There are also regional variations in the nutritive characteristics (Doyle et al. 2001).

Conceptually, the ideal sward would have characteristics, nutrient profile and physical attributes similar to a TMR formulated to provide nutrients in relation to requirements, while having physical characteristics necessary to maintain a stable rumen environment. However, grazed pasture never provides the mix of nutrients provided in a well formulated TMR. For example, the nutritive characteristics of a mixed grass–clover sward compared with a TMR of similar digestibility, had lower concentrations of non-structural carbohydrates and higher concentrations of neutral detergent fibre, in addition to the physical differences of longer particle sizes and lower % DM (Kolver and Muller 1998). Cows grazing this type of sward also consumed less DM, starch and ME, and more crude protein and water, due to the low DM concentration of the herbage. In comparison with TMR, high-producing cows consuming only grazed pasture had lower DM intake, milk production, milk protein and fat concentrations, lost more body condition and had lower liveweight (Kolver and Muller 1998; Bargo et al. 2002).

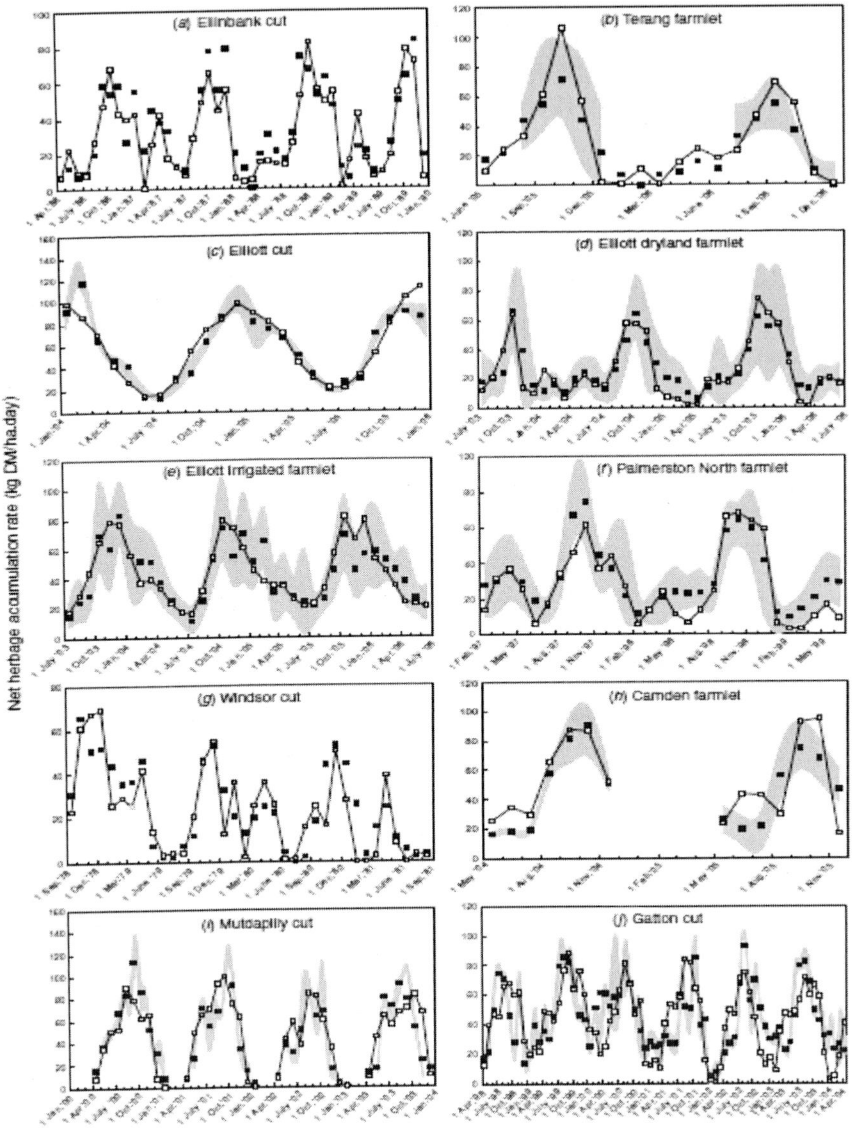

Figure. 1. Measured (■) and modelled (□) monthly mean daily herbage accumulation rates (kg DM/ha.day), with measured variability (grey shaded) for (*a*) Ellinbank, Victoria, Australia, (*b*) Terang, Victoria, Australia, (*c*) Elliott, Tasmania, Australia, (*d*) dryland farmlet, Elliott, Tasmania, Australia, (*e*) irrigated farmlet, Elliott, Tasmania, Australia, (*f*) Palmerston North, North Island, New Zealand, (*g*) Windsor, South Island, New Zealand, (*h*) Camden, New South Wales, Australia, (*i*) Mutdapilly, Queensland, Australia, and (*j*) Gatton, Queensland, Australia. Adapted from Cullen et al. (2008).

Differences in milk production between pasture and TMR was predominantly driven by the lower DM intake of pasture, rather than differences in the proportion or digestibility of individual nutrients. At grazing, DM intake would need to approach 5% of a cow's bodyweight to achieve intakes similar to those observed for cows consuming the TMR. However, Kolver and Muller (1998) concluded that current pasture species are unlikely to provide nutrients in sufficient quantities to achieve milk yields similar to those in cows consuming TMR. As forages are unlikely to provide the ideal mix of nutrients, identifying profitable supplementation strategies will continue to be important.

4. FACTORS LIMITING INTAKE OF GRAZING COWS

In all grazing systems, herbage allowance has the biggest effect on DM intake with characteristics of the sward, such as pasture mass, spatial distribution of plants and their nutritive characteristics also being important (Wales et al. 2005). Additional nutrients are commonly provided by cereal supplements around milking time, with further supplements of pasture silages and hays and maize silage/palm-kernel extract. These types of feeding systems have intake limited to ~4% of liveweight, equivalent to 18 and 24 kg DM for a 450- and 600-kg cow respectively (Doyle et al. 2001). The genetic potential for intake under ideal feeding management may be 5% of liveweight, equivalent to intakes of 23 and 30 kg DM for a 450- and 600-kg cow respectively. Several factors are purported to limit intake in grazing systems and overcoming these limitations remains a key challenge for our industries.

The concept of rumen fill is often brought up as limiting intake, due to breakdown of structural carbohydrates, with neutral detergent fibre (NDF) being rate limiting (Mertens 1987). This concept has been useful when applied to TMR fed to appetite; however, the regulation of intake in grazing systems is more complex because of interactions between signals generated by supply of energy and essential nutrients from the digestive tract and intermediary metabolism (Weston 1982). Rumen fill appears to have a role in intake regulation when digestibility of the diet is less than 75% (Dove 1996) and Ketelaars et al. (1992) suggested that the intake of highly digestible feeds (i.e. >75% DM digestibility) is physiologically determined and physical restrictions to intake are less important. For example, rumen fill was lower on Persian clover (*Trifolium resupinatum*) than on perennial ryegrass (*Lolium perenne*) of similar digestibility but higher NDF (Williams et al. 2005), and in grazing cows consuming different amounts of Persian clover there were no differences in average rumen fill, DM or NDF loads over a range in intake from 5.6 to 20.4 kg DM/cow.day (Williams et al. 2014).

At grazing, factors such as the time available for grazing and rumination (Rook 2000; Gregorini 2012), dietary preferences (Provenza 1995), sensory factors such as palatability (Weston 1982) and management strategies (Gregorini 2012) also play roles in intake regulation. The complexities influencing intake mean that predicting intake from simple relationships based on single factors, such as digestibility or NDF concentration, will not be universally applicable across the extremes of grazing to TMR feeding systems. Therefore, when attempting to increase intake above current barriers, it is necessary to consider sward and plant characteristics and the complex signals between the digestive tract and digestive process, intermediary metabolism and sufficiency of energy and essential nutrients, and behavioural signals.

Another factor, often overlooked, is the depressive effects on intake of diets that are low in DM. For dairy cows, the critical water content was estimated to be 82% (18% DM), with a depression of 0.34 kg DM intake for each percentage increase in water content above this amount (Verite and Journet 1970). When water was added to the rumen *per fistulum*, Cabrera Estrada et al. (2004) showed that intake and eating rate were restricted by internal water of grass, but not by external water, suggesting that the effects of water content may be associated with the large volumes of fresh herbage that need to be processed during digestion. Water content of highly digestible pastures can be as high as 88% water, generally during winter and spring, which would restrict the intake by ~2 kg DM/cow.day.

To maximise intake from pasture, cows must consume plants that degrade quickly and are highly digestible, leading to fast rates of passage through the rumen, with intake determined by the rate of eating (bite mass by bite rate) multiplied by time spent eating (meal duration by number of meals; Allden and Whittaker 1970; Rook 2000). Grazing ruminants vary bite mass, bite rate and grazing time in response to changes in sward conditions (Hodgson 1981; Milne et al. 1982; Penning et al. 1991; Gibb et al. 1997). Bite mass, for example, increases with increased sward height and bulk density, ranging from 0.25 to 4 g DM in cattle offered micro-swards of lucerne (*Medicago sativa*) and paspalum (*Paspalum dilatatum*) (Ungar 1996) and the rate of biting is related to the ease of prehensions, herbage shear force and bite mass, as smaller mouthfuls often lead to an increase in bite rate.

Sward height and sward density also influence hourly intake rate via effects on bite mass (Mayne et al. 1997). Intake of fasted cows in short-term grazing experiments was maximised at between 3.5 and 4 kg DM/h when sward heights were greater than 18 cm (measured with a sward stick). However, intake rate was also high at 3 kg DM/h, when sward height was

15 cm. With shorter swards, bulk density becomes important, with intake rates varying from 1 to 2.5 kg DM/h for swards varying in bulk density from 1.7 to 3.1 kg/m^3. Thus, high intake rates may be achieved by grazing dairy cows offered swards taller than 15 cm or dense swards below 15 cm. However, these high, short-term intake rates are difficult to translate into grazing systems as swards change during grazing events where grazing management and optimum allowance of pasture on offer are important.

5. RECENT RESEARCH ACTIVITIES AIMED AT IMPROVING FEED SUPPLY, WHILE LIMITING ENVIRONMENTAL IMPACTS

Recent feedbase and nutrition research in New Zealand has focussed on the first two themes described by García and Fulkerson (2005) of growing more forage on farm and increasing its utilisation, while also limiting environmental impacts. These solutions result in increases in DM intake and remain relevant to both Australian and New Zealand. In New Zealand, there has been increasing societal concern surrounding the environmental impact of a growing and intensifying agricultural sector (Ministry for the Environment and Statistics New Zealand 2015). In various regions, the dairy industry is facing water-quality regulations, with a strong focus on nitrogen (N) leaching. Consequently, the dairy industry faces increasing pressure to improve efficiency and productivity as well as environmental performance (Monaghan and de Klein 2014). Simply growing more feed on farm can no longer be a sole objective; it must be done with a view to reducing the environmental footprint, with New Zealand farmers increasingly needing to farm within N limits.

5.1 Forages for reduced nitrate leaching

The Forages for Reduced Nitrate Leaching program is a cross-sector approach to developing profitable farming systems that enhances N-use efficiency, and reduce environmental impact for arable, dairy, sheep and beef farms. It has explored optimised pasture mixtures and cropping sequences to meet a research target of 20% reduction in nitrates leached per hectare.

Perennial ryegrass and white clover (*Trifolium repens*) are the most widely used species in New Zealand grazing swards. A promising novel strategy to reduce N leaching is the inclusion of more plant species to these traditional swards (Pembleton et al. 2015). Species such as chicory (*Cichorium intybus*), plantain (*Plantago major*) and lucerne are among the key species being investigated. Such diverse sward mixtures can reduce N leaching by reducing the amount and concentration of N excreted in the urine (Pembleton et al. 2015). This can arise from several mechanisms, including the lower N content of diverse swards, an improved ratio between water-soluble carbohydrates and protein, or the presence of plant secondary compounds that may increase water intake and urination volumes (Totty et al. 2013). In addition, chicory and plantain tend to have higher water contents than do grasses and may act as diuretics, increasing the total volume of urine and, hence, diluting the N concentration of urine patches (Pembleton et al. 2015). Consequently, the inclusion of diverse swards in the diet of dairy cattle has been shown to achieve reductions in the excretion of total urinary N in the order of 20–50% (Woodward et al. 2012). This, together with observed increases in urine volume, has led to 20–60% reductions in N concentrations observed in the urine of cattle grazing diverse swards (Woodward et al. 2012; Totty et al. 2013; Edwards et al. 2015). Plantain appears to be the component of a diverse pasture that drives the reduced urinary N concentrations (Edwards et al. 2015).

In line with these results, system modelling has indicated that reductions in N leaching of 11–19% can be achieved at the farm scale, if 20–50% of the milking platform is allocated to diverse swards (Beukes et al. 2014). Milk production was maintained or increased slightly in this study, and is reflective of forage yields achieved from diverse pastures similar to those from traditional pastures, with diverse-pasture production typically being stronger during summer in non-irrigated areas (Woodward et al. 2013).

Outstanding issues with diverse pastures include sward persistency beyond 3–4 years, and development of optimised management, and the resultant profitability when used by farms that have regulatory N limits.

Other crops and forages have also been shown to be useful in providing good yields and reduced N leaching. Low-N crops such as fodder beet, which can grow more than 24 t DM/ha (Chakwizira et al. 2016), have the potential to reduce urinary N (Edwards et al. 2014). Italian ryegrass (*Lolium multiflorum*), which has high winter growth rates, can also reduce N leached by more than 25% as a result of increased daily N uptake during winter (Malcolm et al. 2014).

The grazing of any winter crop contributes a large proportion of total N leached from the dairy operation. This is due to high crop yields leading to higher stocking densities at a time of year when leaching risk is greater. Following the winter feeding of the forage crop, the land is typically left fallow until a new crop is sown in mid-spring or returned to new pasture. Catch crops, such as oats (*Avena sativa*), can be used to cover fallow ground and take up urinary N deposited during winter grazing. This practice can reduce nitrate leaching by 30%, while boosting annual crop yield by 6–12 t DM/ha (Inside Dairy 2016).

5.2 Pastoral 21

A large national set of farm-system trials, Pastoral 21 Phase 2 (P21), was initiated over a 5-year period (2011–2016) in New Zealand, with the aim of addressing these challenges by developing industry accessible, adoptable, system-level solutions for profitably increasing production, while reducing environmental footprint. A key outcome was to field test the proposed solutions for demonstrable efficacy and value (Chapman et al. 2012).

Results from Waikato and Canterbury region multi-year farm-scale studies showed that large reductions in N leached, in the order of 30–50%, can be achieved with minimal impact on production and profitability (within 5%; Macdonald 2014). In the Waikato, the 4-year farm study of a combination of reduced N fertiliser (46 vs 137 kg N/ha), use of standoff pads at strategic times, limited amounts of a low protein feed (267 vs 0 kg grain/cow), a lower stocking rate (2.6 vs 3.2 cows/ha) and improved cow genetics (199 vs 129 breeding worth, as at 2014) showed that there are feed and management options available to reduce N and P losses. The P21 program is now investigating how to boost profits while retaining the N leaching gains that have been made. Urine spreading and diluting technologies are being tested (Selbie et al. 2015), and may provide a more cost-effective mitigation option.

Similar results (30% reduction in N leached) were achieved over 3 years at a farm scale in Canterbury, with a combination of a lower stocking rate (3.5 vs 5.0 cows/ha), reduced N fertiliser (158 vs 311 kg N/ha), less brought in supplement (0.34 vs 4.95 t DM/ha) and the use of diverse pastures (ryegrass and white clover with chicory, plantain, and red clover (*Trifolium pratense*)) and a catch crop following the winter kale crop (Macdonald 2014). These options are currently being tested on commercial farms.

5.3 Chicory and plantain

Pasture-based dairy farms often face feed shortages during periods of summer–autumn soil moisture deficits, and may require supplementary feed to maintain milk production. Chicory and plantain can produce more DM in summer with superior nutritive characteristics compared with perennial grasses (Li and Kemp 2005; Moorhead and Piggot 2009). Chicory and plantain can also increase milk production (Waugh et al. 1998; Chapman et al. 2008; Pembleton et al. 2016). DairyNZ developed the chicory and plantain research program, which ran from 2010 to 2013, to determine the milk production response to increasing amounts of chicory and plantain in the diet, to develop grazing management guidelines and to understand the impact of chicory and plantain on farm profitability.

In two summer experiments, dairy cows were offered perennial ryegrass-based pasture or ryegrass-based pasture plus either chicory or plantain at 20%, 40% or 60% of the diet (Minnee et al. 2012). When the ME of ryegrass was moderate (10.5 MJ/kg DM), the milk fat+protein yields were similar for all treatments. However, when the ME of the ryegrass dropped to 9.6 MJ /kg DM, cows fed herbs ate ~1 kg DM more per day and milk fat+protein production was 17% higher than for cows fed only ryegrass-based pasture. A third summer experiment included maize silage in the diet with herbs and pasture. At similar feed allowances and intakes, cows fed pasture plus chicory or plantain produced 13% more milk fat+protein than did cows fed only pasture. When maize silage was added to the pasture plus herb diet, milk fat+protein production was increased by 7% (compared with pasture only) but the cows consumed on average 2 kg DM/day more than did cows on the other diets. This implied that the conversion of DM into milk fat+protein was less efficient when maize silage comprised part of the diet, with insufficient intake of crude protein implicated; however, a longer-term study is needed to determine whether this additional feed was allocated to body condition rather than milk.

Grazing management guidelines have been developed from research that indicates that grazing frequency has the greatest impact on yield (Lee et al. 2012). In the first year after spring sowing, grazing chicory and plantain at 25–35 cm height increased herbage production, while in the second year, both species should be grazed at 25 cm height to limit reproductive stem growth and maintain nutritive characteristics.

The value of chicory crops to perennial ryegrass-based dairy farms in the Waikato was assessed using non-linear optimisation models of three farms differing in the degree to which imported supplement was used (Romera et al. 2015). Results indicated that at a milk price of $NZ7/kg fat+protein, farmers were unlikely to receive adequate reward for the additional complexity arising from the utilisation of chicory crops, especially when imported supplement provides more flexibility.

5.4 Future feed system

Predicting future production systems and feeding trends is impossible and the most that can be achieved is a presentation of alternative futures in light of such things as volatile seasonal weather conditions and environmental pressures to reduce nutrient losses and different options farmers may take to achieve targets.

An attempt at what the typical New Zealand dairy farm may look like in 2030–2031 was made by forecasting two future farming scenarios in 2030–2031, on the basis of current trends in types of feed used on-farm, stocking rates, farm growth and milk fat+protein production (DairyNZ 2016b); in the modest milk fat+protein growth scenario, milk production increased by 0.8% per year, and the high-growth scenario assumed milk production would increase at twice this rate, namely 1.7% per year. There were two pathways to the modest-growth scenario, either through holding stocking rates constant and a modest increase in milk fat+protein per cow or by reducing stocking rates and increasing milk fat+protein per cow more to achieve the same overall production. The high-growth forecast scenario also assumed that stocking rates would increase from 2.8 to 3.0 cows/ha.

Annual per cow feed demand was derived on the basis of estimated milk fat+protein per cow and constant average cow liveweights. Two pathways were developed for the high-growth scenario, with either pasture and crop or imported supplementary feed being the drivers of the additional feed consumed, while for the modest-growth scenario, it was assumed that the growth would be driven from increased pasture and crop eaten per cow.

A summary of the key feed use per hectare in 2030–2031 is provided in Table 4. The modest-growth scenario with constant stocking rates shows feed eaten increasing to 14.4 t DM/ha, with the increase occurring from more pasture and fodder beet consumed, while palm-kernel extract imports and maize silage decline. The second modest-growth scenario with reduced stocking rates shows that feed eaten per hectare reaches only 14.1 t DM/ha, with the difference being lower pasture consumed per hectare.

Table 4. Future New Zealand growth scenarios (DairyNZ 2016b)

2030–2031	Modest growth, constant SR	Modest growth, low SR	Fast growth, supplements	Fast growth, pasture and crop
Milk fat + protein (billion kg)	2.1	2.1	2.3	2.3
Milk fat + protein (kg/cow)	403	426	415	415
Milk fat + protein (kg/ha)	1127	1127	1245	1245
Stocking rate (cows/ha)	2.80	2.65	3.00	3.00
Cows milked	5.15	4.88	5.52	5.52
Feed eaten (t DM/ha)	14.40	14.08	15.70	15.70
Pasture eaten (t DM/ha)	11.94	11.62	12.41	12.90
Crop eaten (t DM/ha)	0.97	0.97	0.99	0.96
Harvested supplements eaten (t DM/ha)	0.56	0.56	0.80	0.63
Import supplements eaten (t DM/ha)	0.93	0.93	1.49	1.21

In the fast-growth supplements scenario, total feed eaten increases to 15.7 t DM/ha due to increased milk fat+protein production. Most of the additional feed consumed will be from increased palm-kernel extract, maize silage and barley. In the fast-growth pasture and crop scenario, the increase in feed eaten will be driven by twice as much fodder beet, while the volumes of imported supplements were capped at current levels. Pasture eaten per hectare is 1 t higher than the modest-growth constant stocking-rate scenario.

The modest-growth scenario is the most likely dairy industry pathway into the future, although if more productivity improvements occur over the next decade, the fast growth pasture and crop scenario is considered more likely. Other factors not considered in the analysis, but which will clearly influence the production systems, include such things as consumer acceptance, social licence to operate, natural capital and compliance for market access, emissions, sustainability-framework targets, competing demands on resources and emergence of transformative technologies.

6. RECENT RESEARCH ACTIVITIES AIMED AT IMPROVING FEED EFFICIENCY

In Australia, much of the recent funding from the dairy industry and state governments has supported research focusing on understanding when PMR feeding systems (Australian Systems 3 and 4) are profitable, and fit into the third theme of more efficient use of supplements, as described by García and Fulkerson (2005). This commenced immediately following 10 years of below-average rainfall across Australia, characterised by reduced availability of irrigation water and reduced pasture DM production (Dairy Australia 2011), with research and development supported in Victoria, Queensland and Western Australia. This drought led to an increased reliance on purchased forage and concentrates on at least 18% of Australian dairy farms. In addition, a wider range of supplements was used, including an array of by-products that created challenges in providing a consistent diet with optimised nutrient concentrations.

Feeding systems based on the provision of mixed rations to dairy cows grazing perennial pastures (termed PMR systems) have presented the dairy industry with opportunities for improved intake, production and feed efficiency, but also significant challenges (Wales et al. 2013). There is an expectation that milk production per cow will increase as systems use higher amounts of imported feed. However, results from a survey conducted in 2010 (Dairy Australia 2011) showed that PMR systems were not performing demonstrably better than were those using large amounts of grain offered in the dairy at milking time, despite having opportunities to feed formulated rations. In 2010, the advantage of PMR over high-grain feeding in the bail was 480 versus 471 kg fat+protein/cow.year (~6400 versus 6200 kg milk/cow.year). This lack of a benefit provided the impetus for research to quantify the benefits of PMR systems. To this end, improvements in PMR systems have remained modest, with the latest survey statistics (2014–2015) showing about the same advantage of ~200 kg in average milk production from cows on PMR systems that produced a total of 6600 kg/year.

In Victoria, a series of 11 short-term grazing experiments and six digestibility experiments were conducted in a temperate, high-rainfall environment. The objectives were to define the marginal milk production responses to PMR consumed with limited amounts of pasture, and to quantify the contribution to increased milk production from increased intake and DM digestibility, and hence ME supply. To achieve the objectives, the following four main research questions were addressed: (1) can milk responses be improved by feeding PMR based on cereal and maize grain compared with feeding equivalent ME as cereal grain fed in the dairy and forage in the paddock; (2) can replacement of some wheat with canola meal (*Brassica napus*) in the PMR increase intake and milk production further; (3) can DM intake and milk responses be increased through increasing the herbage allowance; and (4) are there negative impacts on health and welfare of cows in PMR systems compared with systems where supplements are consumed in the dairy?

6.1 Milk responses can be improved by feeding PMR

The PMR component of the diets was formulated to be isoenergetic, with supplements fed in the dairy on the basis of NRC (2001). Overall, 10 kg DM of total supplement (comprising ~70% concentrate and ~30% forage) was required for responses of increased DM intake and increased production of energy-corrected milk (ECM) for cows consuming cereal/maize-based +/- canola meal+forage PMR compared with cereal grain fed in the dairy and forage in the paddock (Auldist et al. 2013, 2014, 2016; Golder et al. 2014). Also, milk production reached a maximum at higher amounts of total supplement intake. For example, Auldist et al. (2013) reported that maximum milk production occurred when cows were consuming 1.5 kg more of supplement DM as a PMR than when they consumed cereal grain in the dairy. These conclusions hold for situations where PMR is offered as two equal amounts after milking, but not for situations where it is offered as a single feed once per day (Wright et al. 2014).

The improved ECM-production responses of PMR were often associated with milk fat concentration not declining to the same extent with an increasing supplement intake, as it did when feeding cereal supplements in the dairy (Auldist et al. 2013, 2014, 2016; Golder et al. 2014). Diet is clearly a major factor in this response (Shingfield et al. 2010; Bauman and Griinari

2003). Diets with low fibre and high starch and/or lipids rich in polyunsaturated fatty acids are often associated with low milk fat concentration (Stockdale et al. 1987). Differences could also be expected in the site, and end products, of digestion in the PMR diets where approximately half of the starch from wheat (*Triticum aestivum*) was replaced with maize grain starch (Khorasani et al. 2001). The transferability of these results to New Zealand systems may be limited as grain supplementation rarely approaches the amounts from which benefits of using a PMR are observed.

6.2 Replacing wheat with canola meal further improves milk responses

Protein supplements have been associated with increased DM intake and milk production (Butler 1998; Broderick 2003; Ipharraguerre and Clark 2005; Olmos Colmenero and Broderick 2006). A series of experiments showed that, under some circumstances, replacing ~20% of the wheat DM with canola meal in a PMR ration fed twice daily to grazing cows can increase intake and milk production. The canola meal provided more CP in the PMR, while maintaining the ME concentration. Cows consuming PMR with canola meal compared with a PMR without canola meal increased DM intake from both pasture (lower residuals or post-grazing biomass) and supplement (less refusals) by 0.5–1.9 kg DM/cow.day (Auldist et al. 2014; Golder et al. 2014). In these spring and autumn experiments (early lactation and late lactation respectively), cows offered the PMR with canola meal produced more ECM (ranging from 1.7 to 4.0 kg ECM/cow.day) than did cows offered the PMR diet without canola meal (Auldist et al. 2014; Golder et al. 2014). These differences occurred when supplement intake was at least 12.0 kg DM supplement/cow.day.

6.3 Increasing pasture allowance further improves DM intake and milk responses

Cows consuming PMRs graze further into swards, resulting in higher DM intakes despite very low pasture allowances (14–15 kg DM/cow.day, measured to ground level) and an equal opportunity to graze compared with cows consuming cereal grain in the dairy (Auldist et al. 2014, 2016). Three experiments were undertaken to quantify the additional DM intake from providing cows with higher allowances of pasture. Increasing the herbage allowance from 15 to 40 kg DM/cow.day (measured to ground level) increased pasture intake by up to 5.4 kg DM/cow.day, when 12 kg DM of supplement PMR was also on offer (Auldist et al. 2016, 2017), with an increase of up to 5 kg ECM/cow.day. The additional pasture intake is a result of increased bite mass and bite rate, with no differences in time spent grazing per day (Wright et al. 2016).

Data from the PMR experiments conducted during early lactation (Auldist et al. 2013, 2016) have been combined to explore relationships between intake and feed efficiency (Fig. 2). Data from experiments where cereal grain was fed in the dairy illustrate the negative relationship between intake and feed efficiency; that is, as intake increases (expressed on % liveweight basis), feed conversion efficiency decreases. Analysis of the series of PMR+canola meal experiments in early lactation showed higher feed efficiencies, such that at equivalent feed conversion efficiency cows consuming the well formulated PMR consumed more DM, and produced more ECM. This demonstrates that the PMR feeding system represents a more optimised diet. These relationships also point out the danger in attempting to optimise a partial measure of productivity, such as feed conversion efficiency. Without context, feed conversion efficiencies of 2 kg ECM/kg intake (% liveweight) appear desirable, but, in this situation, intake was very restricted at 2.6% of bodyweight (equivalent to ~16 kg DM for a 600-kg cow), and body reserves were likely to be supporting a significant amount of milk production, which, over the longer term, is unsustainable and is also associated with negative effects on fertility, health and milk production in the following lactation.

6.4 There are no negative impacts on health and welfare of cows in PMR systems

The final objective was to determine whether there were any negative health and welfare outcomes in cows on PMR systems, with the focus on risk of ruminal acidosis, hoof health and behaviour of cows while on feedpad and at grazing. Auldist et al. (2013) speculated that a well formulated PMR, with slower-degradable starch sources, such as maize grain, could lead to more stable ruminal fermentation than with grain fed in the dairy. Ruminal pH is used as an indicator of acidosis risk. Two metabolism experiments confirmed that ruminal pH was consistently higher in samples from cows consuming pasture and PMR diets than in those from cows consuming pasture and grain diets (Greenwood et al. 2014; see Fig. 3). Similarly, Golder et al. (2014) identified acidosis risk for a range of diets and showed that PMR diets had significantly lower risks of ruminal acidosis than did diets where wheat was fed in the dairy at a range of intakes (Fig. 4). It was further speculated that the reduced ruminal-fluid pH in the cows receiving high amounts of wheat grain in the dairy during milking would compromise the digestion on the diet overall, since ruminal-fluid pH below ~6.0 is known to compromise fibre digestion (Leddin et al. 2010). However, Greenwood et al. (2014) found no difference in whole-tract digestibility of DM, N, starch or NDF between the wheat grain in the dairy and PMR diets, despite the differences in ruminal pH, as described above. Foot health was not affected, with experiments indicating that there was no effect of feeding system on the presence of foot lesions associated with white line disease, paintbrush haemorrhage and traumatic bruising in early lactation (Coombe et al. 2013) or on the concentrations of serum haptoglobin, as an indicator of subacute ruminal acidosis (Coombe et al. 2015).

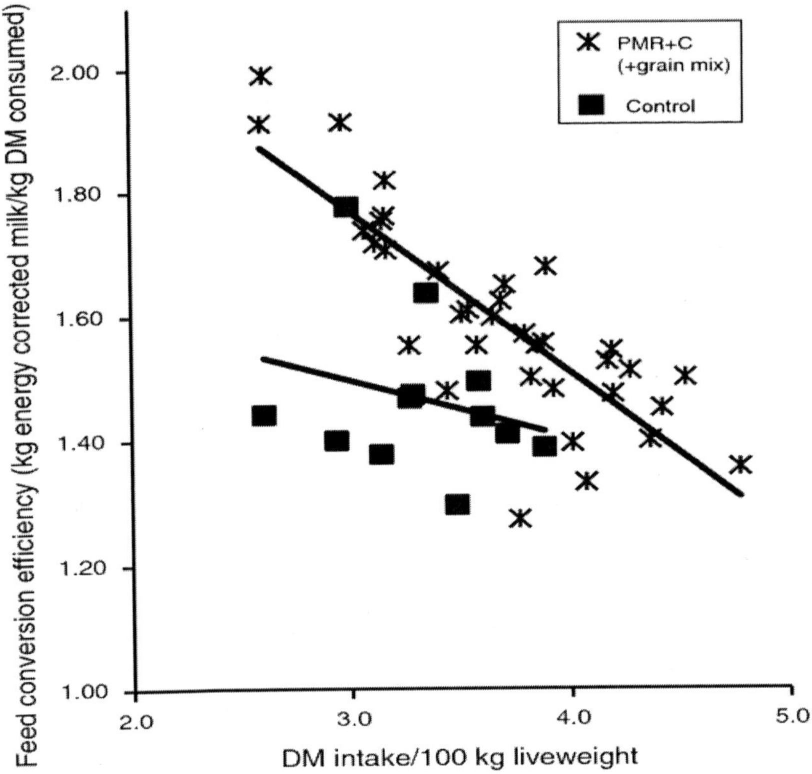

Figure. 2. Relationship between DM intake (% liveweight) and feed conversion efficiency (kg energy-corrected milk/kg DM intake) for cows in early lactation consuming either cereal grain in the dairy at milking time and forage in the paddock (■), or partial mixed ration (PMR), including canola fed twice daily on a feedpad (PMR; ×). Data are from Auldist et al. (2014, 2016).

In terms of feeding and social behaviour, feeding supplements as a PMR for ~1.5 h in the experiment described by Auldist et al. (2013) resulted in a small reduction in grazing time, while rumination and lying time and number of lying bouts were unchanged and, hence, there was no indication of negative consequences for cow welfare (Hetti Arachchige et al. 2013). Aggressive interactions decreased on the feedpad when feeding space was increased from 0.6 to 0.75 to 1.0 m and feeding behaviour improved, with the effects being greatest for subordinate cows (Hetti Arachchige et al. 2014).

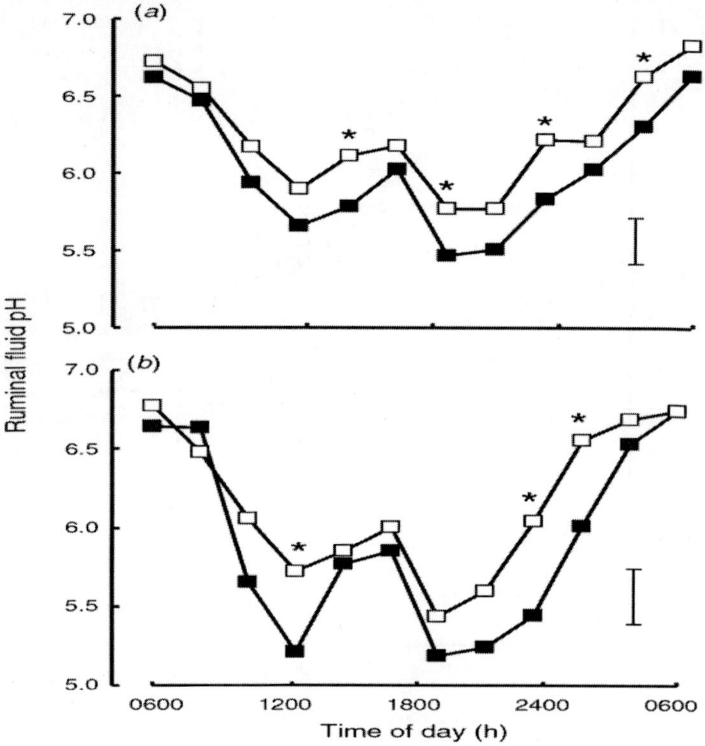

Figure. 3. Diurnal patterns of ruminal fluid pH for cows consuming fresh-cut perennial ryegrass herbage and offered grain and silage either separately (Control; ■) or as a partial mixed ration (PMR; □). Data points represent means of seven cows per diet measured approximately every 2 h over a 24-h period in (*a*) early lactation and (*b*) late lactation. Vertical bars represent least significant differences of means ($P < 0.05$). Asterisks indicate sampling times at which there was a significant ($P < 0.05$) difference between treatment means. Adapted from Greenwood et al. (2014).

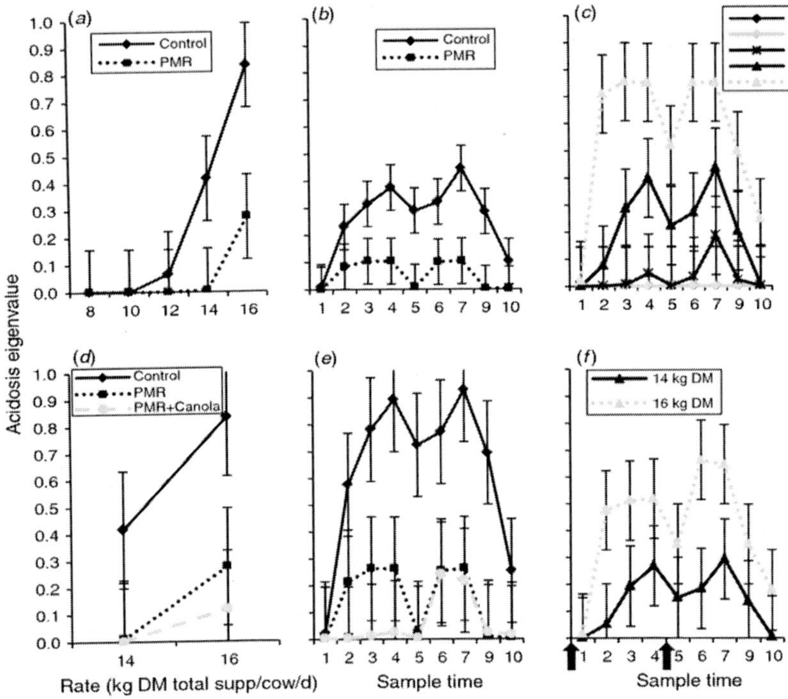

Figure. 4. Mean acidosis eigenvalues for dairy cattle from all feeding groups, showing interactions between (*a*) feeding strategy and feeding rate, (*b*) feeding strategy and sample time and (*c*) feeding rate and sample time. Mean acidosis eigenvectors for dairy cattle from the high feeding-rate groups only (14 and 16 kg of DM of total supplement/cow per day), showing interactions between (*d*) feeding strategy and feeding rate, (*e*) feeding strategy and sample time and (*f*) feeding rate and sample time. Sample times were ~2.4 h apart over a 24-h period. Sample time 1 was ~0820 hours and milking was at 0700 hours and 1500 hours (black arrows). PMR, partial mixed ration; rate, kg of DM of total supplement/cow per day. Source is Golder et al. 2014.

The Queensland subtropical dairy industry produces 5% of the national milk from ~500 dairy farms (Dairy Australia 2015b). These farms are in two regions in the south, one in central and one in northern Queensland, and feeding systems have evolved to exploit the natural advantages of each region. For example, TMR systems dominate in the Darling Downs where concentrates are cheap and grazing and PMR systems are concentrated in the South East Coast region where a more reliable rainfall allows pasture to be grown and grazed, albeit of lower digestibility than that in temperate dairying regions.

A key nutritional challenge has been how to profitably feed dairy cows in environments where home-grown forage supply is highly variable and of lower digestibility than temperate dairying regions. The autumn period of the year is when pasture grazing options are most limited and milk production declines, which is significant in these year-round calving systems. Higher-cost purchased energy and protein concentrate supplements are typically fed to help overcome the nutrient deficit representing ~40% of the diet DM. In a recent survey, average feed-related costs ranged from 29 cents/L for PMR systems to 38 cents/L in TMR systems and margin over feed costs from 25 to17 cents/L (Murphy and Simpson 2014). Reducing costs of production is a key strategy to improve profitability and feed costs make up the largest category of costs.

The dairy industry in Queensland is focussed on development of strategies that reduce the cost of rations, which increases the margin over feed costs through incorporating more highly digestible/high-protein home-grown forages (Findsen 2015). Their approach has been to assemble diets based on grazing and feeding of forages to reduce reliance on purchased concentrate supplements, within the commonly used TMR and PMR feeding systems in the subtropics. The forage options include silages produced from white and forage sorghums (*Sorghum* spp.), maize (high cut), lucerne, soybean (*Glycine max*), lablab (*Lablab purpureus*) and lucerne hay. Well made, these forages provide high-energy and -protein options for incorporation into rations and have reduced the reliance on purchased concentrates to 10% of the total ration (Table 5). To overcome variability in rainfall and forage supply, sufficient silages are produced to provide buffers in line with the risk profile of the farmers. Although this strategy introduces more complexity to the feeding system, the margin over feed costs can potentially be increased by 10 cents/L, due to more optimised supply of nutrients produced on farm, leading to increased milk production and high DM intakes.

Dairy production in Western Australia is conducted in a region with a Mediterranean-type environment, where annual rainfall is ~800 mm, falling in the cool months from May to November when cows graze, followed by a hot, dry period from December to April when pasture growth is minimal and grazing is limited (Bolland and Russell 2010). The predominant feeding strategy is to feed concentrates during milking, and graze during the cooler months and offer conserved forage in the paddock during the hotter months (Australian System 4). An alternative strategy that is being tried on farm is feeding a portion of the concentrate with the forage as part of a mixed ration during the summer period, with the aim of increasing feed efficiency and milk production. A comparison of these two strategies has recently been reported as part of a survey of the physical performance of 10% of dairy farms in Western Australia, conducted over a 15-month period (McDonnell et al. 2016). Their hypothesis was that incorporating concentrate as part of a ration would produce more milk and increase annual feed conversion efficiency because rumen fluctuations would be reduced, in comparison with 'slug feeding' of concentrates in the dairy. Results showed significant variation among farms within each system. Average milk production and annualised feed conversion efficiency were similar (7680 kg/cow and 1.37 kg ECM/kg total intake respectively), and McDonnell et al. (2016) concluded that the alternative feeding systems are currently not optimised. Further research to optimise the alternative strategy is exploring the role of canola meal and slower-degradable starch sources on a commercial farm and preliminary results suggest that milk responses in the order of 3 kg/cow.day are possible through fine tuning proportions of the components of the ration (R. McDonnell, pers. comm.).

Table 5. Queensland feeding systems comparing average performance of partial mixed ration (PMR) system with new total mixed ration (TMR) and PMR strategies that reduce the amount of purchased supplements

Parameter	TMR[A]	PMR[A]	Average PMR[B]
Total DM intake (kg DM/cow.day)	23.8	23.1	–
DM intake grazing (kg DM/cow.day)	0	8.2	–
Milk yield (L/cow.day)	26.0	24.0	20.0
Purchased concentrate (% total DM intake)	10	8	40
Feed conversion ratio (L/kg DM intake)	1.1	1.0	–
Feed-related costs (cents/L)	19	20	29
Margin over feed costs (cents/L)[C]	36	35	26

[A]Data adapted from Findsen (2015).
[B]Data adapted from Murphy and Simpson (2014).
[C]Milk price 55 cents/L.

7. GAPS IN KNOWLEDGE AND FERTILE AREAS FOR FURTHER FEEDBASE AND NUTRITION RESEARCH

7.1 Incorporating high-yielding crops and diverse/alternative forages into systems

The rapid growth of fodder-beet use in New Zealand has demonstrated potentially high yields, and farmers and researchers are understanding the best way to integrate fodder beet into systems, with transition programs being important to maintain animal health. However, for a feed that is currently having such a large impact on the New Zealand feed supply, it has exposed gaps in our understanding, particularly around the health of animals with long-term exposure to fodder beet. An area of concern from an environmental and perceived animal-welfare perspective is winter grazing of crops and exposure of animals to mud.

The development of diverse pastures will likely require further iterations of optimisation before they are proven technologies for farmers from both countries to use with confidence. Most promising is plantain as it is easily incorporated into existing systems, has a reasonable persistency, and has potentially significant environmental-footprint attributes, which could mean that farms can grow productivity and profitability under an imposed N limit.

7.2 Capturing forage-yield benefits from genetic gain

As far as improving the performance of the base pasture, several technologies are in development that could prove to be game-changing over the next 10 years. The challenge will be to demonstrate capture of these benefits at a farm-system level.

The forage-value index (FVI), which combines the performance value and economic value of a range of yields, and potentially nutritive and persistency traits, will help farmers in both countries select pasture cultivars that are expected to deliver greater farm profits. In New Zealand, the difference between the highest-ranked and the lowest-ranked perennial ryegrass cultivar ranges from $556 profit/ha to $863/ha, depending on the region. Since the 1960s, pasture selection has delivered $12–$18/ha.year in extra farm profits (Chapman et al. 2017). The immediate research challenge is to evaluate the FVI in a farm-system context with support from modelling, and to extend the FVI beyond yield evaluation to nutritive and persistence traits of the cultivars.

Currently, there are several exciting genetic options in development that will require evaluation at a farm scale. Genomic selection can be used to predict performance, shorten the breeding cycle and speed up genetic gain, as has been recently demonstrated in dairy cows (Barrett et al. 2015). New DNA tools are allowing identification of ryegrass strains that may be hybridised, with Italian ryegrass demonstrating more than a 20% yield improvement with hybrid vigour (DairyBio 2016). A real step change in Australasian pasture productivity will be achieved if the barriers of photosynthesis and energy levels in pastures can be overcome. AgResearch have recently developed a GM technology to enhance photosynthesis in C3 plants (Barrett et al. 2015; Bryan and Roberts 2015). The high-ME technology has boosted lipid concentrations in ryegrass from 3.5% to 8%, lifting ME levels by 10%, but the real breakthrough has been the 25–40% faster growth rates in glasshouse trials, which are the result of a more efficient photosynthesis. Encouragingly, modelling and glasshouse work conducted so far includes a 12% increase in milk fat+protein production, improvements in animal reproduction, possible increases in animal liveweight, 17% decrease in N_2O emissions, 15–30% decrease in methane emissions, a potentially enhanced water use efficiency and drought tolerance, and reduced nitrate leaching (Bryan and Roberts 2015). Field and animal-nutrition experiments are necessary to confirm the value of these novel forages for Australasia's pastoral industry, leading to recommendations of any change to current supplementary feeding or agronomic or grazing management practices. Current work in New Zealand on condensed tannins in white clover and novel endophytes, and on high-fructan grasses in Australia, will also provide homegrown, grazeable forage options in the future (Barrett et al. 2015).

7.3 Potential of feeding grain mixes to increase milk fat+protein production

The shift towards more complicated feeding systems is at odds with desire from farmers from both countries to manage simple feeding systems that are easy to implement. For example, PMR systems require more investment and expertise to realise the benefits. However, the benefits of PMR over traditional supplementary-feeding systems raised the possibility that feeding the grain portion of the PMR to cows using existing grain-feeding infrastructure in most modern dairies could capture some or all of the milk-production benefit observed with PMR systems, without the need for farmers to purchase a mixer wagon and build a feed pad. Feed companies currently formulate mixes of concentrates and implementing improved formulations would be relatively simple. Feeding a mix of concentrates from a PMR was tested in an experiment with cows grazing a restricted pasture allowance and fed a mix of maize grain, wheat grain and canola meal in the dairy at milking times, with lucerne hay fed in the paddock (Auldist et al. 2016). Milk production was compared with that from a PMR containing the four components fed on a feed pad after milking and wheat grain fed in the dairy (and lucerne in the paddock). Results showed that most of the advantage of the PMR over wheat fed in the dairy was achieved by feeding the grain mix. This indicates that the benefits of the PMR are related to the composition of the ration rather than the method of feeding. This was also supported by a lack of milk response from a simple PMR mix of cereal grain and forage fed to cows on a feed pad (Auldist et al. 2013).

Much of the increased milk response in cows fed a formulated grain mix and PMR strategies was associated with the arrested decline in milk fat concentrations as supplement intake increased, compared with the marked decline in milk fat concentrations observed in the cows consuming equivalent amounts of wheat grain in the dairy. The formulated grain-mix strategy could, therefore, be a much cheaper option than PMR, which would be relevant to a much wider cross-section of pasture-based dairy industries in Australia and New Zealand. Environmental benefits are also possible through the use of supplementary feeding to dilute N intake and increase milk production, which increases the efficiency of N use and potentially reduces N leaching. Beukes et al. (2014) suggested that N-use efficiency of standard pasture only was ~14% and we calculated that diets based on formulated grain mixes can improve the conversion of dietary crude protein intake to milk protein to 26%, on the basis of data presented in Auldist et al. (2016).

Although the recent research has highlighted the potential of using formulated grain mixes, the underlying mechanisms behind the observed milk responses are not fully understood. It is important to understand the reasons for the responses to allow the benefits to be reliably captured in other environments/dairying regions, which may have a different choice of ingredients available. These include the reasons for the consistent milk fat percentage responses and for the increased motivation of cows to graze pasture when canola meal has replaced wheat. Research is needed to quantify the site and the extent of digestion of carbohydrates and protein and the supply of metabolisable protein and adequacy of amino acids, for example, which would provide more precise information on carbohydrate and protein degradation of temperate pastures and local supplements.

7.4 Mechanistic nutrition models to determine complementary supplement mixes in grazing systems

Predicting milk production from information about the diet in grazing systems is difficult due to the complex interactions among the requirement for nutrients by the cow, the characteristics of different feeds and the contributions from body reserves towards the animal's requirements for the milk produced. Historically, we have formulated rations to meet requirements for estimated ME and crude protein, while ensuring that NDF from long forage sources falls within certain bands to ensure that sufficient rumination is occurring. However, there are significant limitations to this approach, particularly with higher-producing dairy cows, where ME intake is often not the limiting nutrient for production. As pasture nutritive characteristics change throughout the growing season, decisions on which and how much supplement (or supplement mixes) to feed are further complicated.

Mechanistic nutrition models, such as the Cornell Net Carbohydrate and Protein System (CNCPS 2000), have been designed to improve the prediction of milk responses. The complex mathematical construct is based on up-to-date literature, and is most applicable to cows in a housed environment consuming TMR to appetite (Fox et al. 2004). Compared with TMR, modelling the performance of dairy cows grazing pasture introduces additional complexity because of the greater variation in the nutritive characteristics of the forage, and, hence, greater variability in nutrient intake among cows.

Further advances in our understanding of nutrient supply from pasture and supplements through targeted research to describe a full suite of nutrients at different times of the year, and the refinement of nutritional models such as the CNCPS, make it possible to improve precision in formulating supplements to complement nutrients from pasture. A good example of this is in the recent research where the formulation of complicated mixes of grain supplements led to increased milk production compared with other dietary treatments of similar ME concentrations (Golder et al. 2014).

7.5 Individualised or differential feeding of concentrates to increase herd milk production

The increased availability of computerised feeding and milk-monitoring technologies has led to farmer interest in individually feeding concentrates to cows during milking on the basis of factors such as milk yield, days in lactation and parity. However, there is currently little evidence that individualised or differential feeding of concentrate supplements to dairy cows increases milk production, particularly in systems where forage is available *ad libitum*, such as those systems where housed cows are provided with silage to appetite and different amounts of the same supplement are offered to different cows (Hills et al. 2015). In these situations, no additional nutrients are brought into the system and the available nutrients are just distributed differently among cows.

The exception may be in systems where cows graze as a herd and receive supplements in the dairy where competition for a restricted pasture resource may limit nutrient intake in cows of higher genetic merit. In these types of systems, there is significant variation in pasture and total DM and nutrient intake of cows, which suggests that individualised feeding could lead to improved marginal milk-production responses. The difference between previous research with housed cows and grazing systems is that nutrient intake can be highly variable in grazing cows because of the effect of the decline in nutrient characteristics with pasture depletion within a grazing event. Hills et al. (2015) concluded that further research must be conducted to better understand the relationship between factors such as grazing and other activity-based behaviours and pasture intake, substitution rates and milk production. To this end, recent research from New Zealand (Dela Rue 2015) concluded that individualised feeding did not increase milk fat+protein production, where more of the same supplement was fed to higher-producing cows and less to lower-producing cows than with flat-rate feeding during early lactation. However, what has not been explored in experiments with grazing cows is identifying the limiting nutrients from grazed pasture and providing these nutrients as supplements. For example, a more energy-dense supplement may be appropriate for a lower-producing cow, while additional protein may be required for cows of higher genetic merit.

7.6 Economic frameworks to reliably assess tactical feeding management options

Research into milk responses to feed in combination with prices and costs has been limited, despite an increased need due to changes to more diverse pasture bases and supplementary feeding across both countries since the 1980s. Milk-response functions to supplementary feeding and production economics can be used to help farmers determine the most profitable amount of an input to use, given input cost and output price. The theory and principles are well established, with Heady (1952) conducting the seminal work in agricultural production economics in the 1950s, particularly in the economics of milk-response functions, with virtually all studies in agricultural production economics since then citing the principles established by Heady.

The experiments used by Heady remain relevant to the USA, but the principles developed are more widely applicable in the analyses of input–output relationships in dairy production in Australia and New Zealand. Some more recent international applications of production economics include those of André et al. (2010) who examined the profit-maximising amount of concentrate to feed cows in confined systems and those of Adduci et al. (2015) who developed a model to evaluate the efficiency of dairy cow rations. A limitation to applying the production-economics approach in grazing systems has been the lack of milk-response functions. Recent Victorian research developed response functions of milk production to cereal-based concentrates (Heard et al. 2017) and this was used in production-economics analyses to find the profit-maximising amount of the supplement (Ho et al. 2017), given a particular situation on-farm and a set of prices and costs.

Dairy farmers often consider supplementary feed, fertiliser and water as substitutes when making decisions about how to feed their cows, and response functions are needed for non-cereal-based supplements and water to understand the trade-offs between such inputs in making short-term feeding decisions. Some of this work is underway, but there are still gaps in other areas. Work is currently being conducted to develop milk-production responses to supplementary forages (C. Ho, pers. comm.). Data about milk-production responses to PMR feeding under a range of conditions are available, which could potentially be used to develop generalised responses to PMR feeding. The milk-production response to non-cereal-based supplements such as protein supplements is lacking, despite their being common supplements to grazing cows. Until research is conducted in this area, economic benefits from addition of protein-based supplements will remain unclear.

Current and previous work investigating the response to supplementary feed has focussed on milk production, but other benefits have not been quantified, for example, the potential benefits to increasing body condition. There is a need to develop analogous response functions, to change in supplement feed intake, for these other benefits. An obvious benefit of better information about response functions is improving the biological relationships used in modelling.

Recent studies by Rogers et al. (2016) defined the relationship between decreased water use and the survival and yield of lucerne and perennial pasture under different strategies. However, the economic costs and benefits, and the subsequent impact on farm profit have not been assessed. A generalised framework is also needed to enable the cost–benefit analysis of other crops and pastures as the water-response functions become available.

The aim is to help farmers make more informed tactical decisions by developing a framework incorporating a range of response functions enabling the exploration of the trade-off among fertiliser, supplementary feed and water. This would allow more informed decisions on how much extra of these inputs could be used at different times of the year, under price and cost combinations, to increase farm profit.

8. INTEGRATING NEW TECHNOLOGIES TO OPTIMISE NUTRIENT INTAKE IN INDIVIDUAL COWS IN REAL TIME

Decisions on what type, and how much, supplement to offer are often based on some estimate of pasture DM intake by the herd and of the average nutrient concentration of the pasture consumed, to achieve a desired total DM intake and milk production, within environmental constraints. This approach cannot optimise milk production because of seasonal changes in the amount and nutrient density of grazed forage, and variation in nutrient intake among cows from both the grazed forage and offered supplements. This contrasts with cows consuming total mixed rations where, in theory, each mouthful contains the optimised mix of nutrients. A better solution would be to optimise nutrient intake in real time at the individual-cow level, which could be achieved by knowledge of the nutrient intake of grazed forage by each cow and a supplement being tailored for each cow to achieve a profitable milk response.

Technology exists to enable some of this to occur, but the greatest challenge will be integrating these elements into a feeding management decision support framework that removes much of the herd manager's input. These technologies include near-infrared sensors that remotely measure canopy spectral reflectance to determine mass and nutrient density of swards (e.g. Starks et al. 2006). Real-time estimation of pasture intake by individual cows in a herd is also now possible with sensor and wireless-sensor networks that relate specific behaviour to intake of pasture. For example, Greenwood et al. (2017) developed reliable algorithms that predict intake of pasture when forage is not limiting, but concluded that further refinement of the algorithms is needed to account for variation within and among pastures and use of different sensor types that enable more specific classification of ingestive behaviours such as bite size, eating time, number of chews and eating rate. With real-time information on nutrient intake, mechanistic nutrition models may provide a supplementary-feed solution allowing optimal milk production that has a reduced impact on the environment. The challenge is to automate the model for each cow in the herd, on the basis of real-time estimates of nutrient intake from grazed forage. The next challenge is to use wireless-sensor networks to inform the feeding systems in the dairy to deliver the supplementary feed to each cow. In theory, the nutrient intake of each cow would approach that of a cow consuming a TMR.

REFERENCES

1. Adduci F, Labella C, Musto M, D'Adamo C, Freschi P (2015) Use of technical and economic parameters for evaluating cow ration efficiency. *Italian Journal of Agronomy* **10**, 202–207
2. Allden WG, Whittaker IAM (1970) The determinants of herbage intake by grazing sheep: the interrelationship of factors influencing herbage intake and availability. *Australian Journal of Agricultural Research* **21**, 755–766.| The determinants of herbage intake by grazing sheep: the interrelationship of factors influencing herbage intake and availability.
3. André G, Berentsen PBM, Van Duinkerken G, Engel B, Oude Lansink AGJM (2010) Economic potential of individual variation in milk yield response to concentrate intake of dairy cows. *The Journal of Agricultural Science* **148**, 263–276
4. Auldist MJ, Marett LC, Greenwood JS, Hannah M, Jacobs JL, Wales WJ (2013) Effects of different strategies for feeding supplements on milk production responses in cows grazing a restricted pasture allowance. *Journal of Dairy Science* **96**, 1218–1231.| Effects of different strategies for feeding supplements on milk production responses in cows grazing a restricted pasture allowance.
5. Auldist MJ, Marett LC, Greenwood JS, Wright MM, Hannah M, Jacobs JL, Wales WJ (2014) Replacing wheat with canola meal in a partial mixed ration increases the milk production of cows grazing at a restricted pasture allowance in spring. *Animal Production Science* **54**, 869–878.
6. Auldist MJ, Marett LC, Greenwood JS, Wright MM, Hannah M, Jacobs JL, Wales WJ (2016) Milk production responses to different strategies for feeding supplements to grazing dairy cows. *Journal of Dairy Science***99**, 657–671.
7. Auldist MJ, Wright MM, Marett LC, Jacobs JL, Wales WJ (2017) Effects of pasture allowance on milk production of dairy cows offered increasing amounts of partial mixed rations in spring. *Animal Production Science* **57**, 1305–1310.
8. Bargo F, Muller LD, Delahoy JE, Cassidy TW (2002) Performance of high producing dairy cows with three different feeding systems combining pasture and total mixed rations. *Journal of Dairy Science* **85**, 2948–2963.
9. Barrett BA, Faville MJ, Nichols SN, Simpson WR, Bryan GT, Conner AJ (2015) Breaking through the feed barrier: options for improving forage genetics. *Animal Production Science* **55**, 883–892.
10. Bauman DE, Griinari JM (2003) Nutritional regulation of milk fat synthesis. *Annual Review of Nutrition* **23**, 203–227.
11. Beever DE, Doyle PT (2007) Feed conversion efficiency as a key determinant of dairy herd performance: a review. *Australian Journal of Experimental Agriculture* **47**, 645–657.
12. Beukes PC, Gregorini P, Romera AJ, Woodward SL, Khaembah EN, Chapman DF, Nobilly F, Bryant RH, Edwards GR, Clark DA (2014) The potential of diverse pastures to reduce nitrogen leaching on New Zealand dairy farms. *Animal Production Science* **54**, 1971–1979.
13. Bolland M, Russell W (2010) Changes in chemical properties of 48 intensively grazed, rain-fed dairy paddocks on sandy soils over 11 years of liming in south-western Australia. *Australian Journal of Soil Research* **48**, 682–692.
14. Broderick GA (2003) Effects of varying dietary protein and energy levels on the production of lactating dairy cows. *Journal of Dairy Science* **86**, 1370–1381.
15. Bryan G, Roberts N (2015) 'Superior forages could result from GM technology.' Technical series online, issue 28, December 2015. pp. 1–5. (DairyNZ: Hamilton, New Zealand) Available at https://www.dairynz.co.nz/media/4439072/technical-series-june-2016-forages.pdf
16. Butler WR (1998) Effect of protein nutrition on ovarian and uterine physiology in dairy cattle. *Journal of Dairy Science* **81**, 2533–2539.
17. Cabrera Estrada JI, Delagarde P, Faverdin P, Peyraud JL (2004) Dry matter intake and eating rate of grass by dairy cows is restricted by internal, but not external water. *Animal Feed Science and Technology* **114**, 59–74.

18. Chakwizira E, de Ruiter JM, Malcolm B, Maley S (2016) Winter growth and quality of fodder beet. Plant & Food Research Report for Seed Force (PFR SPTS No. 12866) pp. 1–13. (The New Zealand Institute for Plant & Food Research Limited: Auckland, New Zealand)
19. Chapman DF, Tharmaraj J, Nie ZN (2008) Milk-production potential of different sward types in a temperate southern Australian environment. *Grass and Forage Science* **63**, 221–233.
20. Chapman DF, Cullen BR, Johnson IR, Beca D (2009) Intrannual variation in pasture growth rate in Australian and New Zealand dairy regions and its consequences for system management. *Animal Production Science* **49**, 1071–1079.
21. Chapman D, Clark D, Beukes P, Edwards G, Dalley D, Pinxterhuis I, Romera A, Shepherd M, Cameron K, Di H, Woodward S, Glassey C, Roach C, Curtis J, Lynch B (2012) Strategies for increasing dairy system productivity while controlling environmental footprint: a common systems research approach in two regions of New Zealand. In 'Proceedings of the 5th Australasian dairy science symposium'. (Ed. JL Jacobs). pp. 417–421 (Australasian Dairy Science Symposium: Melbourne, Australia)
22. Chapman DF, Bryant JR, Olayemi ME, Edwards GR, Thorrold BS, McMillan WH, Kerr GA, Judson G, Cookson T, Moorhead A, Norriss M (2017) An economically based evaluation index for perennial and short-term ryegrass in New Zealand dairy farm systems. *Grass and Forage Science* **72**, 1–21.
23. Clark D, Malcolm B, Jacobs J (2013) Dairying in the Antipodes: recent past, near prospects. *Animal Production Science* **53**, 882–893.
24. CNCPS (2000) 'Cornell Net Carbohydrate and Protein System. [4.0].' Computer program. (Cornell University: Ithaca, NY)
25. Coombe JE, Pyman MF, Mansell PD, Auldist MJ, Anderson GA, Wales WJ, Malmo J, Conley MJ, Fisher AD (2013) The effects on claw health of supplement feeding grazing dairy cows on feed pads. *The Veterinary Journal* **198**, 672–677.
26. Coombe JE, Pyman MF, Mansell PD, Auldist MJ, Anderson GA, Wales WJ, Conley MJ, Manos S, Hannah M, Fisher AD (2015) The effects on ruminal pH and serum haptoglobin after feeding a grain-based supplement to grazing dairy cows as a partial mixed ration or during milking. *Veterinary Journal* **204**, 105–109.
27. Cullen BR, Eckard RJ, Callow MN, Johnson IR, Chapman DF, Rawnsley RP, Garcia SC, White T, Snow VO (2008) Simulating pasture growth rates in Australian and New Zealand grazing systems. *Australian Journal of Agricultural Research* **59**, 761–768.
28. Dairy Australia (2011) 'Situation and outlook.' Available at http://www.dairyaustralia.com.au/Industry-information/About-Dairy-Australia/ ~/ media/ Documents/ Stats%20and%20markets/S%20and%20O/S%20and%20O%20-%20May%202011/Dairy%20Situation%20and%20Outlook%20Report%20May%202011%20V4.pdf
29. Dairy Australia (2015*a*) 'Australian dairy industry in focus 2015.' (Dairy Australia: Melbourne) Available at http://www.dairyaustralia.com.au/~/ media/ Documents/ Stats%20and%20markets/Australian%20Dairy%20Industry%20In%20Focus/Australian%20Dairy%20Industry%20In%20Focus%202015.pdf
30. Dairy Australia (2015*b*) 'Dairy feeding update: briefing notes 2015.' (Dairy Australia: Melbourne) Available at http://www.dairyaustralia.com.au/Industry-information/About-Dairy-Australia/~/media/Documents/Pastures%20and%20feeding/ Feed%20management/Dairy%20Feeding%20Update%20-%20Briefing%20notes.pdf
31. DairyBio (2016) 'Better ryegrass through hybrid breeding.' Available at http://dairybio.com.au/designer-forages/pasture-breeding-tools/ [Verified 10 November 2016]
32. DairyNZ (2008) 'DairyNZ economic survey 2006–07.' (DairyNZ: Hamilton, New Zealand) Available at www.dairynz/publications/ [Verified 10 November 2016]
33. DairyNZ (2010) 'DairyNZ economic survey 2008–09.' (DairyNZ: Hamilton, New Zealand) Available at www.dairynz/publications/ [Verified 10 November 2016]
34. DairyNZ (2012) 'DairyNZ economic survey 2010–11.' (DairyNZ: Hamilton, New Zealand) Available at www.dairynz/publications/ [Verified 10 November 2016]

35. DairyNZ (2014) 'DairyNZ economic survey 2012–13.' (DairyNZ: Hamilton, New Zealand) Available at www.dairynz/publications/ [Verified 10 November 2016]
36. DairyNZ (2015) 'New Zealand dairy statistics 2014–15.' (DairyNZ: Hamilton, New Zealand) Available at www.dairynz/publications/ [Verified 10 November 2016]
37. DairyNZ (2016*a*) 'DairyNZ economic survey 2014–15.' (DairyNZ: Hamilton, New Zealand) Available at www.dairynz/publications/ [Verified 10 November 2016]
38. DairyNZ (2016*b*) Feed use in the NZ dairy industry. Report to Ministry for Primary Industries. DairyNZ, Hamilton, New Zealand.
39. Dela Rue B (2015) 'Individualised supplementation of grazing cows: how does it stack up?' Technical Series in Brief, Issue 28, December 2015. Available at www.dairynz/techseriesdec15
40. Dharma S, Shafron W, Oliver M (2012) 'Australian dairy: farm technology and management practices 2010–11.' (Australian Bureau of Agricultural and Resource Economics and Sciences: Canberra)
41. Dove H (1996) Constraints to the modelling of diet selection and intake in the grazing ruminant. *Australian Journal of Agricultural Research* **47**, 257–275.
42. Doyle PT, Stockdale CR, Wales WJ, Walker GP, Heard JW (2001) Limits to optimising milk production and composition from pastures. In 'Recent advances in animal nutrition in Australia 2001'. (Ed. P Cronje). pp. 9–17 (Animal Science University of New England: Armidale, NSW)
43. Edwards GR, de Ruiter JM, Dalley DE, Pinxterhuis JB, Cameron KC, Bryant RH, Di HJ, Malcolm BJ, Chapman DF (2014) Urinary nitrogen concentration of cows grazing fodder beet, kale and kale–oat forage systems in winter. In 'Proceedings of the 6th Australasian dairy science symposium'. (Ed. JR Roche) pp. 144–147. (Australasian Dairy Science Symposium: Hamilton, New Zealand)
44. Edwards G, Bryant R, Smith N, Hague H, Taylor S, Ferris A, Farrell L (2015) Milk production and urination behavior of dairy cows grazing diverse and simple pastures. In 'Proceedings of the New Zealand Society of Animal Production'. (Ed. J McEwan) pp. 79–83. (New Zealand Society of Animal Production: Dunedin, New Zealand)
45. Findsen C (2015) C4Milk progress report 4. Dairy Australia. Publisher Animal Science, Department of Agriculture and Fisheries, State of Queensland, Gatton, Australia. Available at www.dairyaustralia.com.au
46. Fox DG, Tedeschi LO, Tylutki TP, Russell JB, Van Amburgh ME, Chase LE, Pell AN, Overton TR (2004) The Cornell Net Carbohydrate and Protein System model for evaluating herd nutrition and nutrient excretion. *Animal Feed Science and Technology* **112**, 29–78.
47. García SC, Fulkerson WJ (2005) Opportunities for future Australian dairy systems: a review. *Australian Journal of Experimental Agriculture* **45**, 1041–1055.
48. Gibb MJ, Huckle CA, Nuthal R, Rook AJ (1997) Effect of sward surface height on intake and grazing behaviour by lactating Holstein Friesian cows. *Grass and Forage Science* **52**, 309–321.
49. Golder HE, Denman SE, McSweeney C, Wales WJ, Auldist MJ, Wright MM, Marett LC, Greenwood JS, Hannah MC, Celi P, Bramley E, Lean IJ (2014) Effects of partial mixed rations and supplement amounts on milk production and composition, ruminal fermentation, bacterial communities, and ruminal acidosis. *Journal of Dairy Science* **97**, 5763–5785.
50. Greenwood JS, Auldist MJ, Marett LC, Hannah MC, Jacobs JL, Wales WJ (2014) Ruminal pH and whole-tract digestibility in dairy cows consuming fresh cut herbage plus concentrates and conserved forage fed either separately or as a partial mixed ration. *Animal Production Science* **54**, 1056–1063.
51. Greenwood PL, Paull DR, McNally J, Kalinowski T, Ebert D, Little B, Smith DV, Rahman A, Valencia P, Ingham AB, Bishop-Hurley GJ (2017) Use of sensor-determined behaviours to develop algorithms for pasture intake by individual grazing cattle. *Crop and Pasture Science*
52. Gregorini P (2012) Diurnal grazing pattern: its physiological basis and strategic management. *Animal Production Science* **52**, 416–430.

53. Heady EO (1952) 'Economics of agricultural production and resource use.' (Prentice Hall: Upper Saddle River, NJ)
54. Heard JW, Hannah M, Ho CKM, Kennedy E, Doyle PT, Jacobs JL, Wales WJ (2017) Predicting milk responses to cereal-based supplements in grazing dairy cows. *Animal Production Science* **57**, 746–759.
55. Hetti Arachchige A, Fisher AD, Auldist MJ, Wales WJ, Jongman EC (2013) Effects of different systems of feeding supplements on time budgets of cows grazing restricted pasture allowances. *Applied Animal Behaviour Science* **148**, 13–20.
56. Hetti Arachchige AD, Fisher AD, Wales WJ, Auldist MJ, Hannah MC, Jongman EC (2014) Space allowance and barriers influence cow competition for mixed rations fed on a feed-pad between bouts of grazing. *Journal of Dairy Science* **97**, 3578–3588.
57. Hills JL, Wales WJ, Dunshea FR, Garcia SC, Roche JR (2015) Invited review: an evaluation of the likely effects of individualized feeding of concentrate supplements to pasture-based dairy cows. *Journal of Dairy Science* **98**, 1363–1401.
58. Ho CKM, Heard JW, Wales WJ, Jacobs JL, Doyle PT, Malcolm B (2017) Evaluating the economics of concentrate feeding decisions in grazing dairy cows. *Animal Production Science*
59. Hodgson J (1981) Variation in the surface characteristics of the sward and the short-term rate of herbage intake by calves and lambs. *Grass and Forage Science* **36**, 49–57.
60. Inside Dairy (2016) 'Two wins from using a catch crop.' (DairyNZ: Hamilton, New Zealand) Available at www.dairynz/publications/
61. Ipharraguerre IR, Clark JH (2005) Varying protein and starch in the diet of dairy cows. II. Effects on performance and nitrogen utilization for milk production. *Journal of Dairy Science* **88**, 2556–2570.
62. Jacobs JL (2014) Challenges in ration formulation in pasture-based milk production systems. *Animal Production Science* **54**, 1130–1140.
63. Ketelaars JJ, Tolkamp MH, Tolkamp BJ (1992) Toward a new theory of feed intake regulation in ruminants 1. Causes of differences in voluntary feed intake: critique of different views. *Livestock Production Science***30**, 269–296.
64. Khorasani GR, Okine EK, Kennelly JJ (2001) Effects of substituting barley grain with corn on ruminal fermentation characteristics, milk yield, and milk composition of Holstein cows. *Journal of Dairy Science***84**, 2760–2769.
65. Kolver ES, Muller LD (1998) Performance and nutrient intake of high producing Holstein cows consuming pasture or a total mixed ration. *Journal of Dairy Science* **81**, 1403–1411.
66. Leddin CM, Stockdale CR, Hill J, Heard JW, Doyle PT (2010) Increasing amounts of crushed wheat fed with Persian clover herbage reduced ruminal pH and dietary fibre digestibility in lactating dairy cows. *Animal Production Science* **50**, 837–846.
67. Lee JM, Hemmingson NR, Minnee EMK, Clark CEF (2012) Chicory and plantain management strategies to increase herbage dry matter yield, nutritive value and plant survival. In 'Proceedings of the 5th Australasian dairy science symposium'. (Ed. JL Jacobs). pp. 393–397. (Australasian Dairy Science Symposium: Melbourne)
68. Li G, Kemp PD (2005) Forage chicory (*Cichorium intybus* L.): a review of its agronomy and animal production. *Advances in Agronomy* **88**, 187–222.
69. Macdonald K (2014) Driving production, profit and environment in all-grazed systems. In 'Inside Dairy Technical Series (October) Issue 24'. pp. 1–4. (DairyNZ: Hamilton New Zealand)
70. Malcolm BJ, Cameron KC, Di HJ, Edwards GR, Moir JL (2014) The effect of four different pasture species compositions on nitrate leaching losses under high N loading. *Soil Use and Management* **30**, 58–68.
71. Mayne CS, McGilloway DA, Cushnahan A, Laidlaw AS (1997) The effect of sward height and bulk density of herbage intake and grazing behavior of dairy cows. In 'Proceedings of the XVIII international grassland congress'. Vol. 1, pp. 2-15–2-16. (Winnepeg, Manitoba, Canada)
72. McDonnell RP, Staines MvH, Edmunds BE, Morris R (2016) Feeding management, production and performance of 13 pasture-based dairy farms in a Mediterranean environment. *Animal Production Science*

73. Mertens DR (1987) Predicting intake and digestibility using mathematical models of rumen function. *Journal of Animal Science* **64**, 1548–1558.
74. Milne JA, Hodgson J, Thompson R, Souter WG, Barthram GT (1982) The diet ingested by sheep grazing swards differing in white clover and perennial ryegrass content. *Grass and Forage Science* **37**, 209–218.
75. Ministry for the Environment and Statistics New Zealand (2015) 'Environment Aotearoa 2015.' Available at http://www.mfe.govt.nz/publications/environmental-reporting/environment-aotearoa-2015
76. Minnee EMK, Clark CEF, McAllister TB, Hutchinson KJ, Lee JM (2012) Chicory and plantain as feeds for dairy cows in late lactation. In 'Proceedings of the 5th Australasian dairy science symposium', Melbourne, Australia. pp. 426–428.
77. Monaghan RM, de Klein CAM (2014) Integration of measures to mitigate reactive nitrogen losses to the environment from grazed pastoral dairy systems. *The Journal of Agricultural Science* **152**, 45–56.
78. Moorhead AJE, Piggot GJ (2009) The performance of pasture mixtures containing 'ceres tonic' plantain (*Plantago lanceolata*) in Northland. In 'Proceedings of the New Zealand Grassland Association', Waitangi, New Zealand. pp. 195–199.
79. Murphy R, Simpson G (2014) 'Balancing dairy production and profits in northern Australia, Queensland Dairy Accounting Scheme – 2014.' (Animal Science, Department of Agriculture, Fisheries and Forestry: Gatton, Qld) Available at www.dairyaustralia.com.au [Verified 11 October 2016]
80. National Research Council (2001) 'Nutrient requirements of dairy cattle.' (National Academy Press: Washington, DC)
81. Olmos Colmenero JJ, Broderick GA (2006) Effect of dietary crude protein concentration on milk production and nitrogen utilization in lactating dairy cows. *Journal of Dairy Science* **89**, 1704–1712.
82. Pembleton KG, Tozer KN, Edwards GR, Jacobs JL, Turner LR (2015) Simple versus diverse pastures: opportunities and challenges in dairy systems. *Animal Production Science* **55**, 893–901.
83. Pembleton KG, Hills JL, Freeman MJ, McLaren DK, French M, Rawnsley RP (2016) More milk from forage: milk production, blood metabolites, and forage intake of dairy cows grazing pasture mixtures and spatially adjacent monocultures. *Journal of Dairy Science* **99**, 3512–3528.
84. Penning PD, Parsons AJ, Orr RJ, Treacher TT (1991) Intake and behaviour responses by sheep to changes in sward characteristics under continuous stocking. *Grass and Forage Science* **46**, 15–28.
85. Provenza FD (1995) Postingestive feedback as an elementary determinant of food preference and intake in ruminants. *Journal of Range Management* **48**, 2–17.
86. Rogers ME, Lawson AR, Kelly KB (2016) Lucerne yield, water productivity and persistence under variable and restricted irrigation strategies. *Crop and Pasture Science* **67**, 563–573.
87. Romera AJ, Doole GJ, Khaembah EN (2015) Value of chicory crops to pasture-based dairy farms in the Waikato region of New Zealand. *The Journal of Agricultural Science* **153**, 124–137.
88. Rook AJ (2000) Principles of foraging and grazing behavior. In 'Grass, its production and utilization'. (Ed. A Hopkins) pp. 229–246. (Blackwell Science: Oxford, UK)
89. Selbie D, Buckthought LE, Shepherd MA (2015) The challenge of the urine patch for managing nitrogen in grazed pasture systems. *Advances in Agronomy* **129**, 229–292.
90. Shingfield KJ, Bernard L, Leroux C, Chilliard Y (2010) Role of *trans* fatty acids in the nutritional regulation of mammary lipogenesis in ruminants. *Animal* **4**, 1140–1166.
91. Starks PJ, Zhao D, Phillips WA, Coleman SW (2006) Herbage mass, nutritive value and canopy spectral reflectance of bermudagrass pastures. *Grass and Forage Science* **61**, 101–111.
92. Stockdale CR, Callaghan A, Trigg TE (1987) Feeding high energy supplements to pasture-fed dairy cows, effects of stage of lactation and level of supplement. *Australian Journal of Agricultural Research* **38**, 927–940.

93. Thorrold B, Doyle PT (2007) Nature or nurture: forces shaping the current and future state of dairy farming in New Zealand and Australia. In 'Proceedings of the 3rd Australasian dairy science symposium', Melbourne, Australia. pp. 450–460.
94. Totty V, Greenwood S, Bryant R, Edwards G (2013) Nitrogen partitioning and milk production of dairy cows grazing simple and diverse pastures. *Journal of Dairy Science* **96**, 141–149.
95. Ungar ED (1996) Ingestive behavior. In 'The ecology and management of grazing systems'. (Eds J Hodgson, AW Illius) pp. 185–218. (CAB International: Wallingford, UK)
96. Verite R, Journet M (1970) Influence de la teneur en eau et la deshydration de l'herbe sur la valeur alimentaire por les vaches laitiers. *Annales de Zootechnie* **19**, 255–268.
97. Wales WJ, Stockdale CR, Doyle PT (2005) Plant and sward characteristics to achieve high intake in ruminants: In 'Utilisation of grazed grass in temperate animal systems. Proceedings of a satellite workshop of the XXth international grassland congress', July 2005, Cork, Ireland. pp. 37–47.
98. Wales WJ, Marett LC, Greenwood JS, Wright MM, Thornhill JB, Jacobs JL, Ho CKM, Auldist MJ (2013) Use of partial mixed rations in pasture-based dairying in temperate regions of Australia. *Animal Production Science* **53**, 1167–1178.
99. Waugh CD, Clark DA, Harris SL, Thom ER, Copeman PJA, Napper AR (1998) Chicory for milk production. In 'Proceedings of the New Zealand Grassland Association', Nelson, New Zealand. pp. 33–37.
100. Weston RH (1982) Animal factors affecting feed intake. In 'Nutritional limits to animal production from pastures'. (Ed. JB Hacker) pp. 183–198. (Commonwealth Agricultural Bureaux: Farnham Royal, UK)
101. Williams YJ, Walker GP, Doyle PT, Egan AR, Stockdale CR (2005) Rumen fermentation characteristics of dairy cows grazing different allowances of Persian clover- or perennial ryegrass-dominant swards in spring. *Australian Journal of Experimental Agriculture* **45**, 665–675.
102. Williams YJ, Doyle PT, Egan AR (2014) Diurnal variation in rumen fill of dairy cows grazing Persian clover at different pasture allowances. *Animal Production Science* **54**, 1388–1393.
103. Woodward SL, Waghorn GC, Bryant MA, Benton A (2012) Can diverse pasture mixtures reduce nitrogen losses? In 'Proceedings of the 5th Australasian dairy science symposium', Melbourne, Australia. pp. 463–464.
104. Woodward S, Waugh C, Roach C, Fynn D, Phillips J (2013) Are diverse species mixtures better pastures for dairy farming? In 'Proceedings of the New Zealand Grassland Association', Tauranga, New Zealand. pp. 79–84.
105. Wright MM, Marett LC, Greenwood JS, Hannah M, Jacobs JL, Wales WJ, Auldist MJ (2014) Feeding a partial mixed ration once a day did not increase milk production compared with feeding grain in the dairy and forage in the paddock. *Animal Production Science* 54, 1405–1411.
106. Wright MM, Auldist MJ, Kennedy E, Dunshea FR, Hannah M, Wales WJ (2016) Variation in feeding behavior and milk production among dairy cows when supplemented with 2 amounts of mixed ration in combination with 2 amounts of pasture. Journal of Dairy Science 99, 6507–6518.

CHAPTER 11

Cluster analysis of breeding values for milk yield and lactation persistency in Guzerá cattle

Diego Augusto Campos da Cruz[1], Rodrigo Pelicioni Savegnago[2], Annaíza Braga Bignardi Santana[3], Maria Gabriela Campolina Diniz Peixoto[4], Frank Angelo Tomita Bruneli[4], Lenira El Faro[5] *

[1]*Programa de Pós-graduação em Produção Animal Sustentável, Instituto de Zotecnia/SAA, Nova Odessa, SP, Brasil.*

[2]*Faculdade de Ciências Agrárias e Veterinárias (FCAV), Universidade Estadual Paulista "Júlio de Mesquita Filho" (UNESP), Jaboticabal, SP, Brasil.*

[3]*Universidade Federal de Mato Grosso (UFMT), Rondonópolis, MT, Brasil.*

[4]*Embrapa Gado de Leite, Juiz de Fora, MG, Brasil.*

[5]*Instituto de Zootecnia/SAA, Rodovia Carlos Tonani, Km 94, 14174-000, Sertãozinho, SP, Brasil.*

ABSTRACT

The aim of this study was to explore the pattern of genetic lactation curves of Guzerá cattle using cluster analysis. Test-day milk yields of 5,274 first-lactation Guzerá cows were recorded in a progeny test. A total of 34,193 monthly records were analyzed with a random regression animal model using Legendre polynomials to fit additive genetic and permanent environmental random effects and mean trends. Hierarchical and non-hierarchical cluster analyses were performed based on the EBVs for monthly test-day milk yield, peak yield, lactation persistency, and partial cumulative and 305-day yields. The heritability estimates for test-day milk yields ranged from 0.24 to 0.52. Cluster analysis identified animals in the population that belong to different groups according to milk production level and lactation persistency.

Keywords: breeding values; lactation curve; random regression

1. INTRODUCTION

The persistency of milk production is defined as the capacity of a cow to maintain its milk production after peak lactation. Thus, persistency is related to the shape of the lactation curve as a function of days of production. Persistency is a trait that is directly associated with economic factors of dairy farming since improved persistency may reduce the costs of the production system (TEKERLI et al., 2000; JAKOBSEN et al., 2002), improving reproductive efficiency (MUIR et al., 2004) and reducing the need of concentrate for milk production (SÖLKNER & FUCHS, 1987).

The currently most suitable method for the measurement and representation of persistency takes into account the prediction of breeding values (EBVs) for components of the lactation curve using random regression models (RRM). These models permit to predict EBVs for lactation persistency since each animal will have a set of EBVs for the additive genetic regression coefficients that describe the lactation curve. It is therefore possible to predict EBVs for yields on each day of lactation and consequently for partial cumulative periods (LIN & TOGASHI, 2002). Some countries, especially in the northern hemisphere, already use lactation persistency in their genetic evaluations (GENGLER, 1996; GROSSMAN et al., 1999).

To obtain genetic gains in milk yield, it is desirable to select animals with high EBVs for both milk yield and lactation persistency. Existence of genetic variation permits adding genetically similar cows into groups using multivariate cluster analysis based on EBVs for milk yields on each test day or on components of the lactation curve. Division into groups would permit to select animals based on the genetic curve pattern of milk yield desired by the breeding program. The objective of cluster analysis is to divide similar animals into groups based on certain traits, reducing heterogeneity between groups and increasing intragroup homogeneity (HAIR et al., 2009). There are no reports in the literature that use cluster analysis to explore the genetic profile and shape of the lactation curve in dairy Zebu cows. The aim of this study was to explore the pattern of genetic lactation curves of Guzerá cattle using cluster analysis.

2. MATERIALS AND METHODS

The database used in this study contained 34,193 records of monthly test-day milk yields (TDMY) for first lactations of 5,274 purebred Guzerá cows with calving records comprising the period from 1987 to 2012. Cows were born to 628 sires and belonged to 101 herds. The data were obtained from the National Dairy Guzerá Cattle Breeding Program (PNMGuL), coordinated by Embrapa Gado de Leite in cooperation with the Brazilian Center for the Genetic Improvement of Guzerá Cattle ($CBMG^2$) and the Brazilian Association of Zebu Breeders (ABCZ). The TDMY records were divided into 10 monthly classes ranging from day 6 to day 305 of lactation. EBVs were estimated by random regression using a single-trait animal model that included milk yield on each test day as the response variable. The fixed effects of contemporary group, defined by herd and test-day year and season, age of cow at calving as covariate (linear and quadratic effect), and random direct additive genetic, permanent environmental and residual effects were considered. The fixed lactation curve of the population was modeled using fourth-order orthogonal Legendre polynomials. The random additive genetic and permanent environmental effects were modeled using fourth- and fifth-order orthogonal Legendre polynomials, respectively. Residual variance was considered to be heterogeneous, with seven classes of variance.

This model was chosen after six different models had been tested using Akaike (AIC) and Bayesian Schwarz (BIC) (WOLFINGER, 1993) criteria. The matrix representation of the model is:

$$y = Xb + Za + Wpe + e$$

where: y = vector of TDMY; b = vector of solutions for fixed effects which included the solutions for contemporary group and for the covariates age at calving and days in milk;

a = vector of solutions for additive genetic random effects; pe = vector of solutions for permanent environmental random effects; e = vector of the random residual effect; X, Z and W = incidence matrices for fixed effects and the random effect of animal and permanent environment, respectively.

The following model assumptions were: E(a)=0, E(pe)=0 and E(e)=0, and $V(a)=k_a \otimes A V(pe)=kpe \otimes I_{Nd}$ and V(e)=R, where k_A and k_{PE} are (co)variance matrices between additive genetic and permanent environmental random regression coefficients, respectively; A is the relationship matrix between individuals; I_{ND} is the identity matrix of dimension Nd; $\otimes$is the Kronecker product between matrices, and R is a diagonal block matrix containing residual variances. The variance components were estimated by the restricted maximum likelihood (REML) method using the WOMBAT statistical package (MEYER, 2006).

Additive genetic random regression coefficients predicted by RRM for each animal were used to predict EBVs on each day of lactation. The EBVs for milk yields of animal i on the tth day of lactation were calculated as follows: $EBV_{tj} = (ebv_{0(j)} * f_{int} + ebv1(j) * f_1 + ebv_{2(j)} * f_2 + ebv_{3(j)} * f_3 + ebv_{4(j)} * f_4)$, where ebv_0 to ebv_4 are the EBVs for the random regression coefficients of each animal (j), and f_{int} to f_4 represent the Legendre polynomials referring to each day of lactation (6 to 305 days). The EBVs for milk yield on each day of lactation were used to calculate the cumulative milk yield at 305 days (MRA_{305}), peak yield (P_{Speak}) and partial yields (PS_{100}, PS_{200} and PS_{300} corresponding to early lactation, mid-lactation and late lactation, respectively), as well as three measures of persistency (PS_1, PS_2 and PS_3).

The EBVs for 305-day cumulative milk yield were obtained by the sum of EBVs on each day of lactation of the animal, represented by formula

$$MRA_{305} = \sum_{t=6}^{305} EBV_t \quad MRA_{305} = \sum_{t=6}^{305} EBV_t$$

Peak yield, defined as the period when maximum production is reached (COBUCI et al., 2004), was determined as the average EBV between 6 and 30 days of lactation since the breed exhibits an typical average lactation curve:

$$PS_{PICO} = \sum_{t=6}^{30} EBV_t/25$$

In this study, three measures of lactation persistency were used. The first measure of lactation persistency (PS_1) proposed here was calculated as the mean EBV between 30 and 270 days of lactation:

$$PS_1 = \sum_{t=31}^{270} EBV_i/241$$

Proposed by JAMROZIK et al. (1997), the second measure of persistency (PS_2) uses specific EBVs from two periods of lactation and was calculated as: PS_2 = (EBV_{270}- EBV_{30}) $PS_2 = (VG_{270} - VG_{30})$. The third measure of persistency (PS_3), adapted from Pereira et al. (2012), takes into consideration the sum of deviations of EBVs between 30 and 270 days of lactation in relation to the EBV predicted for peak yield and can be written as follows:

$$PS_3 = \sum_{t=30}^{270} (EBV_t - EBV_{PICO})$$

Measures of the partial periods of lactation correspond to cumulative milk yields in the first (6 to 100 days), second (101 to 200 days) and last third of lactation (201 to 300 days) and were calculated using the following formulas, respectively:

$$PS_{100} = \sum_{t=6}^{100} EBV_t \quad PS_{100} = \textstyle\sum_{t=6}^{100} VG_t ,$$

$$PS_{200} = \sum_{t=101}^{200} EBV_t \quad PS_{200} = \textstyle\sum_{t=101}^{200} VG_t ,$$

And

$$PS_{300} = \sum_{t=201}^{300} EBV_t \quad PS_{300} = \textstyle\sum_{t=201}^{300} VG_t$$

Simple and rank correlations between EBVs for the measures were estimated using the Proc Corr and Spearman procedures of the SAS statistical package (SAS, 2008), respectively.

Hierarchical and non-hierarchical cluster analyses were used to create groups of the animals based on the EBVs obtained for the production measures reported above. Hierarchical cluster analysis was applied to choose the number of clusters into which the population studied could be divided. The Euclidian distance was used as a measure of similarity between animals and Ward's clustering algorithm (1963) to create the groups. After definition of the number of groups, non-hierarchical cluster analysis by the k-means method was performed to explore the genetic profile of milk production. In the cluster analyses, the EBVs for MRA_{305}, PS_{peak}, PS_1, PS_2, PS_3, PS_{100}, PS_{200} and PS_{300}, as well as EBVs at 30, 60, 90, 120, 150, 180, 210, 240, 270 and 305 days of lactation, were used to divide the individuals into groups. In a second step, only the measure of persistency showing the best behavior, in addition to EBVs for PS_{peak}, PS_{100}, PS_{200}, PS_{300} and MRA_{305}, was used to evaluate the genetic profile of the groups. The PROC CLUSTER procedure of the SAS program (SAS, 2008) and the STATISTICA 8.0 program (STATSOFT, Inc., 2008) were used for cluster analysis.

3. RESULTS AND DISCUSSION

Peak yield was observed in the first month of lactation, with a mean milk yield of 9.04±4.62kg. Production continued in the second month of lactation (8.98±4.14kg) and subsequently decreased after the third month until the end of lactation (5.44±2.53kg). The heritabilities of TDMY (Table 1) were higher than those reported by SANTOS et al. (2013) for the same breed. These authors estimated a higher heritability in the second month (0.35) and a lower heritability in the eighth month of lactation (0.18). However, the trajectory and magnitude of the estimates were similar in the two studies after the second test day. Genetic correlations (Table 1) between monthly milk yields ranged from -0.03 to 0.95 and the lowest or negative estimates occurred between yields at the beginning and end of lactation. Similar results have been reported by SANTOS et al. (2013) for Guzerá cattle. Phenotypic correlations (Table 1) were all positive and ranged from 0.18 to 0.81. The same was observed for permanent environmental correlations which ranged from 0.48 to 0.98, with lower correlations between more distant TDMY.

Table 1 Heritability and standard error estimates (diagonal) and genetic (above the diagonal) and phenotypic correlations (bellow the diagonal) between test-day milk yields (TDMY).

	TDMY1	TDMY2	TDMY3	TDMY4	TDMY5	TDMY6	TDMY7	TDMY8	TDMY9	TDMY10
TDMY 1	**0.52** (0.04)	0.95	0.80	0.61	0.47	0.39	0.34	0.29	0.18	-0.03
TDMY2	0.69	**0.38** (0.03)	0.94	0.82	0.70	0.62	0.54	0.44	0.29	0.05
TDMY3	0.60	0.69	**0.33** (0.03)	0.97	0.89	0.82	0.72	0.59	0.40	0.15
TDMY4	0.54	0.65	0.75	**0.32** (0.03)	0.98	0.93	0.84	0.70	0.50	0.25
TDMY5	0.46	0.56	0.68	0.80	**0.29** (0.03)	0.98	0.92	0.79	0.60	0.35
TDMY6	0.41	0.49	0.62	0.76	0.81	**0.26** (0.03)	0.97	0.88	0.72	0.47
TDMY7	0.37	0.44	0.56	0.69	0.76	0.81	**0.24** (0.03)	0.96	0.84	0.61
TDMY8	0.32	0.39	0.49	0.60	0.66	0.73	0.78	**0.24** (0.03)	0.95	0.77
TDMY9	0.26	0.33	0.42	0.50	0.56	0.62	0.69	0.74	**0.27** (0.04)	0.92
TDMY10	0.18	0.23	0.31	0.40	0.45	0.50	0.55	0.61	0.74	**0.38** (0.05)

Simple correlations of EBVs for MRA_{305} with PS_{peak}, PS_{100}, PS_{200} and PS_{300} were positive and high (0.76, 0.90, 0.95 and 0.76, respectively). Higher correlations with milk yields in the first and second third of lactation ($PS_{100}$and PS_{200}) indicated that they are strongly associated with the milk production level during lactation.

Simple correlations between EBVs for measures representative of the beginning of lactation (PS_{peak} and PS_{100}) and the measure representative of the end of lactation (PS_{300}) were considered to be unfavorable. The correlation between PS_{peak} and PS_{300} was 0.22 and the correlation between PS_{100} and PS_{300} was 0.40. PS_{300}expresses the EBVs for milk yields during late lactation. As is known, Zebu breeds exhibit a relatively shorter lactation length than specialized taurine breeds. During this period, many animals have already ceased lactation mainly because of low yields, especially in the case of first lactations. Rank correlations (Spearman) between EBVs for MRA_{305} and partial measures indicated high agreement in animal ranking, with the highest correlation between MRA_{305} and PS_{200} (0.94). Greater errors in animal ranking were obtained when complete lactation was correlated with late lactation (0.69).

According to DEKKERS et al. (1998), measures of persistency that show low genetic correlations with milk yield until 305 days of lactation are the most suitable in selection processes; otherwise, studies designed to select lactation persistency would not be justified since it would be sufficient to select total milk yield and an increase in the level of persistency would consequently be achieved (COBUCI, 2002). Within this context, PS_1 was not a good measure of persistency since the simple correlation of this trait with total milk yield during lactation was close to unity. Measures PS_2 and PS_3 showed simple correlations of -0.55 and -0.42 with MRA_{305}, respectively, indicating an unfavorable association between milk production during lactation and persistency. These results suggest the use of either PS_2 or PS_3 in selection programs since, although calculated differently and showing a low correlation with milk production during lactation, the two measures were highly correlated and practically ranked the same animals. The correlation between these criteria was high because both corresponded to the post-peak period up to 270 days. In studies on Holstein cattle using RRM, the genetic correlations between different measures of persistency ranged from -0.49 to 0.57 (JAKOBSEN et al., 2002; COBUCI et al., 2004).

Only PS_3 was chosen as a measure of persistency in multivariate cluster analysis, taking into consideration its correlation with MRA_{305}, which was the lowest among the three measures (-0.42). However, the antagonistic association between the two measures should be a warning to breeders who only use the increase in cumulative milk yield during lactation as a criterion. Hierarchical analysis using the EBVs of all animals was performed to visualize the genetic profile of the animals based on the selection criteria proposed. This analysis generated a dendrogram (Figure 1), which revealed the possibility of dividing the population into two groups (clusters). Non-hierarchical cluster analysis was then performed on these two groups using all measures of persistency proposed, partial periods, peak yield, total milk yield and measures obtained at intervals of 30 days to certify the results of the correlations discussed above (Figure 2). This analysis demonstrated that it is not necessary to use the persistency measure that takes into consideration mean EBVs between days 31 and 270, with the observation of high similarity between its EBV and the EBV for total milk yield (MRA_{305}) and yields at 30-day intervals. High similarity was also observed between PS_2 and PS_3, indicating that either measure can be chosen for the following discussion. Figure 2 illustrates the results of non-hierarchical cluster analysis obtained by the k-means method using the EBVs for peak yield, partial yields (100, 200, and 300 days), persistency measures, complete lactation, and milk yields at 30, 60, 90, 120, 150, 180, 210, 240, 270 and 305 days of lactation. Regarding this division into two groups, group 1 exhibited EBVs above the average for partial periods of lactation and peak yield, but an EBV below the average for lactation persistency. Group 2 presented negative EBVs for partial periods and peak yield, but an EBV above the average for persistency, indicating that these animals could be more persistent despite the lower production level.

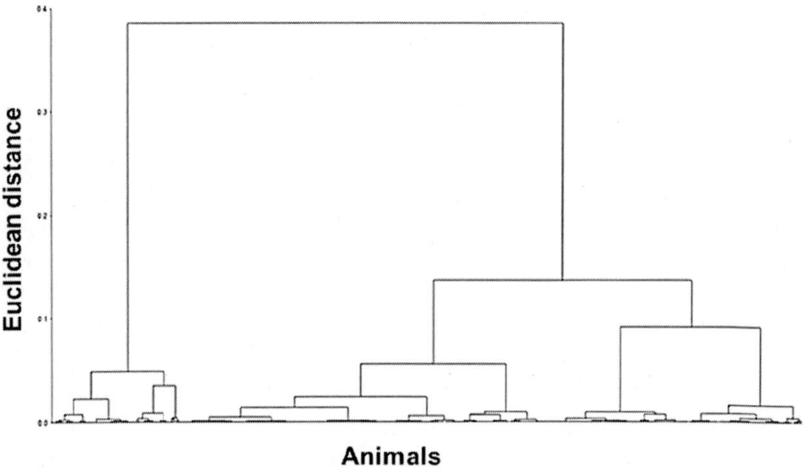

Figure 1 Dendrogram based on predicted breeding values of measures P_{SPICO}, PS_3, PS_{100}, PS_{200}, PS_{300} and MRA_{305}, and milk yields at 30, 60 , 90, 120, 150, 180, 210, 240, 270 and 305 days of lactation obtained by hierarchical cluster analysis by Ward's method.

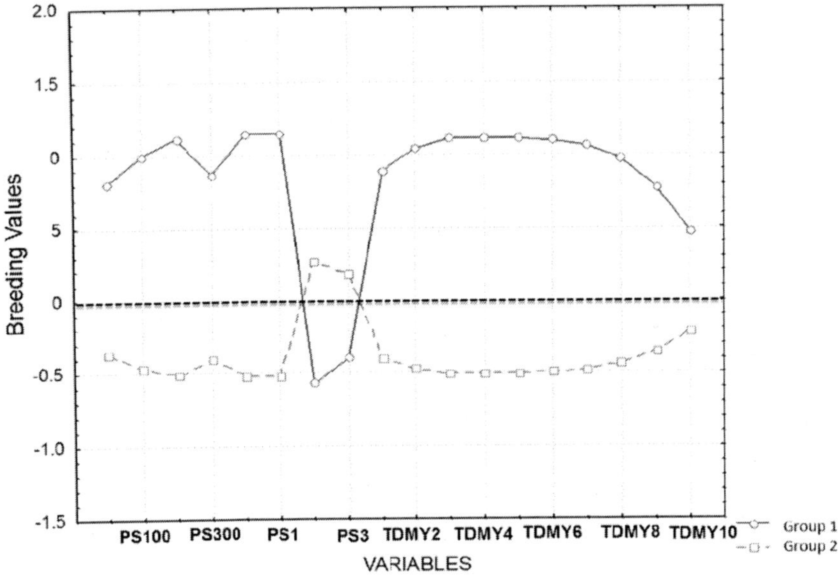

Figure 2 Non-hierarchical cluster analysis obtained by the k -means method, using predicted breeding values for three partial periods (PS100, PS200 and PS300 - representing the initial portion, the middle and the end of lactation, respectively), persistency measures (PS1, PS2 and PS3) and milk production at days in milk 30 (TDMY1), 60 (TDMY2), 90 (TDMY3), 120 (TDMY4), 150 (TDMY5), 180 (TDMY6), 210 (TDMY7), 240 (TDMY8), 270 (TDMY9) and 305 (TDMY10).

With respect to partial measures (Figure 2), the EBVs were higher at the beginning of lactation, decreasing thereafter and reaching the lowest values at the end of lactation. These results support those obtained by correlation analysis in this study, in which higher correlations were observed between measures of the first two-thirds of lactation. The genetic profiles of animals of group 1 obtained by hierarchical cluster analysis indicated the existence of different shapes of the genetic curves within the same group, although these animals exhibited EBVs above the average for partial measures, peak yield and complete lactation. A second multivariate cluster analysis was therefore performed, which only included animals of group 1. The division of animals of group 1 of the first analysis into two subgroups (Figure 3) showed positive values closer to the average in subgroup 1, except for persistency which was negative. The genetic profile of subgroup 2 was characterized by higher values and also a negative EBV for PS_3. However, the EBV for lactation persistency (PS_3) was lower for animals of subgroup 2 compared to subgroup 1. Animals of subgroup 1, which have lower production levels but still positive when compared to the average of the population, can be used in simpler production systems in which the animals are kept on pasture throughout the year. Animals of this subgroup can be used in dual-purpose production systems. Production systems of dairy Zebu breeds, which were not as selected for high milk yields as taurine breeds, should still emphasize the milk production level and animals of group 1, subgroup 2, meet these premises. However, this group of animals had lower EBVs for persistency, a fact resulting in lower total yields at the end of lactation. Thus, lactation persistency should compose a selection index seeking to combine both criteria, MRA_{305} and persistency.

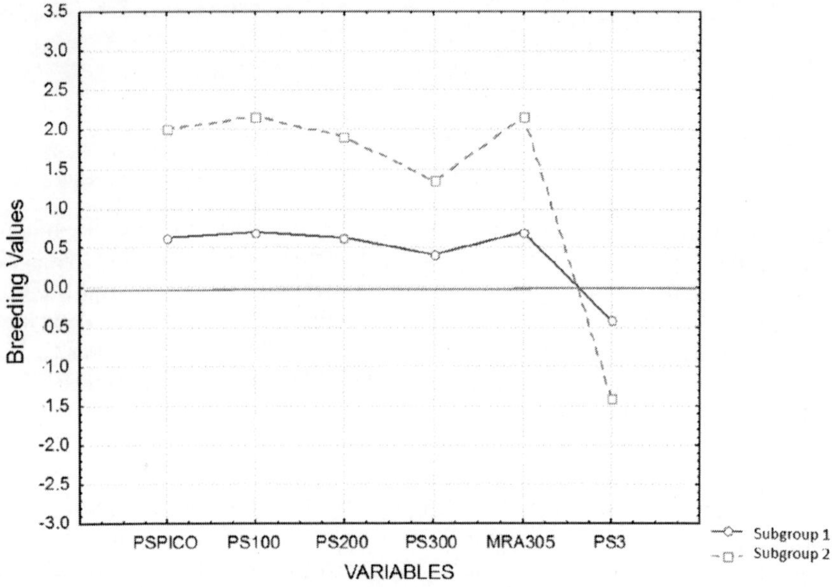

Figure 3 Non-hierarchical cluster analysis obtained by the k-means method using the genetic values of peak lactation (PSPICO) for three partial periods (PS100, PS200 and PS300 - representing the initial portion, the middle and the end of lactation, respectively), persistency of lactation (PS3) and 305-day accumulated milk yield (MRA305).

According to HARDER et al. (2006), cows with smoother production curves can be more persistent than cows exhibiting the same milk production with peak yields, but a sudden decline in production after peak lactation. From an economic point of view, the adoption of these more persistent animals in selection programs would be indicated. According to DEKKERS et al. (1998), cows with higher lactation persistency require less feed and healthcare and reproduction costs are lower. Consequently, these animals are more profitable for the breeder. However, these proposals were made for taurine breeds, animals that have been extensively selected for high milk yield.

In the case of dairy Zebu breeds in which milk production levels are still low, especially the Guzerá breed which has dual-purpose (beef and milk) lineages, milk production level continues to be the main selection criterion. However, certain emphasis is given to persistency of lactation in an attempt to correct long-term problems related to lower lactation persistency and shorter lactation length in animals of this breed. Nevertheless, studies broadening the knowledge of the production systems used by producers of these breeds, in conjunction with this proposal, are needed since they are generally pasture-based production systems.

4. CONCLUSION

The persistency measure (PS_3) was the most suitable for genetic evaluations of lactation persistency, mainly because of its lower correlation with MRA_{305}. Cluster analysis identified animals belonging to different groups according to milk production level and lactation persistency. Study of EBVs of the population using cluster analysis permits to identify superior animals for the desired traits, as well as to distinguish the genetic levels of the population. Selection of animals of this breed should emphasize the level of production, which is still low, including lactation persistency as a selection criterion.

ACKNOWLEDGEMENTS

To Conselho Nacional de Desenvolvimento Científico e Tecnológico (CNPq - grant number 473134/2012-8) to financial resources; L. El Faro held productivity research fellowships from the National Council for Scientific and Technological Development (grant number 301678/2013-8) and D.A.C. Cruz was Granted scholarship by Coordenação de Aperfeiçoamento de Pessoal de Nível Superior (CAPES).

REFERENCES

1. COBUCI, J.A. Uso de modelos de regressão aleatória na avaliação da persistência na lactação de animais da raça holandesa. 2002. 99f. Tese (Doutorado em Zootecnia) - Curso de Zootecnia, Universidade Federal de Viçosa, MG.
2. COBUCI, J.A. et al. Analysis of persistency in the lactation of Holstein cows using test-day yield and random regression model. Revista Brasileira de Zootecnia, v.33, n.2, p.546-554, 2004. Available from: <Available from: http://www.scielo.br/scielo.php?script= sci_ arttext&pid=S1516-35982004000300004 >. Accessed: May 10, 2014. doi: 10.1590/S1516-35982004000300004.
3. DEKKERS, J.C.M. et al. Economic aspects of persistency of lactation in dairy cattle. Livestock Production Science, v.53, p.237-252, 1998. Available from: <Available from: http://www.sciencedirect.com/science/article/pii/S0301622697001243 >. Accessed: May 20, 2014. doi: 10.1016/S0301-6226(97)00124-3.
4. GENGLER, N. Persistency of lactation yields: a review. In: WORKSHOP ON GENETIC IMPROVEMENT OF FUNCTIONAL TRAITS IN CATTLE, 1996, Gembloux, Belgium. v.12, p.97-102. (Interbull Bulletin, 12).
5. GROSSMAN, M. et al. Persistency of lactation yield: a novel approach. Journal of Dairy Science, v.82, p.2192-2197, 1999. Available from: <http://www.journalofdairyscience.org/article/S0022-0302(99)75464-0/abstract > Accessed: May 18, 2014. doi: 10.3168/jds.S0022-0302(99)75464-0.
6. HAIR, J.F. et al. Multivariate data analysis. Upper Saddle River: Prentice Hall, 2009. 816p.
7. HARDER, B. et al. Genetic parameters for health traits and their relationship to different persistency traits in German Holstein dairy cattle., Journal of Dairy Science v.89, p.3202-3212, 2006. Available from: <Available from: http://www.journalofdairyscience.org/article/S0022-0302(06)72595-4/pdf >. Accessed: May 15, 2014. doi: 10.3168/jds.S0022-0302(06)72595-4.
8. JAKOBSEN, J.H. et al. Genetic parameters for milk production and persistency for Danish Holstein estimated in random regression models using REML., Journal of Dairy Science v.85, p.1607-1616, 2002. doi: 10.1590/S1516-35982011000300013.
9. JAMROZIK, J. et al. Genetic evaluation of dairy cattle using test day yields and random regression model., Journal of Dairy Science v.80, p.1217-1226, 1997. Available from: <Available from: http://www.journalofdairyscience.org/article/S0022-0302(97)76050-8/pdf >. Accessed: May 22, 2014. doi: 10.3168/jds.S0022-0302(97)76050-8.
10. LIN, C.Y.; TOGASHI, K. Simultaneous improvement of lactation milk and persistency. In: WORLD CONGRESS ON GENETICS APPLIED TO LIVESTOCK PRODUCTION, 7., 2002, Montpellier. Proceedings... Montpellier: CGALP, 2002. Communication 09-39.
11. MEYER, K. "WOMBAT" - Digging deep for quantitative genetic analyses by restricted maximum likelihood. In: WORLD CONGRESS ON GENETICS APPLIED TO LIVESTOCK PRODUCTION, 8, 2006, Belo Horizonte, MG. Proceedings... Belo Horizonte: CGALP, 2006. CD-ROM.
12. MUIR, B.L. et al. Genetic relationships between persistency and reproductive performance in first-lactation Canadian Holstein., Journal of Dairy Science v.87, p.3029-3037, 2004. Available from: <Available from: http://www.journalofdairyscience.org/article/S0022-0302(04)73435-9/pdf >. Accessed: May 14, 14. doi: 10.3168/jds.S0022-0302(04)73435-9.
13. PEREIRA, R.J. et al. Milk yield persistency in Brazilian Gyr cattle based on a random regression model. Genetics and Molecular Research, v.11, p.1599-1609, 2012. Available from: <Available from: http://www.funpecrp.com.br/gmr/year2012/vol11-2/pdf/gmr 1519. pdf >. Accessed: May 10, 14. doi: 10.4238/2012.June.15.9.
14. SANTOS, D.J.A. et al. Comparison of random regression models to estimate genetic parameters for milk production in Guzerat (Bos indicus) cows., Genetics and Molecular Research v.12, p.143-153, 2013. Available from: <Available from: http://www.

funpecrp.com.br/gmr/year2013/vol12-1/pdf/gmr2025.pdf >. Accessed: May 22, 2014. doi: 10.4238/2013.January.24.6.
15. SAS INSTITUTE INC. SAS/STAT(r) 9.2 User's guide. Cary, NC, 2008.
16. SÖLKNER, J.; FUCHS, W. A comparison of different measures of persistency with special respect to variation of test-day milk yields., Livestock Production Science v.16, n.4, p.305-319, 1987. Available from: <Available from: http://www.sciencedirect.com/ science/ article/pii/0301622687900017 >. Accessed: May 20, 2014. doi: 10.1016/0301-6226(87)90001-7.
17. STATSOFT INC. Statistica. Tulsa, OK, 2008. Version 8.0.
18. TEKERLI, M. et al. Factors affecting the shape of lactation curves of Holstein cows from the Balikesir province of Tukey., Journal of Dairy Science v.6, p.1381-1386, 2000. Available from: <Available from: http://www.journalofdairyscience.org/article/S0022-0302(00)75006-5/pdf >. Accessed: May 23,14. doi: 10.3168/jds.S0022-0302(00)75006-5.
19. WARD. J.H. Hierarchical grouping to optimize a objective function. Journal of the American Statistical Association, v.58, p.236-244, 1963. Available from: <Available from: http://www.csee.wvu.edu/~xinl/library/papers/math/statistics/ward1963.pdf >. Accessed: May 25, 14. doi: 10.1080/01621459.1963.10500845.
20. WOLFINGER, R. Covariance structure selection in general mixed models. Communication In statistics Stat., v.22, p.1079-1106, 1993. doi: 10.1080/03610919308813143.

CHAPTER 12

Artificially Increased Yolk Hormone Levels and Neophobia in Domestic Chicks

Aline Bertin [1,2,3,4,*], Cécile Arnould [1,2,3,4], Chantal Moussu [1,2,3,4], Maryse Meurisse [1,2,3,4], Paul Constantin [1,2,3,4], Christine Leterrier [1,2,3,4] and Ludovic Calandreau [1,2,3,4]

[1] *Physiologie de la Reproduction et des Comportements, Institut National de la Recherche Agronomique (INRA), UMR85, Nouzilly 37380, France*

[2] *Centre National de la Recherche Scientifique (CNRS), UMR7247, Nouzilly 37380, France*

[3] *Université François Rabelais de Tours, Tours 37000, France*

[4] *Institut Français du Cheval et de l'Equitation (IFCE), Nouzilly 37380, France*

ABSTRACT

In birds there is compelling evidence that the development and expression of behavior is affected by maternal factors, particularly via variation in yolk hormone concentrations of maternal origin. In the present study we tested whether variation in yolk hormone levels lead to variation in the expression of neophobia in young domestic chicks. Understanding how the prenatal environment could predispose chicks to express fear-related behaviors is essential in order to propose preventive actions and improve animal welfare. We simulated the consequences of a maternal stress by experimentally enhancing yolk progesterone, testosterone and estradiol concentrations in hen eggs prior to incubation. The chicks from these hormone-treated eggs (H) and from sham embryos (C) that received the vehicle-only were exposed to novel food, novel object and novel environment tests. H chicks approached a novel object significantly faster and were significantly more active in a novel environment than controls, suggesting less fearfulness. Conversely, no effect of the treatment was found in food neophobia tests. Our study highlights a developmental influence of yolk hormones on a specific aspect of neophobia. The results suggest that increased yolk hormone levels modulate specifically the probability of exploring novel environments or novel objects in the environment.

Keywords: yolk hormones; maternal effects; neophobia; fear; behavioral development

1. INTRODUCTION

The reluctance to approach novel objects or novel food resources is observed in a large range of *taxa* and is commonly called neophobia. Neophobia is particularly well described in wild bird populations. It plays a key role in habitat selection, feeding innovation and the capacity to adapt to novel environments or novel resources [1,2]. Neophobia can be defined as the avoidance of an object or other aspect of the environment solely because it has never been experienced and is dissimilar from what has been experienced before. This reaction is associated with fear and the physiological and behavioral correlates of fear responses [3].

Despite domestication, neophobia is also observed in domestic hens (*Gallus gallus domesticus*). In laboratory conditions, this emotional reaction is described as a brief response followed by investigation of the novel food or object ([4] for a review). However, in commercial conditions, incapacity to overcome neophobia and accept novel foods is reported [5]. Domestic hens usually receive food in the form of a complete diet. During the course of development, feed ingredients in the complete diet are changed to meet specific nutritional needs. These changes alter the physical characteristics of feed (e.g., color, odor) in a way that is barely perceptible to humans; however, domestic birds have an acute sensory perception of feed. Birds exhibit neophobic responses when a single sensorial property of their food is changed [6,7]. Moreover, the amplitude of neophobic responses is enhanced when multiple sensorial properties of food are changed simultaneously. For example, the visual and tactile properties of food interact and potentiate the reaction of animals towards novel odors [8]. Thus, under commercial conditions, they sometimes express strong and damaging neophobic responses when facing such changes [9]. The refusal to accept new foods leads to a major reduction in feed intake and, subsequently, in growth and animal welfare [10]. In addition to the expression of fear, food neophobia reduces sanitary conditions with pecking redirected toward litter or feces [11]. For example, turkeys decrease feed intake and increase exploratory behavior upon receiving an unknown feed [12]. The limited research on food neophobia in poultry has focused mainly on how to "deactivate" this behavior at an early age [13]. Currently little is known about early developmental programming effects on this behavior. Evidence for a quantitative genetic basis for different levels of neophobia grew with the variation found between domestic and wild type birds [6,14] and the possibility to divergently select for the capacity to incorporate a novel food (novel insect prey) into the diet (Japanese quail, *Coturnix coturnix japonica*, [15]).

On the other hand, many studies draw attention to the role of developmental environments in shaping the behavior of young domestic birds. In particular, yolk hormones of maternal origin may adjust offspring phenotype according to environmental conditions. In Japanese quail, the environment provided to mothers (human-animal relationship, presence of a shelter, social instability, physical stress) during egg formation engender significant variations in yolk steroid concentrations (testosterone, androstenedione, progesterone). Although the direction of the effects is not predictable, the general emotional reactivity of offspring, including fear responses toward novel objects or humans, was consistently found to be influenced by these hormonal effects [16,17,18,19]. We also found evidence of a link between yolk hormone levels and inherent fearfulness (*i.e.*, propensity to express fear responses) in selected lines of quail. Higher yolk hormone concentration levels were found in eggs of the low inherent fearfulness line compared to eggs of the high inherent fearfulness line. An impact (positive or negative) of maternal androgens on the behavior and body mass of offspring has also been shown by administering physiological doses of hormones (androstenedione, testosterone) into eggs [20,21,22,23].

Far less is known about the implication of yolk hormone levels in the development of fear related behaviors in domestic hens. Janczak *et al.* [24] reported that maternal stress (feed restriction) increased offspring's duration of tonic immobility which indicated increased fearfulness. On the contrary, artificial maternal corticosterone elevation during egg formation has been reported to decrease offspring's duration of tonic immobility and slightly increase activity in an open-field test [25], which indicates lower fearfulness. So far, most studies conducted on domestic chickens focus on the implication of maternal corticosterone levels on offspring's physiology and behavior [26] for a review]. Maternal stress such as unpredictable food access [27] or early experience of social isolation [28] were found to alter the expression of stress-specific genes or genes related to immunity in the offspring. However, as mentioned by the authors, the implication of yolk gonadal hormones remains to be elucidated. It is difficult to identify the pathway of maternal effects. Indeed, maternal stress or increased maternal corticosterone levels also alter other factors such as egg mass [25], which could a play part in the construction of offspring phenotype.

In a previous study we observed a decrease in egg mass and a significant increase in the yolk concentrations of estradiol, testosterone and progesterone in Leghorn hens exposed to moderate heat stress whereas maternal corticosterone levels were not affected [29]. Our hypothesis was that such variations in yolk hormone concentrations could influence body mass and play a part in the propensity to express neophobia in young chicks independently of egg mass. Therefore we experimentally enhanced yolk hormone concentrations prior to incubation and characterized the immediate reaction of chicks to novel items (food, object and environment).

2. EXPERIMENTAL SECTION

2.1. Ethics Statement

All birds were maintained at the Experimental Unit PEAT of INRA (Nouzilly, France). The Experimental Unit is registered by the ministry of Agriculture with the license number B-37-175-1 for animal experimentation. All experiments were approved by the Animal Ethics Committee Val de Loire, CEEA Vdl (permit number 2012-10-6). All experiments were performed in accordance with the European Communities Council Directive 2010/63/UE. All animals were rehabilitated at the end of the experiment.

2.2. Egg Injections

One hundred and forty-five eggs from White Leghorn layers were provided by the experimental unit PEAT of INRA (Nouzilly, France). Two groups were considered: a group of 70 sham control eggs injected with a vehicle solution (C group) and a group of 75 eggs injected with the hormonal solution (H group). Eggs were selected so that the mass of the eggs did not differ significantly between the groups (Mean ± SE, C group: 56.88 ± 0.4 g; H group: 56.9 ± 0.3 g, $F_{1, 143} = 0.001$, $p = 0.97$).

In a previous study we demonstrated that a moderate maternal environmental challenge (moderate heat stress) significantly increased the yolk progesterone (P4), testosterone (T) and estradiol (E2) concentrations in Leghorn eggs but decreased the egg mass [29]. To simulate this increase in yolk hormone levels, we injected H eggs prior to incubation with a hormonal solution containing a mixture of P4, T and E2. We chose to increase the overall yolk concentrations by calculating the difference observed between the control population and the challenged population described by Bertin *et al.* [29]. For example, P4 concentrations were found to be 1823.7 ng/g in the control egg yolks *versus* 2181.9 ng/g in the yolks of the challenged population (which means a difference of 358.2 ng/g). The mean yolk mass was determined with 30 additional eggs (17.8 ± 0.3 g). We used the difference between the two observed concentrations and yolk mass (358.2 × 17.8) to determine the quantity to be injected into each egg. Therefore each egg was injected with 6376 ng of P4, 6.76 ng of T and 16.02 ng of E2 dissolved in 50 μL of vehicle (corn oil, Sigma Life Science).

Artificial increase of yolk hormones was chosen in order to focus specifically on hormone-mediated maternal effects rather than maternal effects induced by changes in egg mass or the general quality of eggs subsequent to maternal challenge. Artificially increasing yolk hormones prior to incubation is commonly used in birds to examine the short- and long-term effects of yolk hormones on the development of phenotypes [30]. Before injection, all eggs were carefully cleaned and disinfected with 70% ethanol and a hole was bored into the eggshell above the air sac using a 25-G drill. The solution was delivered to the yolk using a 100-μL Hamilton syringe mounting a 25-G sterile needle. As fully described by Bertin *et al.* [23], the injection hole was then sealed by gluing a tiny piece of cleaned and disinfected eggshell to the egg immediately after injection. The eggs were left undisturbed for one hour and then placed in alternative rows on each shelf of two incubators maintained at 37.8 °C and 56% relative humidity with automatic and continuous turning. On day 14 of incubation, non-fertile eggs or eggs containing prematurely dead embryos were eliminated (12 for C eggs and 25 for H eggs). Three days before hatching, the rotation was stopped and the temperature was reduced to 37.6 °C.

Artificial increase of yolk hormones was chosen in order to focus specifically on hormone-mediated maternal effects rather than maternal effects induced by changes in egg mass or the general quality of eggs subsequent to maternal challenge. Artificially increasing yolk hormones prior to incubation is commonly used in birds to examine the short- and long-term effects of yolk hormones on the development of phenotypes [30]. Before injection, all eggs were carefully cleaned and disinfected with 70% ethanol and a hole was bored into the eggshell above the air sac using a 25-G drill. The solution was delivered to the yolk using a 100-μL Hamilton syringe mounting a 25-G sterile needle. As fully described by Bertin *et al.* [23], the injection hole was then sealed by gluing a tiny piece of cleaned and disinfected eggshell to the egg immediately after injection. The eggs were left undisturbed for one hour and then placed in alternative rows on each shelf of two incubators maintained at 37.8 °C and 56% relative humidity with automatic and continuous turning. On day 14 of incubation, non-fertile eggs or eggs containing prematurely dead embryos were eliminated (12 for C eggs and 25 for H eggs). Three days before hatching, the rotation was stopped and the temperature was reduced to 37.6 °C.

2.3. Housing Conditions

We kept 70 chicks (40 C and 30 H chicks) that all hatched on the 21st day of incubation. Each chick was identified with a numbered ring on its leg. The 70 chicks were placed in pairs from the same treatment in wire-covered plastic tubs (50 cm × 40 cm × 30 cm; length × width × height) with wood shavings on the floor. The pairs of chicks were equally allocated to two rooms. Chicks were maintained in an 11 h light/13 h dark cycle, with water and food available *ad libitum*. The chicks were fed with a conventional starter mash (PEAT, INRA Val de Loire, France) dispensed in feeding troughs (length 50 cm). The troughs were covered with a metallic roof containing 12 circular holes (diameter, 5 cm); these holes provided the chicks with sufficient access to the feed while avoiding food spillage. Two opaque drinking bottles (1 L) were placed in each cage. The ambient temperature was maintained at 33 ± 1 °C from hatching until the chicks were 8 days old, after which it was decreased by 1 °C per day. The sex of each chick was determined by comb size at 3 weeks of age. The control group was composed of 20 females and 20 males, and the H group was composed of 13 females and 17 males.

2.4. Growth of Chicks

The chicks were weighed at hatching (day 1) and when they were 3 days, 10 days, 17 days and 25 days old.

2.5. Characterization of Neophobia

2.5.1. Food Neophobia

All of the tests used have previously been described by Bertin *et al.* [31]. The chicks are extremely distressed when isolated, and therefore we simultaneously tested two chicks from the same exposure conditions (and the same pen) in all of the tests. Chicks have a very rapid growth rate, therefore, all birds were tested for each test at the same age. This design also controlled for the time during which chicks were in contact with their familiar standard food before each test. We used short-term 3-minute tests to assess the immediate reaction to novel foods. We tested the effect of treatment when chicks were exposed to a change in the olfactory properties of their familiar food (single sensorial modality, Test 1) and a change in multiple sensorial properties of the food (raw cereals). We used cracked corn-wheat (cereals contained in familiar pellets, Test 2) and millet seeds (totally unknown to the birds, Test 3).

For the three tests, each pair of chicks (from the same treatment) was placed in a testing cage, similar to their home cage, after 1 h 30 min of food deprivation. The chicks were transported in a 15 cm × 15 cm × 15 cm container and then deposited and held blind in an enclosure (20 cm × 6 cm × 20 cm) placed at the opposite side of the feeding trough (identical to the familiar trough). After 30 s, the chicks were released and an observer hidden behind a curtain with small observation windows recorded the behavior of one focal bird of each pair (20 pairs of C chicks and 15 pairs of H chicks) for a 3-minute period. Focal birds were chosen randomly beforehand and identified by a blue-colored mark placed on their heads. The experimenter recorded the latency to eat the food and the time spent eating each food (the bird was considered to be eating when mandibulation and neck and throat movements caused by swallowing were observed. Eating behavior was considered to stop when the bird removed its head from the hole of the feeder and no more swallowing was observed for at least 3 s). In Test 1 (4 days of age), the trough contained 200 g of the familiar starter mash odorized with a mixture of artificial scents known to be perceived by chicks. We used a mix of isoamyl acetate (0.03 g), vanilline (0.03 g), and strawberry (0.03 g) in the form of a translucent powder. In Test 2 (7 days of age) we used 200 g of cracked corn-wheat, and in Test 3 (8 days of age) we used 200 g of millet seeds.

2.5.2. Object Neophobia

The protocol was similar to the novel food tests except that the familiar trough was replaced by an unfamiliar green and yellow plastic trough filled with 200 g of the familiar food. The experimenter recorded the latency to eat the food and time spent eating for a 5-minute period. This test was carried out at 18 days of age.

2.5.3. Fear of a Novel Environment

To assess fear of a novel environment we used a classical open-field test. This test involves transferring the birds from the familiar home cage to a new, open, larger and more lighted environment. The chicks were individually placed in the middle of an open cylindrical arena (diameter, 120 cm; height, 35 cm) on a linoleum floor for 5 min. To assess their locomotor activity, two perpendicular 120-cm-long lines were drawn in the arena, dividing it into four equal parts. The latency of the first step and the activity of the birds (number of times a bird crossed a line) were recorded by the experimenter hidden behind a curtain with observation windows. The chicks were tested when they were 15–16 days old.

2.6. Statistical Analyses

We compared the number of chicks hatched out of the fertile eggs using chi-square tests. The mass of chicks was analyzed using ANOVA for repeated measures (age) with treatment and sex as main factors and *post hoc* protected least significant difference Fisher tests.

Behavioral data did not follow a normal pattern of distribution; therefore, the non-parametric Mann-Whitney *U*-test was used for intergroup comparisons. Friedman and Wilcoxon tests were used for intra-group comparisons. We evaluated the intra-individual consistency to approach novel foods and the correlation between novel foods and novel object tests. Therefore we used Spearman rank correlations to correlate the latencies to eat in the different tests. Sex effects are mentioned only when significant. Data are presented as the mean ± standard error (SE). All analyses were performed using the StatView software (SAS Institute Inc., Cary, NC, USA), with significance accepted at $p \leq 0.05$.

3. RESULTS

3.1. Hatching Success and Mass of Chicks

No significant effect of egg treatment was observed on hatching success (C group: 42 chicks hatched out of 58 fertile eggs; H group: 31 chicks hatched out of 50 fertile eggs, Chi2 = 0.89, $p = 0.34$).

We found no significant effect of the treatment on the mass of chicks (Table 1, ANOVA on repeated measures, $F_{1,65} = 0.27$, $p = 0.60$). There was a significant effect of sex but no interaction between sex and treatment (ANOVA, sex effect: $F_{1,65} = 14.25$, $p < 0.01$; sex × treatment: $F_{4,260} = 0.14$, $p = 0.96$). The mass of males was higher than the mass of females.

Table 1. Mean mass (± SE) of C (*n* = 40) and H chicks (*n* = 30) at 1, 3, 10, 17 and 25 days of life.

	Mass (g)				
Groups	day 1	day 3	day 10	day 17	day 25
C chicks	40.2 ± 0.45	41.1 ± 0.71	84.8 ± 1.73	162.5 ± 3.26	269.1 ± 5.41
H chicks	39.2 ± 0.49	40.5 ± 0.6	84.7 ± 1.45	162.7 ± 2.93	267.6 ± 5.49

3.2. Characterization of Neophobia

3.2.1. Food Neophobia

We found no significant difference between C chicks and H chicks in the latency to eat or time spent eating during tests with familiar odorized food or during tests with cracked corn-wheat or tests with millet seeds (Table 2).

Table 2. Mean (± SE) latency to eat (s) and time spent eating (s) during food neophobia tests of C chicks (*n* = 20 pairs) and H chicks (*n* = 15 pairs).

Food Neophobia Test	Parameters Measured	C Chicks	H Chicks	* *p*-Value
Odorized familiar food	latency to eat (s)	72.1 ± 10.9	54.1 ± 10.1	0.48
Test 1	time spent eating (s)	41.7 ± 6	46.4 ± 6.3	0.87
Corn wheat mix	latency to eat (s)	75.3 ± 15.2	63.8 ± 15.5	0.48
Test 2	time spent eating (s)	11.6 ± 2.5	9.6 ± 2.1	0.23
Millet seeds	latency to eat (s)	132.7 ± 15.5	135.1 ± 13.1	0.73
Test 3	time spent eating (s)	5.4 ± 1.9	13.0 ± 5.0	0.75

Within both groups, the latencies to eat differed significantly according to the type of food (Friedman, $p < 0.01$). Within both groups, chicks took longer to eat the millet seeds than to eat the cracked corn-wheat or the familiar odorized food (Table 2, Wilcoxon, $p < 0.01$ for all comparisons). During the test with millet seeds, 18 pairs of chicks refused to eat (nine controls and seven H). The latency to eat the familiar odorized food or the corn-wheat did not differ significantly within both groups (Wilcoxon, $p > 0.05$ for all comparison).

Independent of the treatment, the latency to eat the familiar odorized food was significantly correlated with the latency to eat the cracked corn-wheat ($r = 0.58$, $p < 0.01$) but not with the latency to eat the millet seeds ($r = 0.27$, $p = 0.11$).

3.2.2. Object Neophobia

The latency to eat was significantly shorter in H chicks than in C chicks in the novel-object test (Figure 1). The time spent eating was significantly higher in H chicks than in C chicks.

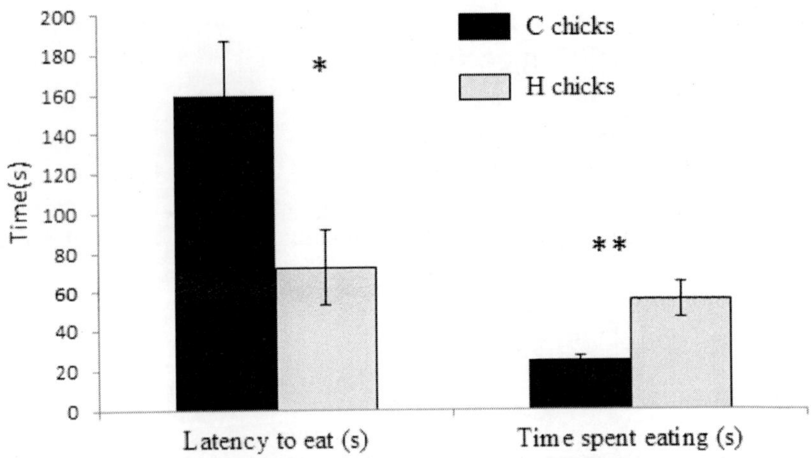

Figure 1. Mean ± SE latency to eat (s) and time spent eating (s) in the novel object test in C chicks (n = 20) and H chicks (n = 15). * Mann-Whitney U-test, $p < 0.05$; ** $p < 0.01$.

The latency to eat during the novel object test did not correlate significantly with the latency to eat the familiar odorized food ($r = 0.03$, $p = 0.83$) nor with the latency to eat the cracked corn-wheat ($r = 0.1$, $p = 0.49$) or the millet seeds ($r = 0.1$, $p = 0.41$).

3.2.3. Fear of a Novel Environment

The number of lines crossed and the number of distress calls were significantly higher in H chicks than in C chicks (Table 3). The latency of first step did not differ significantly.

Table 3. Mean (± SE) latency of first step, number of lines crossed and number of distress calls in C chicks (n = 40) and H chicks (n = 30) in the open-field test.

Parameters Measured	C Chicks	H Chicks	* p-Value
latency of first step (s)	76.3 ± 13.0	47.8 ± 7.93	0.18
number of distress calls	183.5 ± 14.7	227.1 ± 13.7	0.02
number of lines crossed	9.8 ± 1.47	15.0 ± 1.94	0.01

* Mann-Whitney U-tests p-value.

4. DISCUSSION

The aim of this study was to assess whether the propensity of chicks to express neophobia could be shaped by yolk hormone levels. When the maternal population is exposed to a stress, a combination of yolk hormones is commonly affected [16,17,27,28,29]. As a consequence, we did not focus on the action of a specific yolk hormone, as our aim was to mimic the consequences of a maternal stress. Our results showed that chicks exposed *in ovo* to elevated concentrations of estradiol, testosterone and progesterone (simulating a maternal challenge) expressed significantly less fear of a novel object or a novel environment but did not differ from controls regarding food neophobia.

We did not find significant effects of the treatment on latencies to eat or time spent eating each type of food across three testing conditions with different degrees of novelty. This result suggests that yolk hormone levels did not shape the expression of food neophobia specifically. In accordance with our finding in zebra finches (*Taeniopygia guttata*) from eggs with artificially increased testosterone levels that did not differ from controls in their latencies to approach and eat a novel food during their first encounter [32]. Similarly, in canaries (*Serinus canaria*), increased levels of yolk testosterone did not form the main source of variation in the expression of food neophobia [33]. In a previous study, we observed a significant increase in yolk concentrations of estradiol, testosterone and progesterone in Leghorn hens exposed to moderate heat challenge. In comparison with control chicks, offspring from these hens expressed significantly fewer distress calls when facing a novel food, but the latencies to approach and time spent eating were not affected [29]. No distress calls were emitted in the present study, and as maternal challenges modify other properties of the egg (*i.e.*, yolk mass), it remains difficult to attribute the effect on chick vocalization solely to variation in yolk hormone levels.

The significant correlation between the latency to eat the familiar odorized food and the mix of corn wheat (cereals contained in the familiar pellet food) suggests a consistency in responses when foods presented a moderate degree of novelty. Additional work is needed to determine whether food neophobia could be considered a temperament trait in domestic hens, which is suggested in populations of House sparrows (*Passer domesticus*) [34]. We did not find a correlation between the latency to eat the millet seeds and the latencies to eat other types of food, but this lack of correlation could be due to a ceiling effect caused by the short duration of our test and the maximum score (180s) attributed to pairs of chicks refusing to eat the millet seeds.

In contrast to food neophobia tests, very clear effects of the treatment were observed in the novel object test and in the open-field test. This result draws attention to the necessity of considering food neophobia on one hand, and object neophobia and fear of novel environment on the other hand as distinct processes. H chicks approached the novel object significantly faster and spent more time eating than controls, suggesting a reduced emotional response in H chicks compared to controls. Similarly, quail chicks from eggs with increased T levels approached a novel object faster than controls [20], suggesting a more "proactive" profile. However, discrepancies of yolk hormone effects are found in latencies to approach novel objects, as reviewed by Vergauwen *et al.* [33]. Depending on the species and the dose injected, shorter, longer or no effect on latencies to approach a novel object are all observed following increased yolk hormone levels (T or T+A4). We also found no correlation between the latency to approach a novel object and the latency to approach novel foods. In the very few studies investigating the influence of yolk hormones on both food and object neophobia this absence of correlation was also observed [32,33], suggesting two independent behavioral traits in adult birds. In the present study the tests were conducted at different ages and on young birds. As fearfulness changes during development [35] we cannot rule out the possible effect of age.

When tested individually in the open-field test, higher levels of activity (number of lines crossed and number of distress calls) were observed in H chicks compared to controls. In poultry species, fear is associated with behavioral inhibition characterized by both reduced vocalization and motor activity [36]. Quiet and inactive animals are commonly considered to have a higher level of emotional reactivity than active animals [37]. H chicks thus appeared less fearful in the novel environment. Increased levels of activity in novel environments with increased yolk T levels were also found in the pied flycatcher (*Ficedula hypoleuca*) [38]. In the Japanese quail, variation in yolk hormonal contents (both correlational and experimental) was almost consistently found to influence locomotor activity in open-field tests, but with no consensus concerning the direction of the effect [17,18,19,20,21,39,40]. Hegyi and Schwabl [22] have argued that different yolk androgens interact and differentially affect offspring phenotypes. For example, the authors reported that an injection of A4 in the yolk increased the locomotor activity in the open-field test but had no effect on the mass of quail chicks, whereas T injection reduced mass gain and had no effect on the open-field responses. In addition, non-linear dose response effects could also explain the discrepancy observed, as observed for nestling growth in starlings (*Sturnus unicolor*) [41]. Our results show a potential better ability to cope with novelty with increased yolk hormone levels. In the same direction, mothers with increased plasma corticosterone levels were found to produce offspring that were less fearful [25]. However, the mechanisms of action and potential detrimental trade-offs (e.g., immunity) remain to be elucidated.

In our study, no treatment effect was observed on chick mass or hatching success. Although more H eggs were discarded during incubation, it was difficult to determine whether the eggs were non-fertile or contained very early dead embryos. Yolk hormone levels show enhanced, reduced or no effect on growth of young birds in a large range of species ([41,42] for reviews). This lack of consistency highlights our current incomplete understanding of yolk hormones influences in interplay with other egg components and genetic factors. The present study suggests that there are no deleterious effects associated with elevated yolk hormone levels on the growth of chicks.

5. CONCLUSIONS

Our results show that variations in yolk hormone levels modulate the propensity to explore a novel environment and novel objects in the environment. To our knowledge, this developmental influence has not yet been studied in domestic hens. Recent studies in domestic hens have all pointed out that adaptation of mothers to challenging environments translates into maternal effects and shapes the morphology and behavioral development of the offspring [28,29,30,31,32,33,34,35,36,37]. Currently, studies are scarce and leave a gap in knowledge about the nature of these maternal effects. The vast amount of maternal influences during prenatal development can affect the development of fear-related behaviors, which could impair the capacity of young chicks to cope with their environment. We therefore emphasize the importance of studying environmental influences on the early ontogenetic niche of chicks, which is constituted by the egg.

ACKNOWLEDGMENTS

All birds were maintained at the PEAT experimental unit, INRA, Nouzilly, France. We are grateful to all members of the unit, particularly J. M. Brigant and O. Calut for caring for the chicks. This work was supported by the French National Research Agency (ANR); project FeedPhobic ANR-12-JSV7-0011-01.

REFERENCES

1. Greenberg, R.S. The role of neophobia and neophilia in the development of innovative behavior of birds. In *Animal Innovation*; Reader, S.M., Laland, K., Eds.; Oxford University Press: Oxford, United Kingdom, 2003.
2. Sol, D.; Griffin, A.S.; Bartomeus, I.; Boyce, H. Exploring or avoiding novel food resources? the novelty conflict in an invasive bird. *PLoS ONE* **2011**.
3. Greenberg, R.; Mettke-hofmann, C. Ecological aspects of neophobia and neophilia in birds. In *Current Ornithology*; Springer: New York, NY, USA, 2001; pp. 119–178.
4. Marples, N.M.; Kelly, D.J. Neophobia and dietary Conservatism: Two distinct processes? *Evol. Ecol. Res.* **1999**, *13*, 641–653.
5. Picard, M.; Plouzeau, M.; Faure, J.M. A behavioural approach to feeding broilers. *Ann. Zootech.* **1999**, *48*, 233–245.
6. Lecuelle, S.; Leterrier, C.; Chagneau, A.-M.; Laviron, F.; Lescoat, P.; Bastianelli, D.; Bertin, A.; Bouvarel, I. Experience with a variety of feed colours reduces feed neophobia in the turkey. *Appl. Anim. Behav. Sci.* **2011**, *135*, 78–85.
7. Jones, R.B. Food neophobia and olfaction in domestic chicks. *Bird Behav.* **1987**, *7*, 78–81.
8. Siddall, E.C.; Marples, N.M. Better to be bimodal: the interaction of color and odor on learning and memory. *Behav. Ecol.* **2008**, *19*, 425–432.
9. Jones, R.B. Response of domestic chicks to novel food as a function of sex, strain and previous experience. *Behav. Process.* **1986**, *12*, 261–271.
10. Murphy, L.B. Responses of domestic fowl to novel food and objects. *Appl. Anim. Ethol.* **1977**, *3*, 335–349.
11. Yo, T.; Vilarino, M.; Faure, J.M.; Picard, M. Feed pecking in young chickens: New techniques of evaluation. *Physiol. Behav.* **1997**, *61*, 803–810.
12. Lecuelle, S.; Bouvarel, I.; Chagneau, A.M.; Lescoat, P.; Laviron, F.; Leterrier, C. Feeding behaviour in turkeys with a change-over from crumbs to pellets. *Appl. Anim. Behav. Sci.* **2010**, *125*, 132–142.
13. Marples, N.M.; Quinlan, M.; Thomas, R.J.; Kelly, D.J. Deactivation of dietary wariness through experience of novel food. *Behav. Ecol.* **2007**, *18*, 803–810.

14. Desforges, M.F.; Wood-Gush, D.G.M. A behavioural comparison of domestic and mallard ducks. Habituation and flight reactions. *Anim. Behav.* **1975**, *23*, 692–697.
15. Marples, N.M.; Brakefield, P.M. Genetic variation for the rate of recruitment of novel insect prey into the diet of a bird. *Biol. J. Linn Soc.* **1995**, *55*, 17–27.
16. Bertin, A.; Richard-Yris, M.A.; Houdelier, C.; Lumineau, S.; Mostl, E.; Kuchar, A.; Hirschenhauser, K.; Kotrschal, K. Habituation to humans affects yolk steroid levels and offspring phenotype in quail. *Horm. Behav.* **2008**, *54*, 396–402.
17. Guesdon, V.; Bertin, A.; Houdelier, C.; Lumineau, S.; Formanek, L.; Kotrschal, K.; Möstl, E.; Richard-Yris, M.-A. A Place to Hide in the Home-Cage Decreases Yolk Androgen Levels and Offspring Emotional Reactivity in Japanese Quail. *PLoS ONE* **2011**.
18. Guibert, F.; Richard-Yris, M.-A.; Lumineau, S.; Kotrschal, K.; Bertin, A.; Petton, C.; Möstl, E.; Houdelier, C. Unpredictable mild stressors on laying females influence the composition of Japanese quail eggs and offspring's phenotype. *Appl. Anim. Behav. Sci.* **2011**, *132*, 51–60.
19. Guibert, F.; Richard-Yris, M.-A.; Lumineau, S.; Kotrschal, K.; Guémené, D.; Bertin, A.; Möstl, E.; Houdelier, C. Social Instability in Laying Quail: Consequences on Yolk Steroids and Offspring's Phenotype. *PLoS ONE* **2010**.
20. Daisley, J.N.; Bromundt, V.; Mostl, E.; Kotrschal, K. Enhanced yolk testosterone influences behavioral phenotype independent of sex in Japanese quail chicks *Coturnix Japonica*. *Horm. Behav.* **2005**, *47*, 185–194.
21. Okuliarova, M.; Kostal, L.; Zeman, M. Increased testosterone in egg differentially influences behaviour of Japanese quail during ontogeny. *Acta Vet. Brno* **2006**, 325–331.
22. Hegyi, G.; Schwabl, H. Do different yolk androgens exert similar effects on the morphology or behaviour of Japanese quail hatchlings Coturnix japonica? *J. Avian Biol.* **2010**, *41*, 258–265.
23. Bertin, A.; Richard-Yris, M.A.; Mostl, E.; Lickliter, R. Increased yolk testosterone facilitates prenatal perceptual learning in Northern bobwhite quail (Colinus virginianus). *Horm. Behav.* **2009**, *56*, 416–422.
24. Janczak, A.M.; Torjesen, P.; Palme, R.; Bakken, M. Effects of stress in hens on the behaviour of their offspring. *Appl. Anim. Behav. Sci.* **2007**, *107*, 66–77.
25. Henriksen, R.; Rettenbacher, S.; Groothuis, T.G.G. Maternal corticosterone elevation during egg formation in chickens (Gallus gallus domesticus) influences offspring traits, partly via prenatal undernutrition. *Gen. Comp. Endocr.***2013**, *191*, 83–91.
26. Henriksen, R.; Rettenbacher, S.; Groothuis, T.G.G. Prenatal stress in birds: Pathways, effects, function and perspectives. *Neurosci. Biobehav. Rev.* **2011**, *35*, 1484–1501.
27. Nätt, D.; Lindqvist, N.; Stranneheim, H.; Lundeberg, J.; Torjesen, P.A.; Jensen, P. Inheritance of Acquired Behaviour Adaptations and Brain Gene Expression in Chickens. *PLoS ONE* **2009**.
28. Goerlich, V.C.; Nätt, D.; Elfwing, M.; Macdonald, B.; Jensen, P. Transgenerational effects of early experience on behavioral, hormonal and gene expression responses to acute stress in the precocial chicken. *Horm. Behav.* **2012**, *61*, 711–718.
29. Bertin, A.; Chanson, M.; Delaveau, J.; Mercerand, F.; Möstl, E.; Calandreau, L.; Arnould, C.; Leterrier, C.; Collin, A. Moderate Heat Challenge Increased Yolk Steroid Hormones and Shaped Offspring Growth and Behavior in Chickens. *PLoS ONE* **2013**.
30. Groothuis, T.G. G.; Muller, W.; von Engelhardt, N.; Carere, C.; Eising, C. Maternal hormones as a tool to adjust offspring phenotype in avian species. *Neurosci. Biobehav. Rev.* **2005**, *29*, 329–352.
31. Aigueperse, N.; Calandreau, L.; Bertin, A. Maternal Diet Influences Offspring Feeding Behavior and Fearfulness in the Precocial Chicken. *PLoS ONE* **2013**.
32. Tobler, M.; Sandell, M.I. Yolk testosterone modulates persistence of neophobic responses in adult zebra finches, *Taeniopygia guttata*. *Horm. Behav.* **2007**, *52*, 640–645.
33. Vergauwen, J.; Eens, M.; Muller, W. Pre- and postnatal environmental effects as potential sources of variation in neophobic behaviour in canaries. *Behaviour* **2012**, *149*, 849–867.
34. Bókony, V.; Kulcsár, A.; Tóth, Z.; Liker, A. Personality Traits and Behavioral Syndromes in Differently Urbanized Populations of House Sparrows (Passer domesticus). *PLoS ONE* **2012**.

35. Bateson, P. Changes in chicks' responses to novel moving objects over the sensitive period for imprinting. *Anim. Behav.* **1964**, *12*, 479–489.
36. Jones, R.B.; Satterlee, D.G.; Ryder, F.H. Open-field behavior of Japanese quail chicks genetically selected for low or high plasma corticosterone response to immobilization stress. *Poult. Sci.* **1992**, *71*, 1403–1407.
37. Faure, J.; Jones, R.; Bessei, W. Fear and social motivation as factors in open-field behaviour of the domestic chick. A theoretical consideration. *Biol. Behav.* **1983**, *8*, 103–116.
38. Ruuskanen, S.; Laaksonen, T. Yolk hormones have sex-specific long-term effects on behavior in the pied flycatcher (Ficedula hypoleuca). *Horm. Behav.* **2010**, *57*, 119–127.
39. Bertin, A.; Richard-Yris, M.A.; Houdelier, C.; Richard, S.; Lumineau, S.; Kotrschal, K.; Mostl, E. Divergent selection for inherent fearfulness leads to divergent yolk steroid levels in quail. *Behaviour* **2009**, *146*, 757–770.
40. Okuliarova, M.; Skrobanek, P.; Zeman, M. Effect of increasing yolk testosterone levels on early behaviour in Japanese quail hatchlings. *Acta Vet. Brno* **2007**, *76*, 325–331.
41. Muriel, J.; Pérez-Rodríguez, L.; Puerta, M.; Gil, D. Diverse dose–response effects of yolk androgens on embryo development and nestling growth in a wild passerine. *J. Exp. Biol.* **2015**, *218*, 2241–2249.
42. Smiseth, P.T.; Scott, M.P.; Andrews, C. Hormonal regulation of offspring begging and mediation of parent-offspring conflict. *Anim. Behav.* **2011**, *81*, 507–517.

CHAPTER 13

The history of Old World camelids in the light of molecular genetics

Pamela Anna Burger[1]

[1] *Research Institute of Wildlife Ecology, Vetmeduni Vienna, Savoyenstrasse 1, 1160 Vienna, Austria*

ABSTRACT

Old World camels have come into the focus as sustainable livestock species, unique in their morphological and physiological characteristics and capable of providing vital products even under extreme environmental conditions. The evolutionary history of dromedary and Bactrian camels traces back to the middle Eocene (around 40 million years ago, mya), when the ancestors of Camelus emerged on the North American continent. While the genetic status of the two domestic species has long been established, the wild two-humped camel has only recently been recognized as a separate species, Camelus ferus, based on molecular genetic data. The demographic history established from genome drafts of Old World camels shows the independent development of the three species over the last 100,000 years with severe bottlenecks occurring during the last glacial period and in the recent past. Ongoing studies involve the immune system, relevant production traits, and the global population structure and domestication of Old World camels. Based on the now available whole genome drafts, specific metabolic pathways have been described shedding new light on the camels' ability to adapt to desert environments. These new data will also be at the origin for genome-wide association studies to link economically relevant phenotypes to genotypes and to conserve the diverse genetic resources in Old World camelids.

Keywords: Camelus bactrianus, Camelus ferus, Camelus dromedaries, Domestication, Genetic resource management, Genetic diversity

1. INTRODUCTION

Without Old World camelids (Camelini), human survival and prosperity in the marginal agro-ecological zones of African, Arabian, and Asian (semi-) deserts would have been challenging. Since ancient times, camels were exploited as beast of burden, for riding, and as milk and meat source, and today, gain tremendous importance as sustainable livestock species with very specific features (e.g., immunogenes and milk composition). Camels facilitated the trading and cultural dialog between three continents by connecting the Arabian Peninsula with the Sahara and the Levant to the Far East and Asia, with northern Arabia at the crossroads.

Deep historical events like climate change or glacial periods as well as more recent anthropogenic interactions triggered demographic processes in Old World camelids, shaping the patterns of genetic diversity in modern populations. By examining this modern genetic diversity and its global distribution, it is possible to gain insight into past demographic events, which reflect the history of Old Word camelids. This review will give an overview about the evolutionary history of Old World camels and the molecular genetic studies performed over the past years, and summarize the standard molecular genetic methods that underlie these studies. In addition, recent advances in next-generation sequencing (NGS) technologies and their careful implementation in current and future camel research will be discussed.

1.1 Methods for molecular genetic analysis

Depending on the objectives and size of the project, the choice of the marker is critical. The assessment of genetic diversity and population structure might be important if local or regional groups/breeds are investigated, while large-scale phylogenetic studies will be more interested in the global connectivity and demographic history of populations. Association studies between phenotypic traits and underlying genotypes will be necessary if marker-assisted breeding programs are anticipated or if adaptation to a specific environment is addressed.

2. GENETIC MARKERS TO INVESTIGATE DIVERSITY AND POPULATION STRUCTURE

Molecular markers were previously applied in the camelid family for parentage control (Mariasegaram et al. 2002), as well as for the determination of the genetic variability between individual Bactrian camel (Chuluunbat et al. 2014; Silbermayr et al. 2010a; Jianlin et al. 2000, 2004) and dromedary breeds (Mburu et al. 2003), guanacos (Mate et al. 2005), llamas, and alpacas (Kadwell et al. 2001). Molecular evolutionary analysis was used to identify the ancestors of the llama and the alpaca (Kadwell et al. 2001; Stanley et al. 1994).

2.1 Maternal (mitochondrial) and paternal (Y-chromosomal) markers

The most commonly used polymorphic markers in camelid studies are single nucleotide polymorphisms (SNPs) or restriction length polymorphisms (RFLPs) studied in the mitochondrial DNA (mtDNA) reflecting the maternal inheritance (Silbermayr et al. 2010b; Ji et al. 2009; Jianlin et al. 1999). The genetic variation between Mongolian domestic and wild Bactrian camels showed 2.9 % nucleotide difference in the mitochondrial control regions (CR) of the two related species (Silbermayr et al. 2010b; Ji et al. 2009). In addition, mtDNA sequence analysis of ancient DNA (aDNA) proved to be crucial in resolving domestication processes in donkeys (Beja-Pereira et al. 2006), cattle (Troy et al. 2001), and dromedaries (Almathen et al., Ancient and modern DNA reveal dynamics of domestication and cross-continental dispersal of the dromedary, submitted to PNAS (2015-19508R)).

The Y-chromosome is much less variable within species, e.g., Y-SNPs (Lindgren et al. 2004), and Y-chromosome-linked microsatellite analysis detected high homology in different horse breeds (Wallner et al. 2003). Polymorphic Y-microsatellite markers in ruminants are currently only available for small ruminants (Meadows et al. 2006), yak (Xuebin et al. 2005), and cattle (Hanotte et al. 2000).

2.2 Nuclear autosomal markers

The main class of molecular markers used in molecular genetic studies in camelids is microsatellites, short repetitive motives (e.g., ACACAC...) in the nuclear DNA following the co-dominant inheritance (Schlötterer 2004). Autosomal microsatellite loci are important tools to measure population diversity (Chuluunbat et al. 2014; Silbermayr et al. 2010a; Jianlin et al. 2004), genetic variation, and genetic admixture among livestock breeds or between wild and domestic animal species (Bruford et al. 2003). A set of microsatellites established in New World camelids successfully amplified in the Camelini (Mariasegaram et al. 2002) and was applied to study the genetic distance between Mongolian and Chinese domestic Bactrian camels (Jianlin et al. 2004; Charruau 2012).

Nuclear SNPs are single base pair changes that occur approximately each 1000 bps in the mammalian genome. Large-scale nuclear SNP analyses have not been applied in Old World camelids so far. These markers are very useful alternatives to microsatellites and have been employed in many studies about genetic diversity and relevant phenotypic traits in livestock (Goddard and Hayes 2009).

3. POPULATION GENETIC AND PHYLOGENETIC ANALYSES IN OLD WORLD CAMELIDS

The population genetic analyses in Old World camelids reviewed in this study were based on the comparisons of local breeds and addressed population genetic diversity and differentiation. If genetic diversity, population structure, and differentiation are the objectives of a study, the most important analytical steps should involve (1) evaluation of the raw data including confirming single haplotypes, testing for allelic dropout and possible null alleles, and checking for possible relatedness in the sample set to avoid a bias in the genetic diversity estimates, i.e., underestimation of heterozygosity and overestimation of the inbreeding coefficient (FIS) due to related individuals; (2) screening of population structure using Bayesian methods (Pritchard et al. 2000; Corander and Tang 2007) and/or principal component analysis (PCA); (3) estimation of principle population genetic and diversity parameters over all samples and between populations including Hardy-Weinberg equilibrium and heterozygosity (in nuclear markers), nucleotide and haplotype diversities (mtDNA and nuclear sequence data); (4) population differentiation in terms of population pairwise FST values and analysis of molecular variance (AMOVA); (5) phylogenetic and phylogeographic analyses constructing phylogenetic trees (Tamura et al. 2013) and haplotype networks (Bandelt et al. 1999); (6) inferences of demographic history, including pairwise mismatch distributions, calculations of neutrality estimators (e.g., Tajima's D, Fu's FS), and estimation of changes in the effective population size (Ne) (Bouckaert et al. 2014; Beaumont 1999); and (7) individual hypothesis testing.

4. NEXT-GENERATION SEQUENCING AND GENOME-WIDE GENOTYPING TECHNOLOGIES: ADVANTAGES AND PITFALLS

NGS usually refers to the sequencing of whole genomes using massive parallel sequencing technology. This method employs sequencing by synthesis (SBS; Ronaghi et al. 1998) with four fluorescently labeled nucleotides to sequence millions of single-molecule clusters on a flow cell in parallel, e.g., now offered by Illumina®. First draft whole genomes of the Old World camelids have recently been published (Fitak et al. 2015; Wu et al. 2014; Burger and Palmieri 2014; Jirimutu et al. 2012).

Genome-wide genotyping techniques involve high-density SNP chips or methods, which reduce genome complexity with restriction enzymes before massive parallel amplification, e.g., restriction site-associated DNA sequencing (RADseq; Andrews et al. 2016) and genotyping by sequencing (GBS; Hess et al. 2015). NGS and genotyping technologies applied to Old World camelids in the future might include whole-genome optical mapping for improved genome assemblies (Howe and Wood 2015) as well as SNP genotyping with different methods (RADseq, GBS). Furthermore, genome-wide association studies using either full genomes or high-density SNP chips might be performed similar to other livestock (Hayes et al. 2012; Goddard and Hayes 2009). Caveats of these new technologies include the necessity of high computational resources and advanced bioinformatics and statistical knowledge as well as a high sample number with well-defined phenotypes.

5. PHYLOGENETIC ORIGIN: THREE DIFFERENT SPECIES

The origin of the Old World camelids traces back to the Eocene (40–45 mya) when the first ancestors of the camelid family were found in North America. After splitting into New World (Lamini) and Old World (Camelini) camels, the latter migrated via the Bering land bridge to the eastern hemisphere (the Old World). The earliest camel remains in Asia date back to 5 mya (Kozhamkulova 1986). The divergence between one-humped and two-humped camels was estimated between 5 and 8 mya (Wu et al. 2014; Ji et al. 2009), contradicting previous assumptions of the wild Bactrian camel being the common direct ancestor of dromedary and Bactrian camels (Peters and von den Driesch 1997). Furthermore, an earlier observation of the dromedary embryo undergoing a two-humped developmental stage

(Lombardini 1879) has been disproven by demonstrating the prenatal development of a single hump in Arabian camel fetuses (Kinne et al. 2010).

Molecular evolutionary studies estimate the split between Old and New World camels at 11 mya (Kadwell et al. 2001) to 25 mya (Ji et al. 2009). Within the Camelini, the divergence between dromedaries and Bactrian camels has been dated at 5 to 8 mya (Fig. 1) (Wu et al. 2014; Ji et al. 2009). Modern molecular genetic (Silbermayr et al. 2010b; Ji et al. 2009; Jianlin et al. 2004) and genomic (Jirimutu et al. 2012; Wu et al. 2014) studies confirm the presence of three extant Camelini species: Camelus dromedarius, Camelus bactrianus, and Camelus ferus. While the species status of both domestic species had been well established, it was heavily debated if the last wild two-humped camels in the Mongolian Gobi and the Chinese Taklimakan and Lop Nor deserts were feral or an evolutionary unique unit. Wild camels were first described by Przewalski in 1878, and the International Commission of Nomenclature (2003) fixed the first available specific name based on a wild population "Camelus ferus" to the wild camel (Gentry et al. 2004). Molecular genetic analysis of mitochondrial (Silbermayr et al. 2010b; Ji et al. 2009; Jianlin et al. 2004) and nuclear markers (Silbermayr and Burger 2012) demonstrated the divergence between wild and domestic Bactrian camels and estimated the time of separation between 0.7 and 1.5 mya in the Pleistocene, long before domestication (4000 – 6000 ya). This long-term divergence also excludes the wild two-humped camels as direct ancestors of modern domestic Bactrian camels.

The early independent demographic history of the three Old World camel species has now been reconstructed using whole-genome SNP data (Fitak 2014; Wu et al. 2014). The dataset consisted of whole-genome shotgun sequences from the three species of Old World camel (*C. dromedarius* ($n = 9$), *C. ferus* ($n = 9$), and *C. bactrianus* ($n = 7$)) mapped to the wild camel reference genome (Jirimutu et al. 2012) with approximately 15-fold coverage. To model the demographic history, the Pairwise Sequentially Markovian Coalescent model (PSMC; Li and Durbin 2011) was used with 100 bootstraps on the individual genomes masked for repetitive regions (Fitak 2014). PSMC infers Ne at a given time in the past from a single diploid individual using the rates of coalescence events across the genome. The number of coalescent events occurring more recently than 1000 ya is inadequate to accurately infer the demographic history. Figure 2 shows a maximum effective population size (N_e) of 18,000 around 350,000 ya, with a sharp reduction of up to 70 % until the last glacial period (LGP; 100,000–20,000 ya). During the LGP, the N_e continued declining until it increased again in the Bactrian camel species and reached a plateau in the dromedaries after the end of the last glacial maximum (LGM; ~20,000 ya). Similar patterns of demographic changes in large mammals based on climatic conditions during this period have been described (Orlando et al. 2013; Wu et al. 2014).

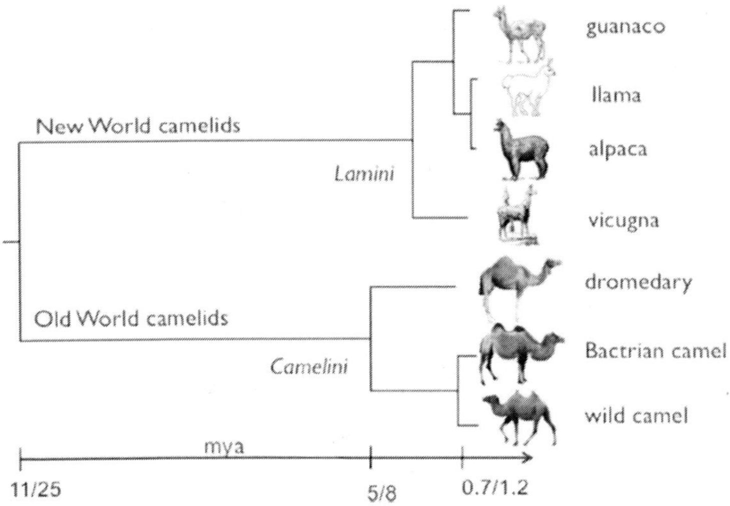

Figure.1 Maximum likelihood tree based on complete mitochondrial genomes of New and Old World camels

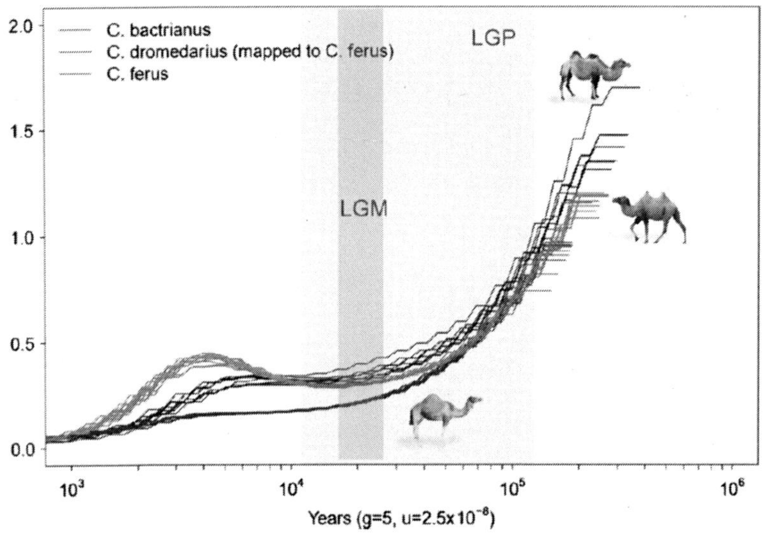

Figure. 2 The historical demography of the three Old World camel species was examined using PSMC v0.6.4

A second, more recent bottleneck starting around 4000–5000 ya further reduced the N_e in the Old World camels (Fig. 2). While this second bottleneck might be related with domestication scenarios of dromedary and Bactrian camels, in the case of the wild camel, it could reflect the beginning (and still ongoing) reduction of habitat and increased hunting activity connected with human population growth in Asia. Overall, a lower Ne in dromedaries than in Bactrian camels has been observed until the recent 3000–4000 ya (Fig. 2).

6. DOMESTICATION

Domestic camels had an important role in the ancient civilizations of the Old World, and today, a number of rural and nomadic societies in North-eastern Africa and in Asia depend on these multipurpose animals in terms of transportation and milk, wool, meat, and dung production. The intensive trading and movements of camels around North Africa and the Arabian Peninsula are reflected in the mitochondrial (maternal) genetic profile of the global camel populations. A panmictic (globally unstructured) dromedary population ranging from North and East Africa to Pakistan and Australia was found (Charruau 2012). At the nuclear level, a more defined global structure was revealed, differentiating East African and South Arabian dromedaries from North African, North Arabian, and South Asian individuals, respectively (Almathen et al., Ancient and modern DNA reveal dynamics of domestication and cross-continental dispersal of the dromedary, submitted to PNAS (2015-19508R); Al-Mathen 2014; Charruau 2012; Mburu et al. 2003; Schulz et al. 2010). Wild dromedaries must have occurred in considerable numbers on the Arabian Peninsula in Holocene Times, as illustrated by the discovery of the second millennium BC hunting sites along the eastern coast yielding large amounts of bones from this species (von den Driesch and Obermaier 2007; Uerpmann and Uerpmann 2002). Zooarchaeological studies suggest that domestication took place in the coastal southeast Arabian Peninsula (Uerpmann and Uerpman 2002, 2012; von den Driesch and Obermaier 2007; Grigson 2012, 2014). Ongoing genetic studies combining modern and ancient DNA confirm this region as one possible place of domestication and suggest a single domestication scenario with recurrent introgression from wild dromedaries into the early-domestic stock (Almathen et al., Ancient and modern DNA reveal dynamics of domestication and cross-continental dispersal of the dromedary, submitted to PNAS (2015-19508R)).

The phylogeographic analysis of modern domestic Bactrian camels from Mongolia, Russia, Kazakhstan, and China revealed that mitochondrial haplotypes are shared between animals and countries (Charruau 2012).

Moreover, a palaeogenetic analysis of 12 Bactrian camel bones from Late Bronze and Early Iron Age sites in Uzbekistan and Siberia showed the same mitochondrial haplotypes as described in modern domestic Bactrian camels, which were genetically distinct from wild Bactrian camels. This reveals that the current Mongolian wild camels are neither the descendants of Late Bronze and Early Iron Age nor the ancestors of modern domestic Bactrian camels. The high homogeneity between these Bronze/Iron Age samples and modern domesticated camels from China and Mongolia led to the conclusion that there was a single domestication for Bactrian camels (Trinks et al. 2012). Until today, the wild relatives of the domestic Bactrian camel still exist in small numbers in Mongolia and China whereas wild dromedaries have become extinct.

7. HYBRIDIZATION

Anthropogenic hybridization between domestic animals and their wild relatives has occurred since the early days of animal breeding. In Old World camels, we find two directions, (i) hybridization between Bactrian camels and dromedaries, which facilitates strong and robust individuals; and (ii) introgression (gene flow from one pool into the other over more generations) from domestic Bactrian camels into wild two-humped camels, and survival of the wild species classified as "critically endangered" by the International Union for Conservation of Nature (IUCN).

8. THE ENCOUNTER OF DROMEDARY AND BACTRIAN CAMELS

Hybridization between Bactrian camels (C. bactrianus) and dromedaries (C. dromedarius) was associated with the transportation of goods along multiple routes of the Silk Road. This practice intended to produce animals with the robustness of the Bactrian camel, the endurance of dromedary, and the ability to tolerate sharply contrasting climatic conditions (Wilson 1984). Today, hybridization facilitates improved milk and wool yield in hybrid Tulu or Nar camels from Middle Eastern and Central Asian countries. Commonly, two hybridizing methods are recognized, Kurt-nar (dromedary female × Bactrian male) and Kez-nar (Bactrian female × dromedary male) followed by F_1-backcrossing with either dromedary for increased milk productivity or Bactrian camel for wool and cold resistance (Faye and Konuspayeva 2012). This heterosis or hybrid vigor effect arises from allelic interactions between parental genomes, potentially leading to increased growth, productivity, and fitness of the hybrids. F_2 hybrids ($F_1 \times F_1$) in Old World camels are usually not favored because of a difficult character and weak progeny performance (Faye and Konuspayeva 2012). A SNP marker set has been developed to identify hybrids and their backcrosses in animals with a cryptic ancestry, or in ancient samples (Ruiz et al. 2015).

9. HYBRIDIZATION AND INTROGRESSION BETWEEN WILD AND DOMESTIC BACTRIAN CAMELS: A MATTER OF CONSERVATION

The hybridization between domestic Bactrian and wild camels is jeopardizing a long process of evolution as it threatens the gene pool of the wild species (Lei et al. 2012; Yadamsuren et al. 2012). With nuclear and mitochondrial markers, different levels of admixture between domestic and wild camels were detected in the Mongolian wild camel population (Charruau 2012; Silbermayr et al. 2010b). This has also been confirmed using whole-genome SNPs as two supposedly pure wild camels showed introgression from domestic camels (Fig. 3; Fitak et al., unpublished data). "Preserving the genetic integrity of the wild camel" was included as a conservation priority to the Mongolian National Conservation Strategy for the wild camel and its desert habitat (2012).

9.1 Ongoing and future research

The genomic resource "Old World camels" has come into focus in the last decade, with a number of scientific groups working in different fields.

10. ADAPTATION TO DESERT ENVIRONMENT AND SELECTION DURING DOMESTICATION

Genes involved in energy metabolism and stress response were described to show accelerated evolutionary rates as compared to New World camels or cattle. Similarly, rapid evolution of genes underlying pathways for salt metabolism and duplications of genes, which play critical roles in sodium reabsorption and water balance in the kidney, were defined to be related to an adaptation to desert environment (Wu et al. 2014; Jirimutu et al. 2012). Comparative analyses between domestic dromedary and Bactrian camels versus wild camels aim to detect signatures of selection during domestication in Old World camels (Fitak et al., unpublished data).

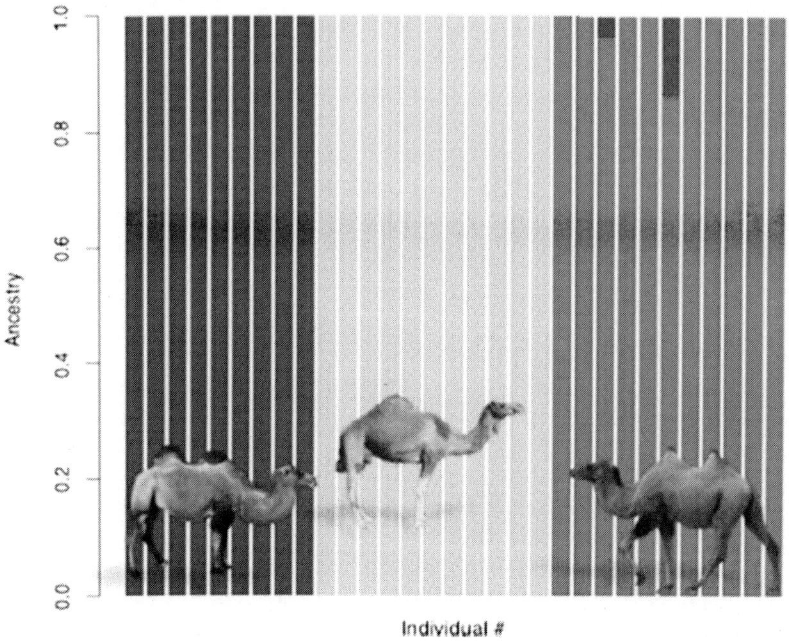

Figure. 3 Admixture analysis using whole-genome SNP data of dromedary (green), Bactrian (blue), and wild (red) camels. Each bar represents the genomic ancestry of a single individual; mixed colors indicate a mixed ancestry between two species

11. IMMUNOGENETICS

The camel immunogenome has been studied so far only to limited extent (e.g., Griffin et al. 2014; Muyldermans et al. 2009). Ongoing studies target the major histocompatibility complex (MHC), a set of immune response (IR) genes playing a crucial role in host-pathogen interactions, to characterize the general organization and molecular evolution of this region in Old World camels (Plasil et al. 2016). Other IR genes, like the T cell receptor, have evolved in the dromedary by mutation in the gamma (TRG) and delta (TRD) genes (Ciccarese et al. 2014). The genomic structure and gene content of these genes are currently investigated (Antonacci et al. 2015).

12. GENETIC ANALYSES OF PRODUCTION TRAITS

The encoding genes for β-casein (Pauciullo et al. 2014) and κ-casein (Pauciullo et al. 2013), two major abundant proteins in camel milk, have been characterized in dromedaries and partially in Bactrian camels. The studies confirm the potential protective role of the camel milk for the human nutrition. The comprehensive characterization of the sequence variability at the MSTN locus coding for myostatin, a negative regulator of skeletal muscle responsible in several livestock species for increased skeletal-muscle mass, in Tunisian dromedaries has been initiated (Muzzachi et al. 2015).

13. NEW CHALLENGES: GENOME-WIDE ASSOCIATION STUDIES AND GENOMIC SELECTION

The identification of genotypes underlying economically relevant phenotypic traits using genome-wide association studies (GWAS) has been applied in other domestic species for a long time (Goddard and Hayes 2009). Selective breeding has tremendously increased market weight in chicken and pigs, and milk production in cattle (Hayes et al. 2012). The trade-off of this intensive selection, however, is loss of genetic diversity, inbreeding and interconnected problems of decreased fertility, and a reduced potential for adaptation to environmental and pathogenic changes. The challenges in this field are numerous, and nowadays many breeding programs shift from only production to include health and fitness traits, like fertility (Hayes et al. 2012).

Genomic selection refers to the application of genome-wide markers to predict the breeding value of selection candidates (Meuwissen et al. 2001). The major improvement over earlier methods of marker-assisted selection, which relied on a small number of causal variants, is the association of complex traits with hundreds or thousands of polymorphisms, each with small effects (Goddard and Hayes 2009). The computation to predict breeding values from SNP genotypes must be estimated from a large sample of animals (around 1000, depending on N_e and linkage disequilibrium in the population), termed the reference population, which have been phenotyped (measured for certain traits) and SNP genotyped. Breeding values for selection candidates solely based on their genotypes can then be calculated with a prediction equation. The candidates are ranked on these estimated breeding values and the best ones are selected to breed the next generation. The advantage of genomic selection is that based on their genomic prediction, individuals can be selected accurately much earlier in life, which results in a reduction of the generation interval and an increase in the rate of genetic gain (Hayes et al. 2012).

In Old World camels, the challenge for future sustainable breeding programs will be to balance the intensification of production traits with the preservation of genetic diversity and local adaptation. The development of high-density SNP chips or potential whole-genome sequencing projects similar to the 1000 bulls genome project (www.1000bullgenomes.com) as a basis for genomic selection need to go along with a thorough and standardized recording of production traits, pedigree, and progeny performance testing. Potential genomic selection projects in camels need to be preceded and complemented by traditional breeding programs based on pedigrees and phenotypes.

14. CONSERVATION OF GENETIC RESOURCES IN WILD AND DOMESTIC CAMELS

The level of genetic variation and as such the ability of a population to evolve and respond to artificial and natural selection is positively associated with the effective population size N_e (Falconer and Mackay 1996). In natural (wild) populations, there are well-known cases of inbreeding and decreased adaptation potential due to reduced genetic variation (Kristensen et al. 2015). Similar to natural populations, inbreeding depression has been observed in domestic species with a small N_e, decreasing fertility and milk yield in cattle (Pryce et al. 2002), and growth rates in sheep (Pedrosa et al. 2010), while increasing diseases (e.g., mastitis; Sørensen et al. 2006). We still record average levels of genetic diversity (nuclear heterozygosity and nucleotide diversity) in domestic Bactrian camels (Chuluunbat et al. 2014) and dromedaries (Almathen 2014; Charruau 2012). However, in wild camels, reduced genetic variation and smaller census size as well as lower Ne have been observed (Charruau 2012; Yadamsuren et al. 2012). Thoughtfully applied genomic tools can contribute to a genetic resource management through monitoring and preserving genetic uniqueness (i.e., C. ferus) and genetic diversity, while improving vitally important production traits in the amazing and precious Old World camels.

ACKNOWLEDGMENTS

Open access funding provided by Austrian Science Foundation (FWF). PB is very grateful to all collaborators, camel owners, and collectors for providing precious samples, to J. Peters for access to the Paleogenetic Core Facility ArchaeoBioCenter, LMU Munich (DE), and to C. Schlötterer, Institute of Population Genetics, Vetmeduni Vienna, for general support. The work was funded by the Austrian Science Foundation (FWF): P24706-B25 to PB, recipient of an APART fellowship (11506) from the Austrian Academy of Sciences.

REFERENCES

1. Almathen F., 2014. Genetic diversity and demographic history of dromedary camel (Camelus dromedarius). Thesis, The University of Nottingham, Nottingham (UK)
2. Andrews K.R., Good J.M., Miller M.R., Luikart G., Hohenlohe P.A., 2016. Harnessing the power of RADseq for ecological and evolutionary genomics. Nat. Rev. Gen., 17, 81–92
3. Antonacci R., Fitak R., Burger P.A., Castelli V., Ciani E., Ciccarese S., 2015. Functional genomics and evolution of the gamma/delta T cell receptor loci in Old World camels. Special Issue of Scientific and Practical Journal Veterinaryia, 42, 344–346
4. Bandelt H., Forster P., Rohl A. (1999) Median-joining networks for inferring intraspecific phylogenies. Mol. Biol. Evol., 16, 37–48
5. Beaumont M.A., 1999. Detecting population expansion and decline using microsatellites. Genetics, 153, 2013–2029
6. Beja-Pereira A., Caramelli D., Lalueza-Fox C., Vernesi C., Ferrand N., Casoli A., Goyache F., Royo L.J., Conti S., Lari M., Martini A., Ouragh L., Magid A., Atash A., Zsolnai A., Boscato P., Triantaphylidis C., Ploumi K., Sineo L., Mallegni F., Taberlet P., Erhardt G., Sampietro L., Bertranpetit J., Barbujani G., Luikart G., Bertorelle G., 2006. The origin of European cattle: evidence from modern and ancient DNA. Proc. Natl. Acad. Sci. U S A., 103, 8113–8118
7. Bouckaert R., Heled J., Kühnert D., Vaughan T., Wu C-H., Xie D., Suchard M.A., Rambaut A., Drummond A.J., 2014. BEAST 2: A Software Platform for Bayesian Evolutionary Analysis. PLoS Computat. Biol., 10, e1003537
8. Bruford M.W., Bradley D.G., Luikart G., 2003. DNA markers reveal the complexity of livestock domestication. Nat. Rev. Gen., 4, 900–909
9. Burger P.A., Palmieri N., 2014. Estimating the population mutation rate from a de novo assembled Bactrian camel genome and cross-species comparison with Dromedary ESTs. J. Hered., 105, 839–846
10. Charruau P., 2012. Insights from evolutionary history and population genetics for domestic and wildlife conservation – cases of the Old World camelids and cheetahs. Thesis, University of Veterinary Medicine, Vienna (Austria)
11. Chuluunbat B., Charruau P., Silbermayr K., Khorloojav T., Burger P.A., 2014. Genetic diversity and population structure of Mongolian domestic Bactrian camels (Camelus bactrianus). Anim. Genet., 45, 550–558
12. Ciccarese S., Vaccarelli G., Lefranc M.P., Tasco G., Consiglio A., Casadio R., Linguiti G., Antonacci R., 2014. Characteristics of the somatic hypermutation in the Camelus dromedarius T cell receptor gamma (TRG) and delta (TRD) variable domains. Dev. Comp. Immunol., 46, 300–313
13. Corander J., Tang J., 2007. Bayesian analysis of population structure based on linked molecular information. Math. Biosci., 205, 19–31
14. Falconer D.S., Mackay T.F.C., 1996. Introduction to Quantitative Genetics. Longman, UK, pp 464
15. Faye B., Konuspayeva G., 2012. The Encounter between Bactrian and Dromedary Camels in Central Asia. In: E.M. Knoll and P.A. Burger (Eds) Camels in Asia and North Africa. Interdisciplinary perspectives on their significance in past and present. Austrian Academy of Sciences, Vienna, 29–36

16. Fitak R., 2014. Assessing the demographic history of Old World camelids through whole-genome sequencing. Scientific report. http://www.esf.org/coordinating-research/research-networking-programmes/life-earth-and-environmental-sciences-lee/current-esf-research-networking-programmes-in-life-earth-and-environmental-sciences/advances-in-farm-animal-genomic-resources-genomic-resources/scientific-activities.html. Accessed on 07 March 2015.
17. Fitak R., Mohandesan E., Corander J., Burger P.A., 2015. The de novo genome assembly and annotation of a female domestic dromedary of North African origin. Mol. Ecol. Res., 16, 314–324
18. Gentry A., Clutton-Brock J., Groves C.P., 2004. The naming of wild animal species and their domestic derivates. J. Archeol. Sci., 31, 645–651
19. Goddard M.E., Hayes B.J., 2009. Mapping genes for complex traits in domestic animals and their use in breeding programmes. Nat. Rev. Gen., 10, 381–391
20. Griffin L.M., Snowden J.R., Lawson A.D., Wernery U., Kinne J., Baker T.S., 2014. Analysis of heavy and light chain sequences of conventional camelid antibodies from Camelus dromedarius and Camelus bactrianus species. J. Immunol. Methods, 405, 35–46
21. Grigson C., 2012. Camels, copper and donkeys in the Early Iron Age of the Southern Levant: Timna revisited. Levant, 44, 82–100
22. Grigson C., 2014. The history of the camel bone dating project. Anthropozoologica, 49, 225–235
23. Hanotte O., Tawah C.L., Bradley D.G., Okomo M., Verjee Y., Ochieng J., Rege J.E., 2000. Geographic distribution and frequency of a taurine Bos taurus and an indicine Bos indicus Y specific allele amongst sub-saharan African cattle breeds. Mol. Ecol., 9, 387–396
24. Hayes B.J., Lewin, H.A., Goddard M.E., 2012. The future of livestock breeding: genomic selection for efficiency, reduced emissions intensity, and adaptation. Trends. Genet., TIG, 1–9
25. Hess J.E., Campbell N.R., Docker M.F., Baker C., Jackson A., Lampman R., McIlraith B., Moser M.L., Statler D.P., Young W.P., Wildbill A.J., Narum S.R., 2015. Use of genotyping by sequencing data to develop a high-throughput and multifunctional SNP panel for conservation applications in Pacific lamprey. Mol. Ecol. Res., 15, 187–202
26. Howe K., Wood J.M.D., 2015. Using optical mapping data for the improvement of vertebrate genome assemblies. Gigascience, 4:10.
27. International Commission on Zoological Nomenclature, 2003. Opinion 2027 (Case 3010): Usage of 17 specific names based on wild species which are pre-dated by or contemporary with those based on domestic animals (Lepidoptera, Osteichthyes, Mammalia): conserved. http://iczn.org/content/biodiversity-studies. Accessed on 8 Feb 2016
28. Ji R., Cui P., Ding F., Geng J., Gao H., Zhang H., Yu J, Hu S., Meng H., 2009. Monophyletic origin of domestic bactrian camel (Camelus bactrianus) and its evolutionary relationship with the extant wild camel (Camelus bactrianus ferus). Anim. Gen., 40, 377–382
29. Jianlin H., Jiexia, Q., Zhenming M., Yaping Z., Wen W., 1999. Rapid communication: three unique restriction fragment length polymorphisms of EcoRI, PvuII, and ScaI digested mitochondrial DNA of Bactrian camels (Camelus bactrianus ferus) in China. J. Anim. Sci., 77, 2315–2316
30. Jianlin H., Mburu D., Ochieng J., Kaufmann B., Rege J.E., Hanotte O., 2000. Application of New World Camelidae microsatellite primers for amplification of polymorphic loci in Old World camelids. Anim. Genet., 31, 404–406
31. Jianlin H., Ochieng J.W., Lkhagva B., Hanotte O., 2004. Genetic diversity and relationship of domestic Bactrian camels (C. bactrianus) in China and Mongolia. J. C. Pract. Res., 11, 97–99
32. Jirimutu, Wang Z., Ding G., Chen G., Sun Y., Sun Z., Zhang H., Wang L., Hasi S., Zhang Y., Li J., Shi Y., Xu Z., He C., Yu S., Li S., Zhang W., Batmunkh M., Ts B., Narenbatu, Unierhu, Bat-Ireedui S., Gao H., Baysgalan B., Li Q., Jia Z., Turigenbayila, Subudenggerile, Narenmanduhu, Wang Z., Wang J., Pan L., Chen Y., Ganerdene Y., Dabxilt, Erdemt, Altansha, Altansukh, Liu T., Cao M., Aruuntsever, Bayart, Hosblig, He F,. Zha-ti A., Zheng G., Qiu F., Sun Z., Zhao L., Zhao W., Liu B., Li C., Chen Y., Tang X., Guo C., Liu W., Ming L.,

Temuulen, Cui A., Li Y., Gao J., Li J., Wurentaodi, Niu S., Sun T., Zhai Z., Zhang M., Chen C., Baldan T., Bayaer T., Li Y., Meng H., 2012. Genome sequences of wild and domestic bactrian camels. Nat. Commun., 3,1202. Erratum in: Nat. Commun., 2013, 4, 3089

33. Kadwell M., Fernandez M., Stanley H.F., Baldi R., Wheeler J.C., Rosadio R., Bruford M.W., 2001. Genetic analysis reveals the wild ancestors of the llama and alpaca. P. Biol. Sci., 22, 2575–2584
34. Kinne J., Wani N.A., Wernery U., Peters J., Knospe C., 2010. Is there a two-humped stage in the embryonic development of the dromedary? Anat. Histol. Embryol., 39, 479–480
35. Kozhamkulova B.S., 1986. The Late Cenozoic two-humped (Bactrian) camels of Asia. Quartärpläontolgie (Abh. Ber. Inst. Quartärpläontolgie Weimar), 6, 93–97
36. Kristensen T.N., Hoffmann A.A., Pertoldi C., Stronen A.V., 2015. What can livestock breeders learn from conservation genetics and vice versa? Front. Genet., 6, 38
37. Lei Y., Hare J., Guoying Y., Yun C., 2012. The Status of the Wild Camel in China. In: E.M. Knoll and P.A Burger (Eds) Camels in Asia and North Africa. Interdisciplinary perspectives on their significance in past and present. Austrian Academy of Sciences, Vienna, 55–60
38. Li H., Durbin R., 2011. Inference of human population history from individual whole-genome sequences. Nat., 475, 493–496
39. Lindgren G., Backstrom N., Swinburne J., Hellborg L., Einarsson A., Sandberg K., Cothran G., Vila C., Binns M., Ellegren H., 2004. Limited number of patrilines in horse domestication. Nat. Genet., 36, 335–336
40. Lombardini L., 1879. Richerche sui camelli. Pisa: Annali Univ. Toscana 2a
41. Mariasegaram M., Pullenayegum S., Jahabar M., Shah S.R., Penedo M.C.T., Wernery U., Sasse J., 2002. Isolation and characterization of eight microsatellite markers in Camelus dromedaries and cross-species amplification in C. bactrianus and Lama pacos. Anim. Genet., 33, 377–405
42. Mate M.L., Bustamante A., Giovambattista G., De Lamo D., Von Thungen J., Zambelli, A., Vidal-Rioja L., 2005. Genetic diversity and differentiation of guanaco populations from Argentina inferred from microsatellite data. Anim. Genet., 36. 316–321
43. Mburu D.N., Ochieng J.W., Kuria S.G., Jianlin H., Kaufmann B., Rege J.E.O., Hanotte O., 2003. Genetic diversity and relationship of indigenous Kenyan camel (Camelus dromedaries) populations: implications for their classification. Anim. Genet., 34, 26–32
44. Meadows J.R., Hanotte O., Drogemuller C., Calvo J., Godfrey R., Coltman D., Maddox J.F., Marzanov N., Kantanen J., Kijas J.W., 2006. Globally dispersed Y chromosomal haplotypes in wild and domestic sheep. Anim. Genet., 37, 444–453
45. Meuwissen T.H., Hayes B.J., Goddard M.E., 2001. Prediction of total genetic value using genome-wide dense marker maps. Genetics, 157, 1819–1829
46. Muyldermans S., Baral T.N., Muyldermans S., Baral T.N., Retamozzo V.C., De Baetselier P., De Genst E., Kinne J., Leonhardt H., Magez S., Nguyen V.K., Revets H., Rothbauer U., Stijlemans B., Tillib S., Wernery U., Wyns L., Hassanzadeh-Ghassabeh G., Saerens D., 2009. Camelid immunoglobulins and nanotechnology. Vet. Immunol. Immunopath., 128, 178–183
47. Muzzachi S., Oulmouden A., Cherifi Y., Yahyaoui H., Zayed M.A., Burger P.A., Lacalandra G.M., Faye B., Ciani E., 2015. Sequence and polymorphism analysis of the camel (Camelus dromedarius) myostatin gene. Emirat. J. Food Agric., 27, 367–373
48. Orlando L., Ginolhac A., Zhang G., Froese D., Albrechtsen A., Stiller M., Schubert M., Cappellini E., Petersen B., Moltke I., Johnson P.L., Fumagalli M., Vilstrup J.T., Raghavan M., Korneliussen T., Malaspinas A.S., Vogt J., Szklarczyk D., Kelstrup C.D., Vinther J., Dolocan A., Stenderup J., Velazquez A.M., Cahill J., Rasmussen M., Wang X., Min J., Zazula G.D., Seguin-Orlando A., Mortensen C., Magnussen K., Thompson J.F., Weinstock J., Gregersen K., Rød K.H., Eisenmann V., Rubin C.J., Miller D.C., Antczak D.F., Bertelsen M.F., Brunak S., Al-Rasheid K.A., Ryder O., Andersson L., Mundy J., Krogh A., Gilbert M.T., Kjær K., Sicheritz-Ponten T., Jensen L.J., Olsen J.V., Hofreiter M., Nielsen R., Shapiro B., Wang J., Willerslev E., 2013. Recalibrating Equus evolution using the genome sequence of an early Middle Pleistocene horse. Nat., 499, 74–88

49. Pauciullo A., Shuiep E.S., Cosenza G., Ramunno L., Erhardt G., 2013. Molecular characterization and genetic variability at κ-casein gene (CSN3) in camels. Gene, 513, 22–30
50. Pauciullo A., Giambra I.J., Iannuzzi L., Erhardt G., 2014. The β-casein in camels: molecular characterization of the CSN2 gene, promoter analysis and genetic variability. Gene, 547, 159–168
51. Pedrosa V.B., Santana M.L., Oliveira P.S., Eler J.P., Ferraz J.B.S., 2010. Population structure and inbreeding effects on growth traits of Santa Ines sheep in Brazil. Small Ruminant Res., 93, 135–139
52. Peters J., A. von den Driesch, 1997. The two-humped camel (Camelus bactrianus): New light on its distribution, management and medical treatment in the past. J. Zool., 242, 651–679
53. Plasil M, Mohandesan E, Fitak R, Musilova P, Kubickova, Burger PA, Horin P, 2016. The Major Histocompatibility Complex in Old World camelids and low polymorphism of its class II genes, BMC Genomics, 17:167
54. Pritchard J.K., Stephens M., Donnelly P., 2000. Inference of population structure using multilocus genotype data. Genetics, 155, 945–959
55. Pryce J.E., Coffey M.P., Brotherstone S.H., Woolliams J.A., 2002. Genetic relationships between calving interval and body condition score conditional on milk yield. J. Dairy Sci., 85, 1590–1595
56. Ronaghi M., Uhlén M., Nyrén P., 1998. A Sequencing Method Based on Real-Time Pyrophosphate. Science, 281, 363–365
57. Ruiz E., Mohandesan E., Fitak R.R., Burger P.A., 2015. Diagnostic single nucleotide polymorphism markers to identify hybridization between dromedary and Bactrian camels. Cons. Genet. Res., 7, 329–332
58. Schlötterer C. 2004. The evolution of molecular markers--just a matter of fashion? Nat. Rev. Genet., 5, 63–69
59. Schulz U., Tupac-Yupanqui I., Martínez A., Méndez S., Delgado J.V., Gómez M., Dunner S., Cañón J., 2010. The Canarian Camel: A Traditional Dromedary Population. Diversity, 2, 561–571
60. Silbermayr K, Burger PA (2012) Hybridization: A Threat to the Genetic Distinctiveness of the Last Wild Old World Camel Species. In: Knoll EM, Burger PA (eds.) Camels in Asia and North Africa. Interdisciplinary perspectives on their significance in past and present. Austrian Academy of Sciences, Vienna (Austria), pp 69–76
61. Silbermayr K., Tero N., Charruau P., Enkhbileg D., Walzer C., Burger P.A., 2010a. Isolation and characterization of nine new microsatellite loci in the domestic Bactrian camel (Camelus bactrianus) and amplification in the wild Bactrian camel (C. ferus). Mol. Ecol. Res., 10, 1106–1108
62. Silbermayr K., Orozco-terWengel P., Charruau P., Enkhbileg D., Walzer C., Vogl C., Schwarzenberger F., Kaczensky P., Burger P.A., 2010b. High mitochondrial differentiation levels between wild and domestic Bactrian camels: a basis for rapid detection of maternal hybridization. Anim. Genet., 41, 315–318
63. Sørensen A.C., Madsen P., Sørensen M.K., Berg P., 2006. Udder health shows inbreeding depression in Danish Holsteins. J. Dairy Sci., 89, 4077–4082
64. Stanley H.F., Kadwell M., Wheeler J.C., 1994. Molecular evolution of the family Camelidae: a mitochondrial DNA study. Proc. Biol. Sci., 22, 1–6.
65. Tamura K., Stecher G., Peterson D., Filipski A., Kumar S., 2013. MEGA6: Molecular Evolutionary Genetics Analysis Version 6.0 Mol. Biol. Evol., 30, 2725–2729
66. Trinks A., Burger P.A, Beneke N., Burger J., 2012. Simulations of populations ancestry of the two-humped camel (Camelus bactrianus). In: Camels in Asia and North Africa. Interdisciplinary perspectives on their significance in past and present, E. Knoll and P. Burger (Eds.), Academy of Science Press, Vienna (Austria), 79–86
67. Troy C.S., MacHugh D.E., Bailey J.F., Magee D.A., Loftus R.T., Cunningham P., Chamberlain A.T., Sykes B.C., Bradley D.G., 2001. Genetic evidence for Near-Eastern origins of European cattle. Nat., 410, 1088–1091

68. Uerpmann M., Uerpman H.P., 2012. Archeozoology of camels in South-Eastern Arabia. In: Camels in Asia and North Africa. Interdisciplinary perspectives on their significance in past and present, E. Knoll and P. Burger (Eds.), Academy of Science Press, Vienna (Austria), 109–122
69. Uerpmann H., Uerpmann M., 2002. The appearance of the domestic camel in south-east Arabia. J. Oman Stud., 12, 235–260
70. von den Driesch A., Obermaier H., 2007. The hunt for wild dromedaries during the 3rd and 2nd millennia BC on the United Arab Emirates coast. Camel bone finds from the excavations at Al Sufouh 2 Dubai, UAE. In: Documenta Archaeobiologiae Bd. 5 – Skeletal series and their socio-economic context. G. Grupe and J. Peters (Eds), Verlag Marie Leidorf GmbH, Rahden (Germany), 133–167
71. Wallner B., Brem G., Müller M., Achmann R., 2003. Fixed nucleotide differences on the Y chromosome indicate clear divergence between Equus przewalskii and Equus caballus. Anim. Genet., 34, 453–456
72. Wilson R.T., 1984. The Camel. Longman, London, 244 pp.
73. Wu H., Guang X., Al-Fageeh M.B., Cao J., Pan S., Zhou H., Zhang L., Abutarboush M.H., Xing Y., Xie Z., Alshanqeeti A.S., Zhang Y., Yao Q., Al-Shomrani B.M., Zhang D., Li J., Manee M.M., Yang Z., Yang L., Liu Y., Zhang J., Altammami M.A., Wang S., Yu L., Zhang W., Liu S., Ba L., Liu C., Yang X., Meng F., Wang S., Li L., Li E., Li X., Wu K., Zhang S., Wang J., Yin Y., Yang H., Al-Swailem A.M., Wang J., 2014. Camelid genomes reveal evolution and adaptation to desert environments. Nat. Commun., 5, 5188.m
74. Xuebin Q., Jianlin H., Lkhagva B., Chekarova I., Badamdorj D., Rege J. E., Hanotte O., 2005. Genetic diversity and differentiation of Mongolian and Russian yak populations. J. Anim. Breed. Genet., 122, 117–126
75. Yadamsuren A., Dulamtseren E., Reading R.P., 2012. The Conservation Status and Management of Wild Camels in Mongolia. In: Knoll EM & Burger PA (Eds). Camels in Asia and North Africa. Interdisciplinary perspectives on their significance in past and present. Austrian Academy of Sciences Press, Vienna, 45–54.

CHAPTER 14

Canonical correlation analysis applied to production and reproduction traits of meat type quails

Jeferson Corrêa Ribeiro[1] *, Luciano Pinheiro da Silva[2], Giovani da Costa Caetano[3], Aline Camporez Crispim[3], Rodrigo de Oliveira Pacheco[3] and Robledo de Almeida Torres[3]

[1]*Departamento de Zootecnia, Instituto Federal Goiano, 75650-000, Morrinhos, GO, Brasil.*

[2]*Departamento de Zootecnia, Universidade Federal do Ceará (UFC), Fortaleza, CE, Brasil.*

[3]*Departamento de Zootecnia, Universidade Federal de Viçosa (UFV), Viçosa, MG, Brasil.*

ABSTRACT

Data from 629 meat type quails were used to study association between two different sets of traits: egg production and reproduction. Traits related to reproduction were: body weight at 42 days (W42), average egg weight (AEW), age at sexual maturity (ASM); and production traits: number of eggs produced from 42 until 182th day considering intervals of 35 days each period (P1, P2, P3 and P4). Canonical correlations were used to describe the association between these two sets of traits. The canonical coefficients indicated that egg production and reproductive traits were moderately related (0.3475). The first partial period of production (P1) and age at sexual maturity were the most relevant traits among the analyzed variables.

Results show that selecting animals with lower age at sexual maturity might lead to an increase on number of eggs produced until 77 days, although it might also lead to a decrease on egg weight. Breeding programs aiming to improve quails production should consider selection strategies to decrease age at sexual maturity and then, as correlated response, increased egg production might be observed.

Keywords: animal breeding; canonical pairs; Coturnix coturnix; egg production; multivariate analysis

1. INTRODUCTION

Quails have been widely used in studies involving genetic selection due to its short interval generation, high fertility and small size (YANG et al., 1998). Moreover, they present physiological and genetic similarity with chicken with the advantage of fast development of specialized lines at specific traits, what would not be possible using animals of ***Gallus*** genus (MINVIELLE, 2009). These points make the use of quail as animal model preferable. Several traits can be included on breeding programs, and a large number of them are directly influenced by environment (BARBOSA et al., 2005), difficulting interpretation through univariate analysis (SAKAGUTI et al., 1996). An alternative in this case is the use of multivariate analysis.

Multivariate analysis allows evaluation of traits simultaneously, taking into account its correlations. Thus, it becomes possible to infer about a set of traits by considering a known significance level (SAKAGUTI et al., 1996), as well as to estimate a better combination of variables that leads to a significant value. GODSHALK & TIMOTHY (1988) describe the use of multivariate analysis techniques as an alternative of selection index in plants, which the main advantage is to provide weights to the traits without needing to estimate genetic parameters on the population of interest.

Canonical correlation is a multivariate analysis technique based on determining two linear combinations, each one for a set of variables, maximizing the correlation coefficient between them. Firstly, the objective is to determine a pair of linear combinations that presents the highest correlation. Therefore, to determine the pair of linear combinations that provide the higher correlation between all the other not correlated pairs and the pair previously selected, and so on. The linear combination pairs are defined as canonical variables and their correlations as canonical correlations. These combinations are canonical variables, and these pairs are denominated as canonical pairs (JOHNSON & WICHERN, 2007).

Among traits with economic relevance on genetic evaluation of breeders, egg production, sexual maturity and average egg weight are well known by their connection with profit determination. Age at sexual maturity, body weight, laying persistence and egg weight can influence egg production. Considering this, canonical correlation analysis may allow inferences about one sets of features that might affect another one. Furthermore, canonical correlation analysis enables to indicate traits that most have influence on other traits.

On a briefly review of literature, canonical correlation analysis have been used on pigs studies (BARBOSA et al., 2005; VENTURA et al., 2011; VENTURA et al., 2012), buffalos (THOMAS & CHAKRAVARTY, 2000), dairy cattle (KESKIN et al., 2004, ÇANKAYA & KAYAALP, 2007, ÇANKAYA et al., 2008a), lambs (SAHIN et al., 2011) and laying hens (YANG et al., 1998; AKBAS & TAKMA, 2005; ROSÁRIO et al., 2006; MENDES & AKKARTAL, 2007; ÇANKAYA et al., 2008b). The aim of this study was to estimate the relation between productive and reproductive traits; to determine which are most relevant on meat type quail selection, leading to higher number off eggs through canonical correlation analysis.

2. MATERIALS AND METHODS

Data from 629 females totalizing four generations of meat type quails (***Coturnix coturnix***), from a poultry breeding program at Universidade Federal de Viçosa were used. Diets for laying quail contained 19% of crude protein and 2900kcal of metabolizable energy per kg. Quails had free access to water and feed and lighting program was 16 hours of light.

Egg production data were collected daily. Age at sexual maturity (ASM) was considered as the age that the bird laid the first egg. The number of eggs produced was recorded starting on 42th until 182th days of age, divided on intervals of 35 days each, totalizing four periods. All birds were weighed at 42th day of age (W42). The average egg weight (AEW) for each period was the average egg weight in a sample of three days record. Animals were weighed on a 0.01 precision scale.

All traits recorded were separated in two sets. The first set with egg production traits, such as number of eggs produced from 43th-77th, 78th-112th, 113th-147th and 148th-182th days of age, relative to first partial period (P1), second partial period (P2), third partial period (P3) and fourth partial period (P4), respectively. The second set with reproductive traits, such as: average egg weigh (AEW), age at sexual maturity (ASM) and female body weight at maturity (W42).

The first step was the estimation of Pearson's correlation among variables. Both Pearson's correlation and canonical correlation analysis were obtained using CANCORR procedure from SAS software (SAS, 2009). Canonical correlation analysis was performed to analyze the relation between the two sets of traits (productive and reproductive). This technique was developed by Hotteling, in 1935, and consists on obtaining the optimal correlation between a linear combination of a set of variables (Y_i) and another linear combination of set of variables (X_i). Therefore, it is possible to define that the linear combination of set of variables, following JOHNSON &WICHERN2007) are:

$$U_1=a_{11}X_1 + a_{12}X_2 + ...+a_{1p}X_p$$

$$V_1=b_{11}Y_1 + b_{12}Y_2 + ...+b_{1q}Y_q$$

Canonical variables U_1 and V_1 belong to the first canonical pair associated with the first canonical correlation, expressed for:

$$r_1 = \frac{\hat{Cov}(U_1, V_1)}{\sqrt{\hat{V}(U_1).\hat{V}(V_1)}}$$

The percentage of variance explained by the canonical variable $\boldsymbol{U}_{2\,x}$ and V^2_y is determined by:

$$U^2_{X_i} = \frac{\sum_{j=1}^{p} a^2_{ij}}{p} \quad V^2_{Y_i} = \frac{\sum_{j=1}^{q} b^2_{ij}}{q}$$

p and q are the number of variables from X and Y, respectively.

To check for significance of canonical correlation, maximum likelihood ratio test was performed, considering Lambda (Λ) from Wilk's whose expressions are presented by KHATTREE & NAIK (2000).

3. RESULTS AND DISCUSSION

Correlation analysis was performed to avoid the severe multicollinearity or linear dependency between the variables, excluding those with correlation higher than 0.80 according to MONTGOMERY & PECK (2012). The third period (P3) from egg production traits was eliminated because it caused severe multicollinearity, which leads to singular matrices or poorly conditioned matrices.

Bivariate correlation among analyzed traits is on table 1. The highest correlation was between P2 and P4 (0.8024), while the lowest was between P1 and ASM (-0.5653). These results indicate that a higher number off eggs produced on the second period results on higher number of eggs on fourth period as well. Furthermore, lower age at sexual maturity imply on higher number of eggs produced until 77 days of age. Similar results involving correlation between sexual maturity and number off eggs were presented by AKBAS & TAKMA (2005) and ÇANKAYA et al. (2008b) on studies with laying hens.

Sets of traits related to egg production and reproduction were not considered independent (Table 2). Only the first canonical correlation was significant ($p<0.01$) and it was 0.3475, showing moderate correlation between reproductive and productive traits. Evaluating traits within first canonical pair there was a high predominance of P1 that is related to egg production and age at sexual maturityin the set of reproductive traits. Proportion of variance in common to the two canonical variables at the first canonical pair, given by squared canonical correlation coefficient ($r^{2)}$ was 12.8%.

Table 1 Bivariate correlation considering productive and reproductive traits.

	P1	P2	P4	ASM	AEW	W42
P1	1.000					
P2	0.7195*	1.000				
P4	0.4720*	0.8024*	1.000			
ASM	-0.5653*	-0.3390*	-0.1811*	1.000		
AEW	0.0566	0.0134	-0.0440	0.0003	1.000	
W42	-0.0541	-0.0589	-0.0747	-0.0011	0.0877	1.000

* = significance ($P<0.001$).

P1 - number of eggs produced at first period; P2 - number of eggs produced at second period; P4 - number of eggs produced at fourth period; ASM - age at sexual maturity; AEW - average egg weight; W42 - body weight at 42 days of age.

After choice of the most appropriate canonical pair, the second step was to analyze the magnitude of canonical coefficients that would be significant on contributions relative with correlated variable. Coefficients indicated effect of reproductive traits on productive traits (Table 3). Through the correlation between the original variable and their canonical correlation, P1 still the most important variable on canonical variable U_1, as well as age at sexual maturity still the most relevant variable on V_1. Thus, lower age at sexual maturity provides higher number of eggs until 77 days.

Table 2 Standardized coefficients, canonical correlation (r), squared canonical correlation (r^2) and F statistical test (F) to canonical pairs relative to productive and reproductive traits.

Traits	Canonical Pairs		
	1st	2nd	3rd
P1	-1.2160	-0.0487	-0.8641
P2	0.3362	-0.7518	2.0595
P4	0.0050	1.5082	-0.8803
ASM	0.9906	-0.0420	0.1302
AEW	-0.1202	-0.7753	0.6279
W42	0.0953	-0.5583	-0.8302
r	0.3475	0.0088	0.0003
F	<0.0001*	0.2827	0.6829
r^2	0.1208	7.84×10^{-5}	9.24×10^{-8}

* = significance (P<0.001).

P1 - number of eggs produced at first period; P2 - number of eggs produced at second period; P4 - number of eggs produced at fourth period; ASM - age at sexual maturity; AEW - average egg weight; W42 - body weight at 42 days of age.

Age at first egg is an important trait, because it characterizes sexual maturity in females, which have influence of several relevant factors on poultry production. According to LIU et al. (1995), the beginning of egg production can be affected by genetics, chronological age and body composition. When the bird reaches sexual maturity it means that reproductive organs developed enough, enabling birds to produce eggs. Early development of reproductive organs may lead to higher egg production and total egg number.

Although age at sexual maturity had been highlighted on the set of reproductive traits, it is important to consider also body weight at maturity. Birds with reasonable weight may persist better laying eggs. According to LEESON & SUMMERS (1997), birds that present a higher body weight at sexual maturity might have better performance and in contrast birds with lower body weight might present later sexual maturity as well as lower performance.

Table 3 Correlation between: productive traits (P1, P2 and P4) and their canonical variables (U); reproductive traits (ASM, AEW and W42) and their canonical variables (V); canonical variables and traits.

	U_1	U_2	U_3		V_1	V_2	V_3
P1	-0.9716	0.1217	0.2031	ASM	0.9893	-0.0397	0.1403
P2	-0.5351	0.4272	0.7289	AEW	-0.1070	-0.8307	0.5463
P4	-0.2981	0.8799	0.3699	W42	0.0742	-0.6346	-0.7693
ASM	0.5832	-0.0037	0.0024	P1	-0.5727	0.0115	0.0035
AEW	-0.0631	-0.0782	0.0095	P2	-0.3154	0.0402	0.0127
W42	0.0438	-0.0597	-0.0134	P4	-0.1757	0.0828	0.0065

P1 - number of eggs produced at first period; P2 - number of eggs produced at second period; P4 - number of eggs produced at fourth period; ASM - age at sexual maturity; AEW - average egg weight; W42 - body weight at 42 days of age

On the other hand, average egg weight had a negative correlation on egg production at first period. RIBEIRO et al. (2012) corroborate these results, when obtained lower egg weights and lower egg mass evaluating the same period of age. Similar results were found by ÇANKAYA et al. (2008).

AKBAS & TAKMA (2005) conclude that age at sexual maturity and body weight at maturity had a positive impact on number of eggs produced on different periods, while egg weight have a negative impact on number of egg. Therefore, it is desirable to select birds that achieve sexual maturity earlier in life, indirectly selecting birds with better egg production.

4. CONCLUSION

Canonical coefficients indicate that the sets of traits related to production and reproductive traits are dependents. Among analyzed variables, egg production until 77 days of age and age at sexual maturity were the most important traits for selection of meat type quails focused on higher egg production. Thus, selection to decrease the age at sexual maturity could lead to an increase of number of eggs.

5. BIOETHICS AND BIOSSECURITY COMMITTEE APPROVAL

The protocol of experiments using animals is in accordance with CONEA and was approved by ethical committee from Universidade Federal de Viçosa, protocol number 73/2014.

ACKNOWLEDGEMENTS

We thank the Conselho Nacional de Desenvolvimento Científico e Tecnológico (CNPq), process 142696/2010-2, Coordenação de Aperfeiçoamento de Pessoal de Nível Superior (CAPES) and Fundação de Amparo à Pesquisa do estado de Minas Gerais (FAPEMIG) for financial support.

REFERENCES

1. AKBAS, Y.; TAKMA, Ç. Canonical correlation analysis for studying the relationship between egg production traits and body weight, egg weight and age at sexual maturity in layers. Czech Journal Animal Science, v.50, n.4, p.163-168, 2005. Available from: <Available from: http://www.cazv.cz/attachments/4-Akbas.pdf >.
2. BARBOSA, L. et al. Association between performance and carcass traits of swine using canonical correlation. Brazilian Journal of Animal Science, v.34, n.6, p.2218-2224, 2005. Available from: <Available from: http://www.scielo.br/scielo.php?pid=S1516-35982005000700008&sccript=sci_arttext >. Accessed: Jan. 10, 2015. doi: 10.1590/S1516-35982005000700008.
3. ÇANKAYA, S.; KAYAALP, G.T. Estimation of relationship between live weights and some body measurements in german farm x hair crossbred by canonical correlation analysis. Journal of Animal Production, v.48, n.2, p.27-32, 2007. Available from: <Available from: http://www.zooteknidernegi.org/dergi/icerik/makale/2007_48_2_27-32.pdf >.
4. ÇANKAYA, L.S. et al. Canonical correlation analysis for estimation of relationship between some body measurements at the birth and 6 month periods in holstein friesian calves. Journal Animal and Veterinary Advances, v.7, n.8, p.953-958, 2008a. Available from: <Available from: http://medwelljournals.com/abstract/?doi=javaa.2008.953.958 >.
5. ÇANKAYA, S. et al. Canonical correlation analysis for estimation of relationships between sexual maturity and egg production traits upon availability of nutrients in pullets. Asian-Australasian Journal Animal Science, v.21, n.11, p.1576-1584, 2008b. Available from: <Available from: http://dx.doi.org/10.5713/ajas.2008.80106 >.
6. GODSHALK, E.B.; TIMOTHY, D.H. Factor and principal component analyses as alternatives to index selection. Theoretical and Applied Genetics, v.76, p.352-360, 1988. Available from: <Available from: http://link.springer.com/article/10.1007/BF00265334 >.
7. JOHNSON R.A.; WICHERN D.W. Applied multivariate statistical analysis. 6.ed. Upper Saddle River: Pearson, 2007. 773p.
8. KESKIN, I. et al. Canonical correlation analysis for studying the relationship between reproductive traits and milk yield traits of brown Swiss herd raised at the state farm of konuklar in konya province. Journal of Animal and Veterinary Advances, v.3, n.12, p.797-

799, 2004. Available: <Available: http://medwelljournal.com/abstract/?doi=javaa. 2004.797799 >. Accessed: Nov. 08, 2014.

9. KHATTREE, R.; NAIK, D.N. Multivariate data reduction and discrimination with SAS software. New York: SAS Institute, 2000. 584p.
10. LEESON, S.; SUMMERS, J.D. Commercial poultry nutrition. 2.ed. Ontario: University Books, 1997. 350p.
11. LIU, G. et al. Correlated responses to long-term divergent selection for eight-week body weight in chickens: growth, sexual maturity and egg production. Poultry Science, v.74, p.1259-1268, 1995. Available from: <Available from: http://ps.oxfordjournals.org/content/74/8/1259.full.pdf >.
12. MENDES M.; AKKARTAL, E. Canonical correlation analysis for studying the relationships between pre- and post-slaughter traits of Ross 308 broiler chickens. Archives Geflügelk, v.71, n.6, p.267-271, 2007. Available from: <Available from: http://www.european-poultryscience.com/artikel.dll/m07-04mk_NDIxODIxMA.PDF >. Accessed: Nov. 07, 2014.
13. MINVIELLE, F. What is quail good for in a chicken-focused world? World's Poultry ScienceJournal, v.65, p.601-608, 2009. Available from: <Available from: http://dx.doi.org/10.1017/S0043933909000415 >.
14. MONTGOMERY, D.C.; PECK, E.A. Introduction to linear regression analysis. 5.ed. New York: Willey, 2012. 672p.
15. RIBEIRO, J.C. et al. Genetic evaluation for egg mass in partial periods and complete period in meat quails. Revista Brasileira de Zootecnia, v.41, n.5, p.1158-1162, 2012. Available from: <Available from: http://www.scielo.br/pdf/rbz/v41n5/12.pdf >. Accessed: Jan. 12, 2015. doi: 10.1590/S1516-35982012000500012.
16. ROSÁRIO, M.F. et al. Canonical discriminant analysis applied to broiler chicken performance. Animal, v.2, n.3, p.419-424, 2008. Available from: <Available from: http://www.lce.esalq.usp.br/tadeu/Millor_Silva_Coelho_Savino_Tadeu.pdf >. Accessed: Jan. 10, 2015. doi: 10.1017/S1751731107001012.
17. SAHIN, M. et al. Canonical correlation analysis for estimation of relationships between some traits measure dat weaning time and six-month age in merino lambs. Bulgarian Journal of Agricultural Science, v.17, n.5, p.680-686, 2011. Available from: <Available from: http://www.agrojournal.org/17/05-15-11.pdf >. Accessed: Dec. 02, 2015. doi: 10.1017/S1751731107001012.
18. SAKAGUTI, E.S. et al. Evaluation of tests of multivariate analysis of variance in experiments with unbalanced data. Brazilian Journal of AnimalScience, v.25, p.437-448, 1996. Available from: <Available from: http://www.revista.sbz.org.br/artigo/visualizar.php?artigo=238 >. Accessed: Jan. 10, 2015.
19. THOMAS P.R.; CHAKRAVARTY, A.K. Canonical correlation analysis for studying the association of breeding efficiency and breeding values with growth and reproductive traits of murrah buffaloes. Indian Jounal AnimalResearch, v.34, n.2, p.100-103, 2000. Available from: <Available from: http://www.indianjournals.com/ijor.aspx?target= ijor:ijar1&volume=34&issue=2&articl e=003 >.
20. VENTURA, H.T. et al. Canonical correlation analysis of the association between carcass and ham traits in pigs used to produce dry-cured ham. Genetics and Molecular Biology, v.34, n.3, p.451-455, 2011. Available from: <Available from: http://dx.doi.org/10.1590/S1415-47572011000300014 >.
21. VENTURA, H.T. et al. Use of multivariate analysis to evaluate genetic groups of pigs for dry-cured ham production. Livestock Science, v.148, p.214-220, 2012. Available from: <Available from: http://dx.doi.org/10.1016/j.livsci.2012.06.007 >.
22. YANG, N. et al. Forty generations of bidirectional selection for mating frequency in male japanese quail. Poultry Science, v.77, p.1469-1477, 1998. Available from: <Available from: http://ps.oxfordjournals.org/content/77/10/1469.full.pdf >.

CHAPTER 15

The nutritive value of marula (*Sclerocarya birrea*) seed cake for broiler chickens: nutritional composition, performance, carcass characteristics and oxidative and mycotoxin status

Doctor Mziwenkosi Nhlanhla Mthiyane[1], Bhekumusa Sabelo Mhlanga[1,2]

[1] *Department of Animal Science, Luyengo Campus, Faculty of Agriculture, University of Swaziland, P.O. Box M205, Luyengo, Swaziland*

[2] *Crane Feeds (Pty) Ltd., P.O. Box 1974, Matsapha, Swaziland*

ABSTRACT

This study was aimed at investigating the nutritive value of marula seed cake (MSC) as an alternative protein source for broilers. In a completely randomised design involving six replicate pens of five chickens assigned to each of five treatments, body weight gain (BWG), feed intake (FI), feed conversion efficiency (FCE) and carcass characteristics were measured in an experiment in which 150 28-day-old broilers were fed maize-based diets containing, respectively, 0, 5, 10, 15 and 20% MSC at finisher phase. The results showed MSC to be remarkably high in CP (470.0 g/kg DM) and EE (343.5 g/kg DM), with moderate CF (58.2 g/kg DM), ash (54.3 g/kg DM), Ca (1.1 g/kg DM) and P (11.0 g/kg DM). Whilst very poor in lysine, MSC was found to be rich in methionine, cyst(e)ine, arginine and glutamic acid; it also contains good levels of valine, glycine, threonine, isoleucine, leucine, histidine, phenylalanine, serine, proline and alanine. Also, it contained 85.24% oleic (OA), 9.65% palmitic and 5.11% stearic acids, with a high peroxide value and low levels of mycotoxins deoxynivalenol (DON) and T-2 toxin. BWG, FI and FCE of broiler chickens significantly decreased ($P < 0.001$) as the dietary level of MSC increased. Further, dietary MSC significantly decreased bird live weight at slaughter ($P < 0.001$), plucked weight ($P < 0.001$), dressed weight ($P < 0.001$) and weights of the liver ($P < 0.001$) and neck ($P < 0.05$). The results therefore demonstrate MSC to be a good source of CP, fat, Ca, P, amino acids (except lysine) and OA that can replace soya bean meal (SBM) in broiler diets. However, its use is currently limited by lipid peroxidation and presence of mycotoxins.

Keywords: Marula seed cake, Broilers, Chemical composition, Performance, Carcass characteristics, Mycotoxins

1. INTRODUCTION

The rapidly increasing human population particularly in the developing world has increased the demand for animal protein, especially for pork and poultry products, for human nutrition (World Bank Group 2015). This is an opportunity for smallholder farmers in Africa to increase household income and improve their livelihoods by connecting with the livestock and poultry value chain. Poultry production, in particular, has a potentially great role to play in terms of poverty alleviation in most developing countries. In Swaziland, for example, poultry are the most widely owned farm animal species, with more than 90% of rural households rearing poultry in the country (Rajiur Rahman 2012; Siyaya and Masuku 2013). However, smallholder farmers in sub-Saharan Africa (SSA) often lack access to good quality feed with sufficient energy and a balanced amino acid profile that is needed to ensure satisfactory animal performance (Martens et al. 2012).

SBM is a widely used feed supplement with high energy content and an amino acid profile close to ideal. The protein quality of SBM is high for poultry, and SBM is a particularly good source of lysine, arginine and tryptophan, but it is deficient in methionine plus cysteine, threonine and valine (Baker 2000). Globally, however, SBM is predominantly used in highly industrialised production systems for pigs and poultry, and there is little surplus SBM for smallholder farmers. Thus, many livestock farmers in most developing countries cannot afford the exorbitant costs of importation of this protein supplement. Therefore, to meet the nutritional requirements of poultry in SSA, there is a need to identify alternative low-cost protein sources (Martens et al. 2012; Acheampong-Boateng et al. 2016).

MSC is a by-product of oil extraction from the dry seeds of the ripe fruits of marula (*Sclerocarya birrea* A. Rich.), an indigenous fruit tree distributed throughout most of SSA from 17° 15′ N in the Aïr Mountains of Niger to 31° 00′ S near Port Shepstone in South Africa (Hall et al. 2002; Chirwa and Akinnifesi 2008). It is a novel locally available low-cost alternative protein supplement for livestock and poultry in Swaziland. In previous studies, it has been found to be a good dietary protein supplement for beef cattle (Mlambo et al. 2011a), dairy cattle (Mdziniso et al. 2016) and goats (Mlambo et al. 2011b). However, its nutritional value for poultry has hitherto not been investigated. This study was therefore undertaken to investigate the nutritional composition and oxidative and mycotoxin status of MSC and performance and carcass characteristics of broilers fed maize-based diets supplemented with graded levels of MSC at finisher phase.

2. MATERIALS AND METHODS

2.1 Location of study and source of materials

The study was conducted in a small-scale commercial broiler house outside Luve, a small town in central Swaziland that is located on the MR5 route between Mpisi and Mliba, 25 km northeast of Manzini city. The broiler house was made of wooden materials with a metal roof of corrugated iron sheets and removable tarpaulin side curtains.

All the feed materials, including MSC, as well as day-old broiler chicks, were supplied by Crane Feeds (Pty) Ltd. (Swaziland). MSC was purchased from Swazi Secrets, a local small-scale marula oil extracting firm at Mpaka (Swaziland). The oil extraction procedure has previously been described by Mlambo et al. (2011a). The unsexed day-old chicks of the Cobb 500 strain were purchased from National Chicks Swaziland at 1 day of age.

2.2 Chemical assay

Chemical analyses of MSC samples were performed at the Central Analytical Laboratories (Roodepoort, South Africa) and AminoLab at Evonik Africa (Pty) Ltd. (South Africa). The DM (930.15), ash (942.05), CP (954.01), EE (920.39) and CF were determined according to procedures of the AOAC (2000). CP was calculated using N × 6.25. Ca and P composition was determined according to VDLUFA (2007). Amino acid analysis, preceded by acid hydrolysis, were performed (method 982.30) using an amino acid analyser (SY-KAM, Erising, Germany) following modifications by Mills et al. (1989). Analysis of fats and fatty acids was performed using a gas chromatography following the revised (method 996.06) procedures of the AOAC (2002). Mycotoxin analysis was performed using high pressure liquid chromatography-mass spectrophotometry (HPLC-MS) and the Vicam Myco6in1® LC/MS/MS method. Peroxide values were determined using the Titrino apparatus following the American Oil Chemist Society (AOCS) official method (AOAC 1993). Lipid peroxidation was determined by measuring the formation of thiobarbituric acid reactive substances (TBARS)/malondialdehyde using an HPLC method adapted from Moselhy et al. (2013).

2.3 Experimental design, birds and diets

A total of 150 Cobb-500 28-day-old broiler chickens, fed a maize and soya bean-based commercial starter diet until they were 28 days old, were used. On day 29, they were assigned to stainless steel pens (each with five birds) so that the initial body weights (8.0 kg/pen) were equal among treatments. A completely randomised design was used with six replicate pens of five chickens assigned to each of five dietary treatments. The chickens were maintained on natural light during the day and on continuous artificial light at night. They were allowed ad libitum access to feed and water from day 29 to day 44 and were inspected daily for any health-related problems.

Iso-energetic and iso-nitrogenous maize and SBM-based diets, in mash form formulated to meet the nutritional requirements of finisher birds as recommended by the NRC (1994) and to which graded levels of MSC were sequentially added in replacement of SBM (Table 1), were used. MSC was incorporated into the diets to produce dietary treatments containing, respectively, 0 (control), 5, 10, 15 and 20% MSC.

2.4 MEASUREMENTS

Chickens were group weighed by pen at the beginning (day 29) and then daily thereafter until the end of the experiment on day 44. Also, feed offered and left was weighed per pen daily in order to calculate average FI per bird per day. Feed conversion ratio was calculated from average BWG per bird per day (g) divided by average FI per bird per day (g).

Table 1 Ingredients and nutrient composition (g/kg) of experimental finisher diets (as-fed basis)

	Dietary MSC level (%)				
Ingredient	0	5	10	15	20
Yellow maize meal	639	676	684	656	597
Soya bean meal (47% CP)	190	112	74	23	–
Full-fat soya bean	64.3	100	22	–	–
MSC	–	50	100	150	200
Wheat bran	–	–	–	–	96.5
Sunflower seed meal	29.6	–	61	125	64
Fishmeal (65% CP)	–	20	20	–	–
Limestone	14.2	15	15	18	17.5
Monocalcium phosphate	5.3	11	10	11.2	10
Mineral/vitamin premix[a]	3.5	3.5	3.5	3.5	3.5
Soya bean oil	29.6	3.2	–	–	–
Salt	3.3	2.9	2.9	3.4	3.3
Lysine	2.6	2.4	3.4	4.7	4.6
Sodium bicarbonate	2.0	2.0	2.0	2.0	2.0
DL-Methionine	2.5	1.9	1.6	1.4	1.4
Threonine	0.7	0.6	0.7	0.8	0.8
Calculated composition					
ME (MJ/kg)	12.5	12.5	12.5	12.5	12.5
CP	18.0	18.0	18.0	18.0	18.0
CF	3.0	3.0	4.0	5.3	5.0
EE	6.6	6.6	6.8	8.0	9.7
Ash	4.2	4.2	4.3	4.4	4.5

Ash	4.2	4.2	4.3	4.4	4.5
NFE	56.6	56.6	54.9	51.9	50.5
Ca	0.85	0.85	0.85	0.90	0.85
P	0.64	0.64	0.68	0.73	0.76
Na (mg/kg)	0.19	0.19	0.19	0.19	0.19
Cl (mg/kg)	0.25	0.25	0.25	0.25	0.24
K (mg/kg)	0.77	0.77	0.72	0.71	0.75
Lysine	1.1	1.1	1.1	1.08	1.1
Methionine	0.50	0.50	0.49	0.47	0.46
Cysteine	0.32	0.32	0.34	0.37	0.39
Methionine + cysteine	0.82	0.82	0.83	0.84	0.85
Threonine	0.74	0.74	0.75	0.76	0.77
Tryptophan	0.21	0.21	0.20	0.21	0.21
Valine	0.02	0.02	0	0	0
Arginine	1.30	1.30	1.48	1.69	1.85
Isoleucine	0.72	0.72	0.69	0.66	0.63
Xantophyll (ppm)	13.5	13.5	13.7	13.1	11.9

[a]The mineral/vitamin premix supplied the following per kilogram of complete feed: vitamin A (retinol acetate), 2250 mg; vitamin D3 (cholecalciferol), 82.5 mg; vitamin E (d-α-tocopherol), 40,000 mg; vitamin K3 (menadione), 1300 mg; vitamin B1 (thiamine mononitrate), 1300 mg; vitamin B2 (riboflavin), 3300 mg; vitamin B3 (nicotinic acid), 33,000 mg; calcium panthonate, 8000 mg; vitamin B6 (pyridoxine chlorhydrate), 2000 mg; folic acid, 1300 mg; vitamin B12 (cyanocobalamin), 6.5 mg; biotin, 70 mg; antioxidant, 125,000 mg; Mn, 70,000 mg; Fe, 25,000 mg; Zn, 70,000 mg; Cu, 6500 mg; Co, 350 mg; I, 1300 mg; Se, 200 mg; choline, 200,000 mg

On day 44, all the birds were weighed again in order to determine live weight at slaughter and then killed by stunning with electric shock and slitting the throat of each bird with a sharp knife. After plucking off the feathers, the carcasses were weighed in order to determine plucked weights. Carcasses were then dressed and weighed in order to determine dressed weight. During evisceration, the internal organs were removed and weighed.

2.5 Statistical analyses

Data on growth performance and carcass/organ characteristics of birds were analysed using the GLM procedure of Minitab (2000). Statistical significance was accepted based on the 0.05 level of probability.

3. RESULTS AND DISCUSSION

Our results showed MSC to have a remarkably high content of CP (470.0 g/kg DM) and EE (343.5 g/kg DM), with moderate CF (58.2 g/kg DM) and ash (54.3 g/kg DM) contents (Table 2). The results corroborate previous findings by Mlambo et al. (2011a) and Mdziniso et al. (2016) who reported similar CP (470.0–470.21 g/kg DM), EE (289.6–394 g/kg DM) and ash (55.5–66.1 g/kg DM) values in MSC.

Table 2 The proximate, mineral and amino acid composition of MSC

Component	Amount
DM (g/kg)	946.6
CP (g/kg DM)	470.0
EE (g/kg DM)	343.5
CF (g/kg DM)	58.2
Ash (g/kg DM)	54.3
Ca (g/kg DM)	1.1
P (g/kg DM)	11.0
Na (g/kg DM)	ND
Amino acids (% of DM)	
Lysine	0.88
Methionine	0.79
Cystine	1.18
Methionine + cystine	1.96
Threonine	0.98

Isoleucine	1.77
Leucine	2.66
Phenylalanine	1.91
Histidine	1.13
Arginine	6.36
Valine	2.14
Alanine	1.40
Aspartic acid	3.51
Glycine	2.11
Serine	1.91
Proline	1.44
Glutamic acid	10.78
Total (without ammonia)	41.23
Ammonia	1.11
Total	42.34

ND not detected

The CP content of MSC evidently surpasses that of most alternative protein sources for non-ruminants such as sunflower oilcake (341 g/kg DM), canola oilcake (339 g/kg DM), cotton seed cake (403 g/kg DM) and tropical leguminous forages (≤360 g/kg DM) (Ramachandran et al. 2007; Martens et al. 2012). In this connection, MSC is indeed a good source of CP at par with SBM (Banaszkiewicz 2011).

Also, whilst very poor in lysine, MSC was found to be rich in sulphur amino acids methionine and cyst(e)ine, with an abundance of arginine and glutamic acid (Table 2). It was also found to contain rather good levels of valine, glycine, threonine, isoleucine, leucine, histidine, phenylalanine, serine, proline and alanine that are similar to typical values for SBM (Goldflus et al. 2006; Carrera et al. 2011; Nahashon and Kilonzo-Nthenge 2011). These findings are in agreement with previous observations that the protein in the kernel of marula seed is limiting in lysine yet a good source of leucine and sulphur-containing amino acids methionine and cysteine (Mojeremane and Tshwenyane 2004; Mariod et al. 2005; Muhammad et al. 2011a), with a predominance of glutamic acid and arginine (Mariod and Abdelwahab 2012). The results suggest that

MSC can be used to supply much needed protein and amino acids in resource-poor farming systems in developing countries. Also, formulation of broiler diets involving MSC as a protein source in the future would not necessarily require any fortification with synthetic amino acids except for lysine.

The EE content of MSC (343.5 g/kg DM) is too high relative to other conventional protein sources such as SBM (155–247 g/kg DM; Banaszkiewicz 2011). However, this level is now rather lower than what has been previously reported (394 g/kg DM; Mlambo et al. 2011a) and is an indication that the oil extraction process at the firm that produces the MSC has improved over the years.

The mineral analysis showed the Ca content of MSC (Table 2) to be 4× lower than values previously found in marula seed kernels in Kenya (Wairagu et al. 2013). However, it was interestingly comparable to that of SBM (1.3 g/kg DM; Ramachandran et al. 2007). On the other hand, the P content of MSC was found to be approximately 7× higher than that previously found in marula seed kernels in Kenya (Wairagu et al. 2013), probably due to sub-species or genetic variation (Moganedi et al. 2011; Nyoka et al. 2015). A comparable value of P in the seed kernel (9.0 g/kg DM) has previously been observed by Akinnifesi et al. (2008). However, of great interest is that the P content of MSC was found to be 1.6× higher than that of SBM (6.9 g/kg DM; Ramachandran et al. 2007).

In terms of fats and fatty acid composition, MSC was found to contain 85.24% monounsaturated (MUFAs) and 14.76% saturated (SFAs) fatty acids, with neither polyunsaturated (PUFAs) nor trans (TFAs) fatty acids. Interestingly, all the MUFAs were composed of OA, a fatty acid less sensitive to oxidation relative to PUFAs, whilst the SFAs were only palmitic (9.65%) and stearic (5.11%) acids (Table 3). In previous studies, the oil in marula seed kernels has been reported to be 10× more stable to oxidation and to contain a similar fatty acid composition to olive oil (Mariod and Abdelwahab 2012), with as high as 67.2% OA, 5.9% linoleic acid, 14.1% palmitic acid and traces of linolenic acid (Mariod 2005). Hence, we suspect that lipid peroxidation might have occurred resulting in the depletion of PUFAs that might have been present in the MSC used in our study. Therefore, with such a high MUFA content in the form of OA, and if it can be protected from lipid peroxidation during processing and storage thereafter, MSC incorporation into broiler diets would be expected to increase the level of MUFAs, and possibly PUFAs, in the meat and, consequently, improve health benefits to consumers. Considering the hypocholesterolemic properties of OA (Kurushima et al. 1995), a substantial increase in the content of this fatty acid in broiler meat is highly desirable. Also, with the oil in marula seed kernels being highly stable to oxidation, it is further hypothesised that MSC incorporation into broiler diets would enhance the shelf life of meat.

Table 3 Total fats and fatty acid content of MSC

Fats (% of total)	
Monounsaturated fats	85.24
Polyunsaturated fats	0.00
Saturated fats	14.76
Trans fatty acids	0.00
Fatty acids (g/100 g)	
C16:0 (palmitic acid)	9.65
C18:0 (stearic acid)	5.11
C18:1 cis-9 (oleic acid)	85.24

Whilst MSC is evidently a good protein, energy, mineral, amino acid and OA source for poultry in terms of its nutritional composition, its dietary substitution for SBM appears to cause remarkably deleterious effects on broiler performance. In this connection, BWG, FI and FCE were significantly decreased ($P < 0.001$) by dietary inclusion of MSC (Table 4). Similar depressive effects of MSC on FI, BWG and FCE in broilers have previously been observed at our university research farm (Magagula 2013; Mamba 2013). Also, a significant decrease in body weights in contrast to an increase in serum total protein, albumin, bilirubin, transaminases, creatinine, urea, uric acid and electrolytes was observed in rats orally administered with 3000 and 4000 mg/kg body weight of marula fruit kernel extract (Muhammad et al. 2011b).

The deleterious effects of MSC on broiler performance were also mirrored in the live weights as well as the carcass characteristics of the birds at slaughter. In this connection, a significant decrease was observed in the live weight ($P < 0.001$), plucked weight ($P < 0.001$), dressed weight ($P < 0.001$), liver weight ($P < 0.001$) and neck weight ($P < 0.05$) of broilers as the level of MSC increased in the diet (Table 5). Although the trend was generally similar to the above carcass characteristics, there were however no significant effects ($P > 0.05$) of dietary MSC on the weights of the gizzard, heart, intestine, head, abdominal fat, feet and collective viscera. In a previous study at our institution's research farm, Magagula (2013) also observed that dietary MSC supplementation significantly decreased the live weight, dressed weight, dressing percentage, length of feathers, as well as weights of the wing, breast, thigh, drumstick, shank, abdominal fat, liver, gizzard, heart and intestines in broilers.

Table 4 Body weight gain, feed intake and feed conversion efficiency in broilers fed diets containing graded levels of MSC at finisher phase

	MSC dietary level (%)						
Items	0	5	10	15	20	SEM[a]	*P* value
BWG (g/bird/day)	81.33	62.90	64.52	37.05	36.87	3.295	$P < 0.001$
FI (g/bird/day)	183.17	147.50	146.50	135.30	120.0	1.911	$P < 0.001$
G/F (g/g)	0.44	0.43	0.44	0.27	0.31	0.012	$P < 0.001$

G/F gain/feed ratio

[a] Standard error of the mean (SEM) based on pooled estimate of variation ($n = 6$)

Table 5 Carcass characteristics (g) of broilers fed diets containing graded levels of MSC at finisher phase

	MSC dietary level (%)						
Items	0	5	10	15	20	SEM[a]	*P* value
Live weight	2838.50	2482.50	2557.80	2171.00	2169.70	63.44	$P < 0.001$
Plucked weight	2439.80	2056.20	2156.00	1842.80	1844.80	69.60	$P < 0.001$
Dressed weight	1948.40	1608.00	1735.40	1451.80	1418.10	64.14	$P < 0.001$
Liver	48.73	43.67	45.83	38.00	32.33	2.52	$P < 0.001$
Gizzard	35.20	35.50	36.17	34.33	36.77	1.45	$P > 0.05$
Heart	13.43	10.83	14.50	10.33	10.53	1.45	$P > 0.05$
Intestine	103.47	106.67	113.03	99.67	99.83	8.15	$P > 0.05$
Neck	58.37	51.33	53.10	47.50	50.17	2.31	$P < 0.05$
Head	48.73	45.67	48.00	47.50	46.50	1.62	$P > 0.05$
Abdominal fat	27.93	25.50	24.87	19.67	29.03	5.12	$P > 0.05$
Feet	85.83	75.50	84.37	75.33	78.27	3.54	$P > 0.05$
Viscera	420.33	400.17	419.03	370.00	379.50	14.51	$P > 0.05$

[a] Standard error of the mean (SEM) based on pooled estimate of variation ($n = 6$)

The negative effects of MSC on performance and carcass parameters are unlikely to be due to poor palatability of MSC as marula kernels are generally consumed and cherished by people in rural areas across the continent of Africa (ANU Forestry 2001; Gouwakinnou et al. 2011) with neither bitterness nor unpleasant odour being reported. Notwithstanding, this may be related to extensive lipid peroxidation of MSC as indicated by the high peroxide value (Table 6). This is corroborated by the absence of PUFAs in MSC (Table 3) which contrasts previous studies that found the oil in marula seed kernels to contain 5.9% linoleic acid and traces of linolenic acid (Mariod 2005). It would appear that all the linoleic and linolenic acids that might have been present in the marula kernels were completely oxidised during MSC processing or thereafter, resulting in a product with a high peroxide value. Lipid peroxidation results in stale, rancid flavour in food (Kerler and Grosch 1996), and feeding of oxidised lipids has been demonstrated to decrease body weight gain (Enberg et al. 1996), cause poor feed conversion efficiency (Cabel et al. 1988; McGill et al. 2011) and reduce meat quality (Racanicci et al. 2008; Delles et al. 2014) in broilers. Thus, consumption of oxidised lipids in MSC might have increased the absorption of lipid-derived radicals into the bloodstream, propagating oxidative reactions throughout various tissues of the birds, thereby impairing their productive performance and carcass characteristics.

Table 6 Oxidative and mycotoxin status of MSC

Item	Amount
Oxidative status	
Peroxide value (mEq/kg)	164.02
Thiobarbituric acid (mg/kg)	0.15
Mycotoxins (ppb)	
Aflatoxin B_1	ND
Aflatoxin B_2	ND
Aflatoxin G_1	ND
Aflatoxin G_2	ND
Deoxynivalenol	66.31
T-2 toxin	68.75
HT-2 toxin	ND
Ochratoxin A	ND
Zearalenone	ND
Fumonisin B_1	ND
Fumonisin B_2	ND

ND not detected

Another plausible mechanism underlying the deleterious effects of MSC entails infestation of this protein-rich product with mycotoxins. In this regard, we found MSC to be infested with DON and T-2 toxin, though at low levels (Table 6). Other mycotoxins might have been present in MSC; however, their levels might have been too low to be detected by analytical methods used in our study. Prolonged dietary exposure of animals to DON generally induces impaired feed intake, decreased weight gain and altered immune function (EFSA 2004). On the other hand, T-2 toxin is a highly toxic trichothecene that primarily affects the immune and haematopoietic systems (Fink-Gremmels 1999). It can also cause weight loss, liver damage, reproductive toxicity and neurotoxicity (Schuhmacher-Wolz et al. 2010). DON and T-2 toxin induce free radical production that serves as the mechanism underlying membrane and DNA damage during oxidative stress caused by these toxins (Wu et al. 2014). DON is also a well-known inhibitor of protein synthesis. It binds to peptidyl transferase (Feinberg and McLaughlin 1989), inhibiting the synthesis of RNA and DNA via binding to the ribosome. Cells and tissues with high protein turnover rates, such as the small intestine, the liver and the immune system, are most severely affected by DON intoxication (Doll et al. 2003). Although both DON and T-2 toxin were detected at rather low levels in MSC, their effects on broiler performance and carcass characteristics would appear to have been more pronounced when these toxins were present in combination and during prolonged exposure to them. Indeed, Kubena et al. (1989) found that total body weight gains and final body weights in broiler chicks were significantly reduced by the DON/T-2 toxin combination but were not significantly affected by the toxins singly. Also, whilst the incidence and severity of oral lesions induced by T-2 toxin were increased in the DON/T-2 toxin combination, several parameters not altered by DON or T-2 toxin singly were significantly affected by the combination (Kubena et al. 1989), indicating that the combination may pose a potentially greater problem to the poultry industry than either of the mycotoxins individually.

In conclusion, our results demonstrate MSC to be rich in CP, EE, Ca, P, essential (except for lysine) and non-essential amino acids, as well as OA. However, its potential use in broiler diets is limited by its rancidity and presence of mycotoxins. The use of dietary antioxidant additives and mycotoxin binders when preparing feeds may lead to improvement of the nutritional value of MSC for poultry in Africa and beyond.

ACKNOWLEDGEMENTS

We are grateful to Crane Feeds (Swaziland) for the supply of all dietary ingredients used in the study. Also, we express our gratitude to the Central Analytical Laboratories (Roodepoort, South Africa), AminoLab at Evonik Africa (Pty) Ltd. (South Africa) and Kemin® (Sterkfontein, South Africa) for the chemical and microbiological assays and Mr. Ben J. Holtzhausen of Kemin® for diet formulations.

COMPLIANCE WITH ETHICAL STANDARDS

The care and use of broiler chickens were performed following the ethical guidelines of the University of Swaziland Department of Animal Science Board that approved the protocol used in the experiment.

REFERENCES

1. Acheampong-Boateng, O., Bakare, A.G. and Mbatha, K.R., 2016. The potential of replacing soyabean oil cake with macadamia oil cake in broiler diets, Tropical Animal Health and Production DOI 10.1007/s11250-016-1057-y
2. Akinnifesi, F.K., Leaky, R.R.B., Ajayi, O.C., Sileshi, G., Tchoundjeu, Z., Matakaala, P. and Kwesigga, F.R., 2008. Indigenous fruit trees in the tropics: domestication, utilisation and commercialisation, (CAB International, Wallingford, UK), pp. 210–212 & pp. 288–300.
3. ANU Forestry, 2001. Marula fruit. <http://www.anu.edu.au/Forestry/wood/ nwfp/toesu/Toesu2.html (Accessed 15 April 2016).
4. Association of Official Analytical Chemists (AOAC)., 1993. Official Methods and Recommended Practices of the American Oil Chemist's Society. D. Firestone (Ed.), AOAC Press, Illinois.
5. Association of Official Analytical Chemists (AOAC)., 2000. Official Methods of Analysis. Vol. I. 17th Ed., (Arlington, Washington DC, USA).
6. Association of Official Analytical Chemists (AOAC)., 2002. Official Method 996.06. Fat (Total, saturated, and unsaturated) in Foods. (Arlington, Washington DC, USA).
7. Baker, D.H., 2000. Nutritional constraints to use of soy products by animals. In: J.K. Drackley (Ed), Soy in Aminal Nutrition, (Federal Animal Science Society, Savoy, IL), pp. 1–12.
8. Banaszkiewicz, T., 2011. Nutritional value of soybean meal, soybean and nutrition. In: Prof. Hany El-Shemy (Ed), ISBN: 978–953–307-536-5, InTech, Available from: http://www.intechopen.com/books/soybean-andnutrition/nutritional-value-of-soybean-meal.
9. Cabel, M.C., Waldroup, P.W., Shermer, W.D., and Calabotta, D.F., 1988. Effects of ethoxyquin feed preservative and peroxide level on broiler performance. Poultry Science, 67: 7125–1730.
10. Carrera, C.S., Reynoso, C.M., Funes, G.J., Martínez, M.J., Dardanelli, J. and Resnik, S.L., 2011. Amino acid composition of soybean seeds as affected by climatic variables, Pesquisa Agropecuária Brasileira, 46 (12), 1579–1587.

11. Chirwa, P.W. and Akinnifesi, F.K., 2008. Ecology and biology of *Uapaca kirkina, Strychnos cocculoides* and *Sclerocarya birrea* in Southern Africa. In: F.K. Akinnifesi, R.R.B. Leakey, O.C. Ajayi, G. Sileshi, Z. Tchoundjeu, P. Matakala and F.R. Kwesiga (Eds), Indigenous Fruit Trees in the Tropics: Domestication, Utilisation and Commercialisation, pp. 322–340 (CABI International, London).
12. Delles, R.M., Xiong, Y.L., True, A.D., Ao, T. and Dawson, K., 2014. Dietary antioxidant supplementation enhances lipid and protein oxidative stability of chicken broiler meat through promotion of antioxidant enzyme activity. Poultry Science, 93: 1561–1570. doi:10.3382/ps.2013-03682.CrossRefPubMedPubMedCentral
13. Doll, S., Danicke, S., Ueberschar, K.H., Valenta, H., Schnurrbusch, U., Ganter, M., Klobasa, F. and Flachowsky, G., 2003. Effects of graded levels of Fusarium toxin contaminated maize in diets for female weaned piglets. Archives of Animal Nutrition, 57 (5): 311–334.
14. Enberg, R.M., Lauridsen, C., Jensen, S.K., and Jakobsen, K., 1996. Inclusion of oxidized vegetable oil in broiler diets: its influence on nutrient balance and on the antioxidative status of broilers. Poultry Science, 75: 1003–1011.
15. European Food Safety Authority (EFSA)., 2004. Opinion of the Scientific Panel on Contaminants in the Food Chain on a request from the Commission related to deoxynivalenol (DON) as undesirable substance in animal feed. EFSA Journal, 73: 1–42.
16. Feinberg, B. and McLaughlin, C.S., 1989. Biochemical mechanism of action of trichothecene mycotoxins. In: V.R. Beasley (Ed.), Trichothecene mycotoxicoses: pathophysiologic effects, pp. 27–35 (Vol. 1, CRC Press, Boca Raton).
17. Fink-Gremmels, J., 1999. Mycotoxins: their implications for human and animal health. Veterinary Quarterly, 21: 115–20.
18. Goldflus, F., Ceccantini, M. and Santos, W., 2006. Amino acid content of soybean samples collected in different Brazilian States—Harvest 2003/2004, Brazilian Journal of Poultry Science, 8 (2), 105–111.
19. Gouwakinnou, G.N., Lykke, A.M., Assogbadjo, A.E., and Sinsin, B., 2011. Local knowledge, pattern and diversity of use of *Sclerocarya birrea*, Journal of Ethnobiology and Ethnomedicine, 7: 8.
20. Hall, J.B., O'brien, E.M. and Sinclair, F.L., 2002. *Sclerocarya birrea*: a monograph. School of Agricultural and Forest Sciences Publication No. 19, (University of Wales, Bangor), p. 157.
21. Kerler, J., and Grosch, W., 1996. Odorants contributing to warmed-over flavour (WOF) of refrigerated cooked beef. Journal of Food Science, 61: 1271–1274.
22. Kubena, L.F., Huff, W.E., Harvey, R.B., Phillips, T.D., Rottinghaus, G.E., 1989. Individual and combined toxicity of deoxynivalenol and T-2 toxin in broiler chicks. Poultry Science, 68: 622–626.
23. Kurushima, H., Hayashi, K., Toyota, Y., Kambe, M. and Kajiyama, G., 1995. Comparison of hypocholesterolemic effects induced by dietary linoleic acid and oleic acid in hamsters, Atherosclerosis, 114 (2): 213–221.
24. Magagula, B.S., 2013. The effect of substituting soyabean meal with cold pressed marula (*Sclerocarya birrea*) seed cake as a protein source in broiler starter diet. B.Sc. Agric (Animal Science) Project Report, Department of Animal Science, Faculty of Agriculture, University of Swaziland.
25. Mamba, S.S., 2013. The effect of replacing soyabean meal with marula (*Sclerocarya birrea*) seed cake as a protein source in broiler finisher diet. B.Sc. Agric (Animal Science) Project Report, Department of Animal Science, Faculty of Agriculture, University of Swaziland.
26. Mariod, A.A., 2005. Investigations on the oxidative stability of some unconventional Sudanese oils, traditionally used in human nutrition, *PhD theis,* Münster University.
27. Mariod, A.A. and Abdelwahab, S.I., 2012. *Sclerocarya birrea* (Marula), an African tree of nutritional and medicinal uses: a review, Food Reviews International, 28, 375–388.
28. Mariod, A.A., Ali, A.O., Elhussein, S.A. and Hussien, I.H., 2005. Quality of proteins and products based on *Sclerocarya birrea* (Marula) seed, Sudan Journal of Science and Technology, 6, 184–192.
29. Martens, S.D., Tiemann, T.T., Bindelle, J., Peters, M. and Lascano, C.E., 2012. Alternative plant protein sources for pigs and chickens in the tropics—nutritional value and

constraints: a review, Journal of Agriculture and Rural Development in the Tropics and Subtropics, 113 (2), 101–123.

30. McGill, J., McGill, E., Kamyab, A., and Firman, J.D., 2011. Effect of high peroxide value fats on performance of broilers in a normal immune state. International Journal of Poultry Science, 10 (3): 241–246.
31. Mdziniso, P.M., Dlamini, A.M., Khumalo, G.Z and Mupangwa, J.F., 2016. Nutritional evaluation of marula (*Sclerocarya birrea*) seed cake as a protein supplement in dairy meal, Journal of Applied Life Sciences International, 4 (3), 1–11, Article no. JALSI.23273.
32. Mills, P. A., Rotter, R. G. and Marquardt, R. R., 1989. Modification of the glucosamine method for the quantification of fungal contamination, Canadian Journal of Animal Science, 69, 1105–1106.
33. Minitab, 2000. Minitab Statistical Software, Release 13.1 for Windows (Minitab Incorporation™, USA).
34. Mlambo, V., Dlamini, B.J., Ngwenya, M.D., Mhazo, N., Beyene, S.T. and Sikosana J.L.N., 2011a. *In sacco* and *in vivo* evaluation of marula (*Sclerocarya birrea*) seed cake as a protein source in commercial cattle fattening diets, Livestock Research for Rural Development, 23 (5). http://www.lrrd.org/lrrd23/5/mlam23121.htm.
35. Mlambo, V., Dlamini, B.J., Nkambule, M.T., Mhazo, N. and Sikosana J.L.N., 2011b. Nutritional evaluation of marula (*Sclerocarya birrea*) seed cake as a protein supplement for goats fed grass hay, Tropical Agriculture (Trinidad), 88 (1), 35–43.
36. Moganedi, K., Sibara, M., Grobler, P. and Goyvaerts, E., 2011. An assessment of genetic diversity among marula populations using the amplified fragment length polymorphism (AFLP) technique, African Journal of Agricultural Research, 6 (4), 790–797.
37. Mojeremane, W., and Tshwenyane, S.O., 2004. The role of morula (*Sclerocarya birrea*): a multipurpose indigenous fruit tree of Botswana, Journal of Boilogical Sciences 4 (6): 771–775.
38. Moselhy, H.F., Reid, R.G., Yousef, F. and Boyle, S.P., 2013. A specific, accurate, and sensitive measure of total plasma malondialdehyde by HPLC, Journal of Lipid Research, 54: 852–858.
39. Muhammad, S., Hassan, L.G., Dangoggo, S.M., Hassan, S.W., Umar, K.J. and Aliyu, R.U., 2011a. Nutritional and anti-nutritional composition of *Sclerocarya birrea* seed kernel, Studia Universitatis "Vasile Goldiş", Seria Ştiinţele Vieţii, 21 (4), 693–699.
40. Muhammad, S., Hassan, L.G., Dangoggo, S.M., Hassan, S.W., Umar, K.J. and Aliyu, R.U., 2011b. Acute and sub-chronic toxicity studies of kernel extract of *Sclerocarya birrea* in rats, Science World Journal, 6 (3), 11–14.
41. Nahashon, S.N. and Kilonzo-Nthenge, A.K., 2011. Advances in soybean and soybean by-products in monogastric nutrition and health, Soybean and Nutrition, Prof. Hany El-Shemy (Ed.), ISBN: 978-953-307-536-5, InTech, http://www.intechopen.com/books/soybean-and-nutrition/advances-in-soybeanand-soybean-by-products-in-monogastric-nutrition-and-health.
42. National Research Council (NRC), 1994. Nutrient requirements of poultry. 9th edition. (National Academic Press, Washington, DC).
43. Nyoka, B.I., Chanyenga, T., Mng'omba, S.A., Akinnifesi, F.K. and Sagona, W., 2015. Variation in growth and fruit yield of populations of *Sclerocarya birrea* (A. Rich.) Hochst, Agroforestry Systems, 89, 397–407.
44. Racanicci, A.M., Menten, J.F.M., Regitano-D'arce, M.A.B., Torres, E.A.F.S., Pino, L.M. and Pedroso, A.A., 2008. Dietary oxidized poultry offal fat: broiler performance and oxidative stability of thigh meat during chilled storage. Brazilian Journal of Poultry Science, 10: 29–35.
45. Rajiur Rahman, S.M., 2012. Documentation of good practices for family poultry development in Swaziland. Report prepared for the IFAD funded FAO Project "Smallholder Poultry Development" (GCP/INT/197/IFA). http://www.fao.org/3/a-aq624e.pdf (Accessed 02 June 2016).
46. Ramachandran, S., Singh, S.K., Larrhoche, C., Soccol, C.R. and Pandey, A., 2007. Oilcakes and their biotechnological applications—a review, Bioresource Technology, 98, 2000–2009.

47. Schuhmacher-Wolz, U., Heine, K., Schneider, K., 2010. Report on toxicity data on trichothecene mycotoxins HT-2 and T-2 toxins. CT/EFSA/CONTAM/2010/03. Available at: http://www.efsa.europa.eu/en/supporting/doc/65e.pdf. Accessed March 04, 2017.
48. Siyaya, B.J. and Masuku, M.B., 2013. Determinants of profitability of indigenous chickens in Swaziland, Business and Economic Research 3 (2): 205–2017.
49. VDLUFA, 2007. Verband Deutscher Landwirtschaftlicher Unterschungs-und Forschung-sanatalten. Handbuch der Landwirtschaftlichen Versuchs-und Untersuchungsmethodik (VDLUFA Methodenbuch), Bd. III. Die Chemische Untersuchung von Futtermitteln. VDLUFA-Verlag. Darmstadt.
50. Wairagu, N.W., Kiptoo, J., and Githiomi, J.K., 2013. Nutritional assessment of *Sclerocarya birrea* (amarula) fruit from Kenya, International Journal of Current Research, 5 (5), 1074–1078.
51. World Bank Group., 2015. Ending poverty and hunger by 2030: an agenda for the global food system. 2nd edition. http://www-wds.worldbank.org/external/default/ WDSContent Server/WDSP/IB/2015/06/03/090224b082eed2bb/2_0/Rendered/PDF/Ending0povert y0e0global0food0system.pdf (Accessed 02 June 2016).
52. Wu, Q., Wang, X., Yang, W., Nussler, A.K., Xiong, L., Kucˇa, K., Dohnal, V., Zhang, X., Yuan, Z., 2014. Oxidative stress-mediated cytotoxicity and metabolism of T-2 toxin and deoxynivalenol in animals and humans: an update. Archives of Toxicology, 88: 1309–1326.

CHAPTER 16

ECOALIM: A Dataset of Environmental Impacts of Feed Ingredients Used in French Animal Production

Aurélie Wilfart[1*], Sandrine Espagnol[2], Sylvie Dauguet[3], Aurélie Tailleur[4], Armelle Gac[5], Florence Garcia-Launay[6]

1 UMR SAS, INRA, Agrocampus Ouest, Rennes, France

2 IFIP, Institut du porc, Le Rheu, France

3 Terres Inovia, Pessac, France

4 ARVALIS-Institut du veÂgeÂ tal, La Chapelle Saint Sauveur, France

5 Institut de l'Elevage, Monvoisin, Le Rheu, France

6 UMR PEGASE, INRA, Agrocampus Ouest, Saint-Gilles, France

ABSTRACT

Feeds contribute highly to environmental impacts of livestock products. Therefore, formulating low-impact feeds requires data on environmental impacts of feed ingredients with consistent perimeters and methodology for life cycle assessment (LCA). We created the ECOALIM dataset of life cycle inventories (LCIs) and associated impacts of feed ingredients used in animal production in France. It provides several perimeters for LCIs (field gate, storage agency gate, plant gate and harbour gate) with homogeneously collected data from French R&D institutes covering the 2005–2012 period.

The dataset of environmental impacts is available as a Microsoft® Excel spreadsheet on the ECOALIM website and provides climate change, acidification, eutrophication, non-renewable and total cumulative energy demand, phosphorus demand, and land occupation. LCIs in the ECOALIM dataset are available in the AGRIBALYSE® database in SimaPro® software. The typology performed on the dataset classified the 149 average feed ingredients into categories of low impact (co-products of plant origin and minerals), high impact (feed-use amino acids, fats and vitamins) and intermediate impact (cereals, oilseeds, oil meals and protein crops). Therefore, the ECOALIM dataset can be used by feed manufacturers and LCA practitioners to investigate formulation of low-impact feeds. It also provides data for environmental evaluation of feeds and animal production systems. Included in AGRIBALYSE® database and SimaPro®, the ECOALIM dataset will benefit from their procedures for maintenance and regular updating. Future use can also include environmental labelling of commercial products from livestock production.

1. INTRODUCTION

The Food and Agriculture Organization (FAO) stated that livestock farming systems are involved in 14.5% of human-induced greenhouse gas (GHG) emissions [1]. Recent studies investigated ways to provide protein to feed the increasing world population by 2025 while also decreasing environmental impacts of agricultural production. Billen *et al.* [2] stated that a change in human diets in several regions of the world and moderate agricultural intensification in all regions of the world may reduce international trade, improve food sovereignty and reduce nitrogen (N) pollution. In contrast, the FAO argued that no technically or economically viable alternative exists for intensive animal production to feed the world [3]. Although there is a clear debate on how to provide an adequate supply of nutrients to the world population by 2050, the scientific community agrees on the need to decrease pollutant emissions related to livestock production; consequently, it focuses on estimating environmental impacts of livestock production reliably enough to identify mitigation options [4].

Life cycle assessment (LCA) is the most popular and recognised methodology, with an ISO standard [5], to estimate environmental impacts of animal products [6, 7]. In pig and poultry systems, feed production contributes greatly to environmental impacts of animal products. In particular, it accounts for 50–85% of climate change impact, 64–97% of eutrophication potential, 70–96% of energy use and nearly 100% of land occupation [8–13], with differences among animal products (e.g. broiler vs. layer), type of production systems (e.g. industrial, backyard) and geographic location [14]. In dairy production, methane from enteric fermentation is the main GHG-emission source, but feed production accounts for approximately 36% of GHG emissions, with differences among geographic regions of the world [15]. In beef production, feed production can contribute up to 55% of GHG emissions [16]. Therefore, given the large contribution of feed production to the environmental load of livestock products, it is necessary to implement mitigation strategies based on robust and accurate data on environmental impacts of feed ingredients.

Environmental impacts of feeds are influenced mainly by their ingredients. Some of these ingredients account for more than 10% of feed composition and have relatively high impacts on resulting rations. For instance, soya bean meal imported to France is usually incorporated at levels of 10% and 22% in pig [17] and poultry growing feeds [18], respectively, and has relatively high potential climate change impacts, related to deforestation in the past 20 years in the soya bean production area of west-central Brazil [19]. Other feed ingredients are incorporated in small amounts into feeds but have high environmental impacts per kilogram, e.g. amino acids and monocalcium phosphate [17]. From a nutritional viewpoint, feed ingredients are classified into cereals and cereal co-products (mainly to supply energy), protein-rich crops and oilseeds (mainly to supply protein), by-products from industry (mainly to supply fibre), and additives (mainly to achieve balance among amino acids and improve phosphorus (P) digestibility). Normally, feed formulation aims to reduce input costs under nutritional constraints, without considering environmental impacts of the feed ingredients incorporated. Feed formulation based on environmental criteria can promote substitutions among feed ingredients within these categories, with a potential difference in feed cost and environmental impacts.

To formulate low-environmental impact feeds, it is necessary to incorporate data on environmental loads of feed ingredients into the feed formulation program. One alternative is to use a dataset based on consistent perimeters and methodologies for LCA. Bertoluci *et al.* [20] emphasised that combining data from heterogeneous studies with different assumptions and perimeters can compromise the consistency of the dataset and the relevance of conclusions. Currently, few international databases include environmental impacts of feed ingredients [21, 22]. The ecoinvent® database relies mainly on Swiss data and practices for the life cycle inventories (LCIs) of agricultural production and provides only a few feed-ingredient LCIs. The European database Agri-Footprint [21] contains standardised data (LCIs and environmental impacts) for several countries but not for all feed ingredients (e.g. feed ingredients processed in France, amino acids). Using it to formulate feeds in France would require incorporating data from other sources for missing feed ingredients, thus increasing the uncertainty in results. Agri-Footprint is also mainly intended as a source of secondary or background data for LCI or LCA of animal products from the perspective of environmental labelling of food products [21]. It is thus more appropriate for comparing animal production systems and mitigation options at the level of the animal unit subsystem. Consequently, an LCA database dedicated to feed formulation is needed, which requires greater data accuracy and model complexity for LCAs of feed ingredients [23].

Feed formulation in France involves mostly French feed ingredients, which result from several cropping systems under a variety of soil and climate conditions and cropping practices. A previous study [18] highlighted the need for homogeneously developed data on environmental impacts of feed ingredients to formulate low-impact feeds and investigate mitigation options.

The ECOALIM dataset is the result of a collaborative project between research and extension institutes, with the participation of animal-nutrition companies. ECOALIM collects the most accurate and representative data available to date for LCIs of French feed ingredients. This dataset can be used in France and in countries importing French feed ingredients. This article presents the methodological choices and perimeters implemented to develop the ECOALIM dataset, as well as its potential application.

2. MATERIALS AND METHODS

LCA is a methodology which calculates environmental impacts of a specific product by including all the resources needed to produce it and all pollutant emissions associated with its life cycle (from extraction of raw materials to recycling of the product's disposal or recycling). An LCA follows 4 steps defined by an ISO standard [5]: goal and scope definition, inventory analysis, impact assessment, and interpretation of results.

2.1 Goal and scope definition

For the ECOALIM dataset, several system perimeters were defined (Fig 1): field gate, storage-agency gate, plant gate, and harbour gate. The field gate is relevant for assessing impacts of on-farm feed production (i.e. crops produced and directly used on-farm). Feed formulation by feed companies requires additional perimeters: plant gate for the co-products of cereals, oilseeds and protein crops (e.g. meals) as well as for industrial products (e.g. amino acids) processed in France; storage-agency gate for cereals (perimeter includes grain drying); and harbour gate for imported feed ingredients.

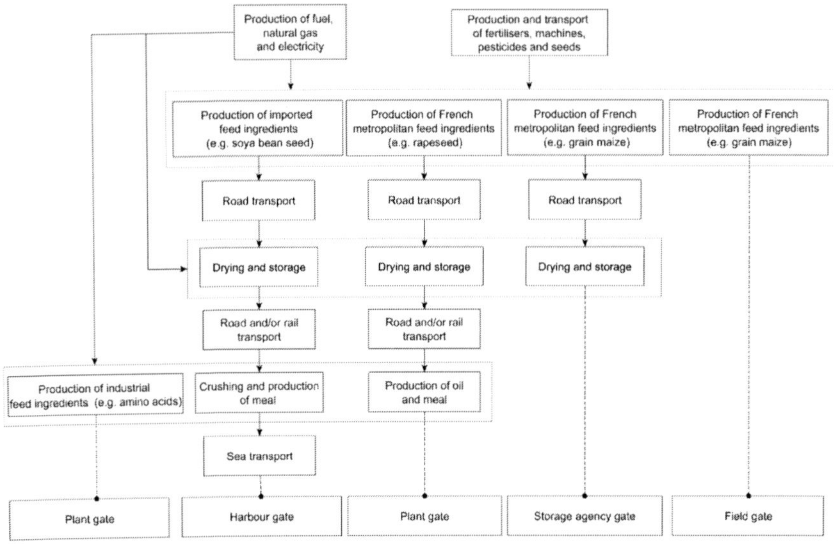

Figure 1. Flow diagram for production of the feed ingredients included in the ECOALIM dataset, with main processes for crop-input production, crop production and feed-ingredient production.

The impacts considered were climate change with (CCLUC) or without land-use change (CC), eutrophication (EU), acidification (AC), land occupation (LO), non-renewable (CEDNR) and total energy demand (CEDTOT) and P demand (PD). PD was included to account for the non-renewable P resource incorporated in fertilisers and feeds. The other impact categories are usually included in agricultural LCAs.

When a production process generates multiple final co-products, it is necessary to allocate the process's impacts to the co-products. In ECOALIM, impacts of co-products of cereals and maize, as well as oils and meals, were calculated using economic allocation. To perform this allocation based on the relative economic value of each co-product, the Olympic five-year (2008–2012) average price of each co-product was calculated. The functional unit used is kilogram of feed ingredient at the reference water content, which is the usual functional unit for feed ingredients.

2.2 Life cycle inventories

2.2.1 French crops

Average LCIs for the following French crops were included in the ECOALIM dataset: maize, wheat, barley, oat, triticale, sorghum, rapeseed, sunflower, faba bean, flaxseed, pea, and soya bean (Table 1).

Table 1. Main inputs used and dry matter yield for French crops used as ingredients in animal feed.

Name \ Unit	N mineral	N^a Manure	P_2O_5 (mineral + manure)	K_2O (mineral + manure)	Seed	Pesticide (active ingredient)	Diesel	Agricultural machinery	Irrigation water	Moisture content at harvest	Yield (dry matter)
	kg/ha	kg/ha	kg/ha	kg/ha	kg/ha	kg/ha	kg/ha	kg/ha	m^3/ha	%	kg/ha
Wheat FR	163	11	25+7	42+12	137	1.87	77	6.7	15	15	7100
Grain maize FR	176	37	45+25	40+54	28	2.03	83	8	645	28	10,672
Barley FR	135	15	33+4	33+14	128	2.5	72	7	9	15	6722
Silage sorghum FR	15	45	18+13	18+65	9	2.3	135	12.2	120	100	10,700
Grain sorghum FR	60	0	48+10	27+0	9	2.3	80	7.2	120	15	5800
Oat FR	110	30	30+7	28+0	97	1.7	63	6	0	15	4900
Triticale FR	120	11	23+41	11+23	120	1.24	62	6	0	15	5200
Pea FR	0	9	22+7	19+10	219	4.2	92.5	8.6	50	15	3910
Faba bean FR	0	8	46+20	31+30	20.6	3.05	81	7.9	0	15	4309.5
Rapeseed FR	161.6	17	38+10	28+0	2.5	1.98	78.3	8.2	0	9	2951
Sunflower FR	40.5	11	25+7	25+0	4.3	1.67	77.4	7.7	24	9	2169
Soya bean FR	1.5	9	24+9	28+0.3	91.2	1.39	62	6	750	14	2683
Flaxseed FR	84	6	33+2	24+20.7	35	0.25	81.5	7.3	0	10	1823

[a] amount of N from manure after allocation among crops in rotation

LCIs for French crops were based on those in the French national database, Agribalyse®, for the main agricultural products [24], with updated emission factors for ammonia [25] and references for agricultural practices, when available. One major improvement compared to Agribalyse® is the distribution of the impacts associated with P fertilisation and nitrate leaching among the crops in the same crop rotation (according to crop requirements, P uptake and nitrate leaching among crops).

For all French crops, average LCIs (yields; amounts of fertilisers, pesticides and seeds) were obtained from French agricultural data [26–28] (Table 2).

Table 2. Data references for average life cycle inventories for the main French crops.

Crop	Inventory source	Yield / average national inventory	Organic fertilisation	Other inputs (mineral fertiliser, seeds, agricultural machinery, etc.)
Rapeseed	AGRIBALYSE®	Agreste[a] 2006–2010	Agreste survey 2006[b]	Terres Inovia surveys 2008–2010 and expert knowledge
Sunflower	AGRIBALYSE®	Agreste[a] 2006–2010	Agreste survey 2006[b]	Terres Inovia surveys 2006 and 2009
Pea	AGRIBALYSE®	Statistical data[c] from UNIP[d]-ARVALIS[e]2005–2009	Agreste survey 2006[b] + expertise	Agreste survey 2006[b] and expert knowledge
Faba bean	AGRIBALYSE®	Statistical data[c] from UNIP[d]-ARVALIS[e] 2005–2009	Expert knowledge	Expert knowledge
Flaxseed	AGRIBALYSE®	Agreste[a] 2005–2009	Agreste survey 2006[b] and expert knowledge	2013 National survey by Terres Inovia and expert knowledge
Soya bean	ECOALIM	Agreste[a] 2008–2012	Agreste survey 2006[b] and expert knowledge	2012 National survey by Terres Inovia and expert knowledge
Soft wheat	AGRIBALYSE®	Agreste[a] 2005–2009	Agreste survey 2006[b]	Agreste survey 2006[b]
Maize	AGRIBALYSE®	Agreste[a] 2005–2009	Agreste survey 2006[b]	Agreste survey 2006[b]
Barley	AGRIBALYSE®	Agreste[a] 2005–2009	Agreste survey 2006[b]	Agreste survey 2006[b]
Sorghum	ECOALIM	Agreste[a] 2008–2012	Expert knowledge	Expert knowledge, Arvalis experimental farm network 2008–2012
Oat	ECOALIM	Agreste[a] 2008–2012	Expert knowledge	Expert knowledge, Arvalis experimental farm network 2008–2012
Triticale	AGRIBALYSE®	Agreste[a] 2005–2009	Expert knowledge	Expert knowledge

[a]Online database (in French) available at: http://www.agreste.agriculture.gouv.fr/enquetes/statistique-agricole-annuelle-saa/

[b]Agreste, 2006. Enquête sur les pratiques culturales en 2006. Available from: http://agreste.agriculture.gouv.fr/publications/chiffres-et-donnees/article/enquete-sur-les-pratiques

[c]Online database (in French) available at: http://www.unip.fr/marches-et-reglementations/statistiques-france/surfaces-rendements-et-productions.html

[d]UNIP: French agricultural technical institute dedicated to legume crops

[e]ARVALIS: French agricultural technical institute dedicated to crops

Table 3. Main inputs used and dry matter yield for European and non-European crops serving as feed ingredients for animal feed (BR: Brazil, US: United States of America, UK: United Kingdom, MY: Malaysia, PK: Pakistan, UA: Ukraine).

Crop \ Unit	N mineral	N manure	P_2O_5 (Mineral +manure)	K_2O (Mineral +manure)	CaO	Seed	Pesticide (active ingredient)	Diesel	Agricultural machinery	Irrigation water	Moisture content at harvest	Yield (dry matter)
	kg/ha	kg/ha	kg/ha	kg/ha	kg/ha	kg/ha	kg/ha	kg/ha	kg/ha	m^3/ha	%	kg/ha
Soya bean BR	5.5	1.3	80 +0.5	80+0.5	518	53	1.7	76.0	18.0	0	18	2708
Soya bean US	18.5	0	52+0	88+0	0	72.5	5.4	53.4	4.9	5	14	2411
Wheat UK	188	0	17+0	23+0	0	200	8.6	62.6	5.6	0	15	6715
Sorghum US	81.6	0	31+0	24+0	0	1.3	4.0	79.8	7.3	838	14	3354
Maize US	156	0	66+0	93+0	0	20.3	4.1	80.4	7.5	871	28	6732
Palm oil tree MY	157	0	32	236	43	-	1.7	26	67,119.0	448	47	24,978
Sugarcane PK	125	29	130+19	0+24	0	-	8.25	98.2	-	11616	70	15,090
Sunflower UA	28.8	0	15+0	15+0	0	3.84	2.04	75.1	7.1	0	22	1432

For maize, wheat, barley, rapeseed and sunflower, specific LCIs were constructed for systematic cover cropping, systematic organic fertilisation, and introduction of a protein crop in the crop rotation, based on data from the farm network of the French agricultural institute dedicated to cereals (ARVALIS-Institut du végétal), to assess variability in the results due to crop management.

2.2.2 Non-French crops

Average LCIs of the following non-French crops were included in the dataset: Brazilian soya bean; US soya bean, sorghum, and maize; UK wheat; Malaysian palm oil tree; Pakistani sugarcane; and Ukrainian sunflower (Table 3). Data for resources came from scientific publications [19, 22, 29] and statistics from national databases [30–33] or FAOSTAT [34]. Resources and emissions in LCIs were calculated according to previously published methodologies [19, 35].

2.3 Industrial products

Industrial products in the dataset include products derived from crops and products that result from chemical synthesis, fermentation, mining extraction, and animal by-product processing. Co-products directly derived from crops are oils, meals, Dried Distilled Grains with Solubles (DDGS), molasses, pulps, and flours associated with extrusion processes, crushing, milling, distillation, etc. Products secondarily derived from crops are starch and gluten feeds. Processing factories were surveyed to collect the relevant underlying data in 2013–2014, except for French oils and meals and non-French feed ingredients [36]. For other industrial products, LCIs are from Garcia-Launay *et al.* [17] for amino acids, confidential industrial data for minerals, the French Technical Centre for Meat (ADIV) data for animal by-products, and ecoinvent data adapted to the French context for vitamins [36].

2.4 Calculation of emissions

Calculations of gaseous emissions (NH_3, NO_x, N_2O, and CO2) and emissions to soil and water (NO_3^-, PO_4^{3-}, heavy metals) were based on the AGRIBALYSE® method [24]. Briefly, NH_3emissions came from EMEP/EEA 2013 Tier 2 [25] and N_2O emissions from IPCC Tier 1 [37]. Phosphate emissions were calculated based on previously developed models for soil loss [38] parametrised for France and the SALCA-P model [22] for P emissions into water (i.e. leaching, run-off, erosion). NO_x emissions were calculated according to EMEP/EEA 2013 Tier 1 [25]. Nitrate emissions were calculated according to a spatial grid of leaching risk based on the crop and the regional area from COMIFER [39] adapted by Tailleur *et al.*[40]. Gaseous emissions from agricultural machinery combustion were estimated with the method of Nemecek and Kägi [22]. CO_2 emissions from urea fertilisation and lime application were estimated from IPCC Tier 1 [38]. Heavy metal emissions were estimated from the SALCA-SM model adapted to the French context [41, 42].

2.5 Impact assessment

ECOALIM uses the ILCD characterisation method recommended by the Joint Research Centre [43], as well as the CML-IA characterisation method [44], which is the most popular in agricultural LCA. Energy demand is calculated according to the CED 1.08 method [44].

CC and CCLUC are expressed in kg CO_2-eq, AC in both molc H^+-eq (AC_{ILCD}) and kg SO_2-eq (AC_{CML}), terrestrial EU in molc N-eq, freshwater EU in kg P-eq, marine EU in kg N-eq, EU_{CML}in kg PO_4^{3-} eq, LO in m^2.y, and CEDNR and CEDTOT in MJ. PD is expressed in kg P and includes all P and phosphate inputs throughout the life cycle. Calculations were performed using SimaPro® software v8.0.5.13 (PRé Consultants, Amersfoort, The Netherlands) and the attributional ecoinvent v3.1 database for background data [36].

2.6 Statistical analysis

The dataset was subject to a series of multivariate analyses using R statistical software [45] to describe variability in the impacts associated with each type of feedstuff. From all of the impact categories available in the dataset, we selected CCLUC, AC_{CML}, EU_{CML}, CEDNR, PD, and LO after examining the correlation coefficient between each pair of impacts and excluding impacts that were highly correlated. To address the high skewness of impact distributions, the quantitative variables CCLUC, AC_{CML}, EU_{CML}, CEDNR, PD, and LO were converted into qualitative variables with levels corresponding to the 4 quartiles of the distribution of each impact. The feed-ingredient typology was constructed with the active qualitative variables CCLUC, AC_{CML}, EU_{CML}, CEDNR, PD, and LO and the illustrative qualitative variable “Feedstuff Type” using multiple correspondence analysis (MCA), followed by hierarchical clustering (HC) on the 5 first dimensions of the MCA. This method classified feed ingredients that are similar to each other but different from others by maximising intra-group homogeneity and inter-group diversity. After plotting the inertia of the classifications as a function of the number of classes, we chose a number of classes that did not result in a gap in inertia. Finally, levels of active variables, the illustrative variable, as well as the classes of the typology, were plotted on the factor map.

3. RESULTS AND DISCUSSION

The objectives of this collaborative work among institutes were to develop a dataset of LCIs and environmental impacts of feed ingredients used for French livestock nutrition. This section addresses the content of the dataset, illustrates the typology of the categories of environmental impacts of feedstuffs available in it, and discuss its potential uses.

3.1 Dataset description

The ECOALIM dataset contains LCIs for 149 average feed ingredients and 16 feed ingredients from specific crop-management practices (Table 4). It offers a wide range of feedstuffs used in several kinds of livestock production (cattle, pig and poultry) and different perimeters for LCA, for feed formulation by feed manufacturers, farm cooperatives and farmers (on-farm). A Microsoft® Excel file is available online (Fig 2) which shows the partners involved and provides an overview of the environmental impacts available by default (CCLUC, AC_{ILCD}, EU_{CML}, CEDNR, PD, LO_{CML}) in the dataset.

Figure 2. Overview of the ECOALIM impact dataset in Microsoft® Excel, freely available at http://www6.inra.fr/ecoalim_eng/ and from the Dryad Digital Repository: http://dx.doi.org/10.5061/dryad.14km1.

Table 4. Number of life cycle inventories available in the ECOALIM dataset according to the type of feed ingredient and perimeter.

Type of feed ingredient	Perimeter			
	At field	At French harbour	At mill or processing plant	At storage agency
Cereals	8	2	0	17
Co-products from food industry	0	0	2	0
Co-products of maize	0	1	5	0
Co-products of wheat	0	0	5	0
Fats	0	6	22	0
Industrial amino acids	0	0	5	0
Minerals	0	0	10	0
Oil seeds and protein crops	6	0	4	13
Oil meals	0	7	22	0
Other co-products of animal origin	0	0	2	0
Other co-products of plant origin	1	1	4	0
Silage	5	0	0	0
Vitamins	0	0	1	0

3.2 Typology of the feed ingredients

The MCA and HC provided a general overview of the dataset. In the MCA, the 1st and 2nd dimensions distinguish the 1st and 4th quartiles of the following environmental impacts: CCLUC, AC_{ILCD}, EU_{CML}, CEDNR, PD, LO_{CML} (Fig 3A). The 1st quartiles of these impacts are close to each other on the factor map, which indicates that some feed ingredients in the dataset have low values for all environmental impacts. The 4th quartiles of 4 of these impacts (AC_{CML}, EU_{CML}, CEDNR and PD) are also close to each other on the factor map, which indicates that some feed ingredients in the dataset have high values for these 4 impacts. The 1st quartiles of impacts tend to be observed for co-products from the food industry, wheat co-products, co-products of animal origin, other co-products of plant origin, and minerals (Fig 3B). The 4th quartiles of impacts correspond to industrial amino acids, fats and vitamins (Fig 3B).

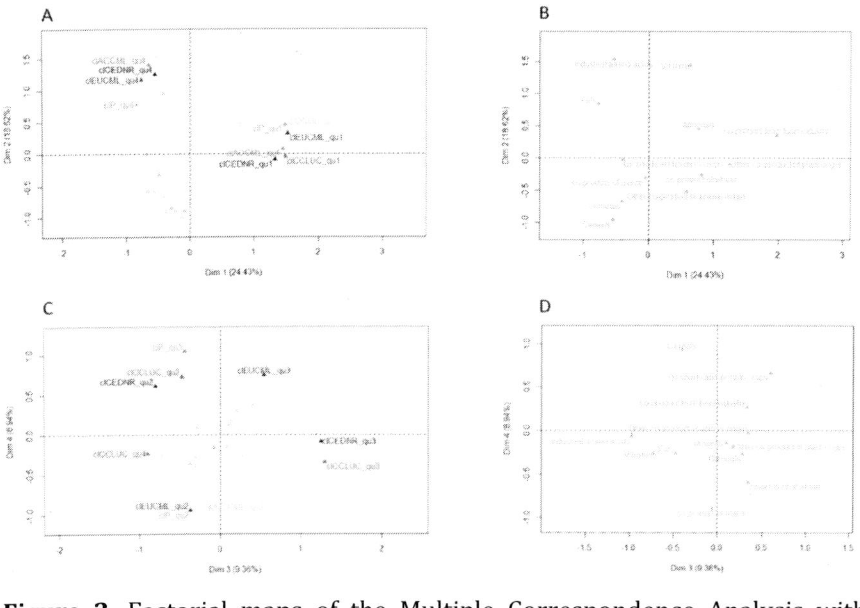

Figure 3. Factorial maps of the Multiple Correspondence Analysis with plotting of the active variable levels contributing more than 10% of the first 4 dimensions (A and C) and plotting of the feed-ingredient categories in the first 4 dimensions (B and D).

CCLUC, AC_{CML}, EU_{CML}, CEDNR, PD, and LO_{CML} are climate change with land-use change, acidification, eutrophication, non-renewable cumulative energy demand, phosphorus demand, and land occupation, respectively. Levels qu1 to qu4 correspond to the 1st to 4th quartiles of each variable.

The 3rd and 4th dimensions of the MCA distinguish quartiles of CCLUC, EU_{CML}, CEDNR, PD, and LO_{CML} (Fig 3C). The 4th dimension is positively linked to the 2nd quartile of CCLUC, EU_{CML}, and CEDNR, and the 3rd quartile of PD, both of which characterise cereals and oilseeds and protein crops (Fig 3D).

The HC built 6 classes that are separated on the 1st to 4th dimensions (Fig 4A and 4B). Class 1 contains 14 feed ingredients with the highest impacts for CCLUC, AC_{CML}, EU_{CML}, CEDNR, PD, and LO_{CML}: industrial amino acids, fats and vitamins. This finding is consistent with previous LCAs on industrial amino acids [46, 47] and fats [48, 49]. Class 2 contains 14 feed ingredients with the lowest impacts for CCLUC, AC_{ILCD}, EU_{CML}, CEDNR, PD, LO_{CML}: minerals, co-products of plant origin and co-products from the bread industry. These low impacts are partly due to the economic allocation of impacts among co-products. Class 3 contains 14 feed ingredients with intermediate impacts for AC_{CML}, EU_{CML}, PD, LO_{CML}, (2ndquartile) and CCLUC and CEDNR (3rd quartile). This class contains 85% of French feed ingredients, including pea, monocalcium phosphate, sunflower meal, DDGS and maize co-products. Class 4 contains 11 feed ingredients which are cereals and imported feed ingredients with intermediate values of CCLUC, AC_{CML}, EU_{CML}, CEDNR, and LO_{CML}. Class 5 contains 10 feed ingredients that are oilseeds and protein crops as well as cereals with intermediate values of CCLUC, CEDNR, PD, and LO_{CML}. Class 6 contains oil meals and oilseeds with high PD values and 3rd quartile values of CCLUC, EU_{CML} and CEDNR which come mainly from Brazil and the US.

The feed ingredient types in the ECOALIM dataset (Fig 3B and 3D) have different levels of environmental impacts. Analysis of the dataset gives some highlights for feed formulation:

- Some types have high impacts regardless of the ingredient (Class 1, e.g. amino acids) and are essential for formulating feeds. Their impacts are influenced by high energy inputs for industrial processes, such as those for amino acids, vitamins, and soya bean oil. Since these products are always incorporated in low quantities, they are not a mechanism for reducing feed impacts, except if amino acids are combined with cereals to replace soya bean meal associated with deforestation [50].
- Other types, such as co-products of plant origin and from milling and bread industries, have low impacts because of economic allocation of impacts and the low-input processes used to produce them. They can be incorporated at higher percentages into feeds to reduce the latter's environmental burdens. Recent studies have investigated nutritional benefits of co-products [51, 52] and the subsequent potential reduction in impacts [53].

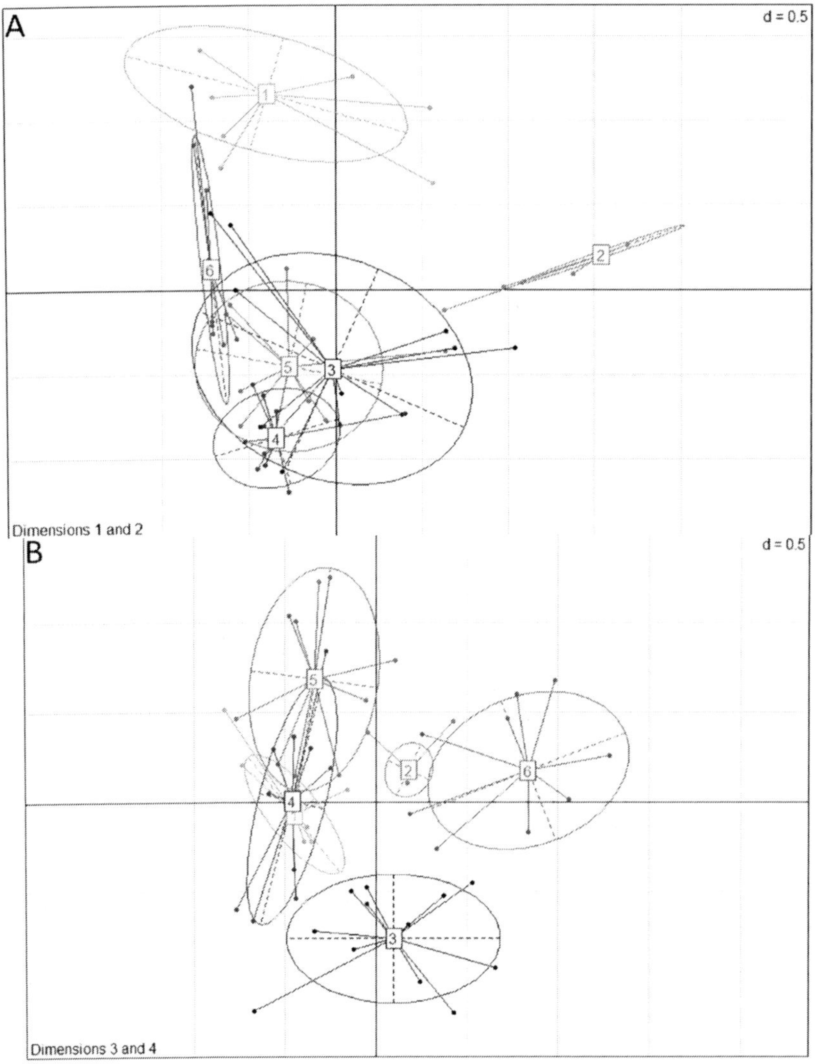

Figure 4. Factorial maps from Multiple Correspondence Analysis and Hierarchical Clustering, with representation of point classes of feed ingredients: Class 1—highest impacts for CCLUC, AC_{CML}, EU_{CML}, CEDNR, PD, and LO_{CML}; Class 2 with the- lowest impacts for CCLUC, AC_{CML}, EU_{CML}, CEDNR, PD, and LO_{CML}; Class 3—intermediate impacts for EU_{CML}, PD, LO_{CML}, 2nd quartile impacts for $AC_{CML}$and 3rd quartile impacts for CCLUC and CEDNR; Class 4—intermediate impacts for CCLUC, AC_{CML}, EU_{CML}, CEDNR, and LO_{CML}; Class 5—intermediate impacts for CCLUC, CEDNR PD, and LO_{CML},; Class 6—high PD impacts and 3rd quartile impacts for CCLUC, $EU_{CM}$and CEDNR $_L$.

- Feed ingredient types that are incorporated in large quantities, such as cereals, oil meals and oilseeds, are distributed among classes with different impact levels (Classes 4, 5 and 6). The main processes driving their impacts include N and P fertilisation, diesel use and combustion, and deforestation for soya bean meal. Therefore, formulating feeds width constraints on potential environmental impacts should lead to substitutions among similar types of feed ingredients, which should decrease impacts.

3.3 Comparison with other databases

To our knowledge, only two other databases include French feed ingredients or crops: Agri-Footprint and ecoinvent®. We focused comparison on the former (Fig 5) because the latter relies mainly on Swiss data and practices for the LCIs of agricultural production and provides only a few feed-ingredient LCIs. Between ECOALIM and Agri-Footprint, rankings of CCLUC impacts of cereals, oil seeds and protein crops differ (Fig 5A) among crops (e.g. faba beans, oat, winter barley). Nevertheless, some consistency exists: rapeseed has the highest impact and sunflower and triticale have similar values in both datasets. Regarding CEDNR (Fig 5B), except for the lowest impact observed for faba beans in both datasets, ECOALIM and Agri-Footprint show different values for most crops: winter barley, wheat, maize grain, oat, rapeseed, and sunflower. Consequently, the ranking of the crops differs between the two datasets.

Note that freshwater eutrophication has two axes, which correspond to (left) ECOALIM and (right) Agri-Footprint.

The values of freshwater EU (Fig 5C) are much higher (x100) in ECOALIM than in Agri-Footprint, and the ranking of crops is completely different. P leaching is calculated as a fixed proportion of P applied in Agri-Footprint, whereas the SALCA-P model used in ECOALIM accounts for the P applied (mineral and organic), crop duration (time between harvest of the considered crop and the preceding crop), soil erosion, and allocation of fertilisation among crops in the same rotation. AC impact (Fig 5D) is consistent between both datasets. In each case, rapeseed has the highest value, and oat and triticale have the lowest ones values. The other crops have moderate differences in value between datasets, rendering their rankings different.

When formulating feeds based on environmental impacts, these databases would yield different feeds. Feed formulation is an optimisation problem that relies on linear programming with nutritional constraints. Consequently, when minimising environmental impacts of feeds at a given nutritional level, incorporation of feed ingredients will follow the impact ranking of feed ingredients. Therefore, feed manufacturers need the most accurate data on environmental impacts that account for specific crop-management practices, data provided for the French context by the ECOALIM dataset.

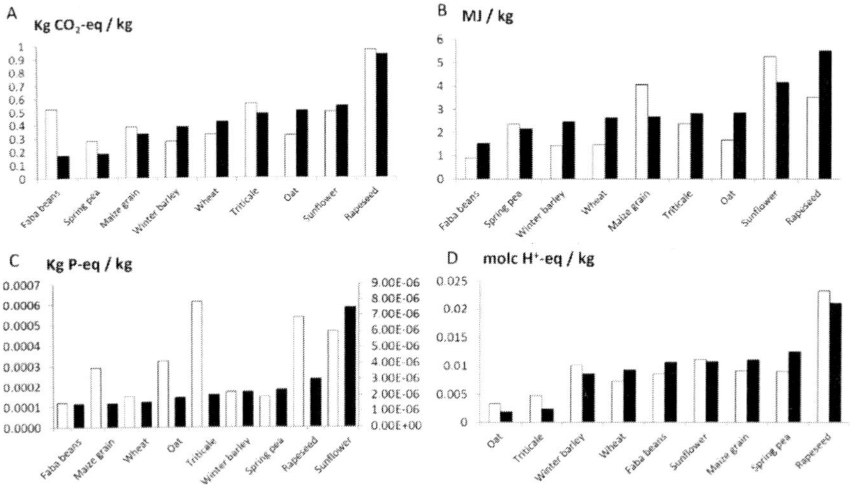

Figure 5. Comparison of impact assessment results per kg of feed ingredient for cereals, legumes, and oil and protein crops in the ECOALIM and Agri-Footprint databases for (A) climate change ILCD (kg CO_2-eq), (B) non-renewable energy (MJ), (C) freshwater eutrophication ILCD (kg P-eq) and (D) acidification ILCD (molc H^+-eq).

3.4 Use, maintenance and updating of the dataset

The ECOALIM dataset is the animal-nutrition subset of the AGRIBALYSE® database. Therefore, it is also available in the Agribalyse® v1.3 database in SimaPro®, with the following name format for processes:

- for unprocessed feed ingredients: "<Feed ingredient name>, <type of production>, <crop-management practice>, animal feed, <LCA perimeter> /FR U", e.g. Feed barley grain, conventional, national average, animal feed, at farm gate/FR U.
- for processed feed ingredients: "<Feed ingredient name>, animal feed, <LCA perimeter> /FR U", e.g. Sunflower meal, with low dehulling, animal feed, at transformation plant/FR U.

The ECOALIM dataset can be used to estimate environmental impacts of feeds, animal production systems and mitigation options (Fig 6).

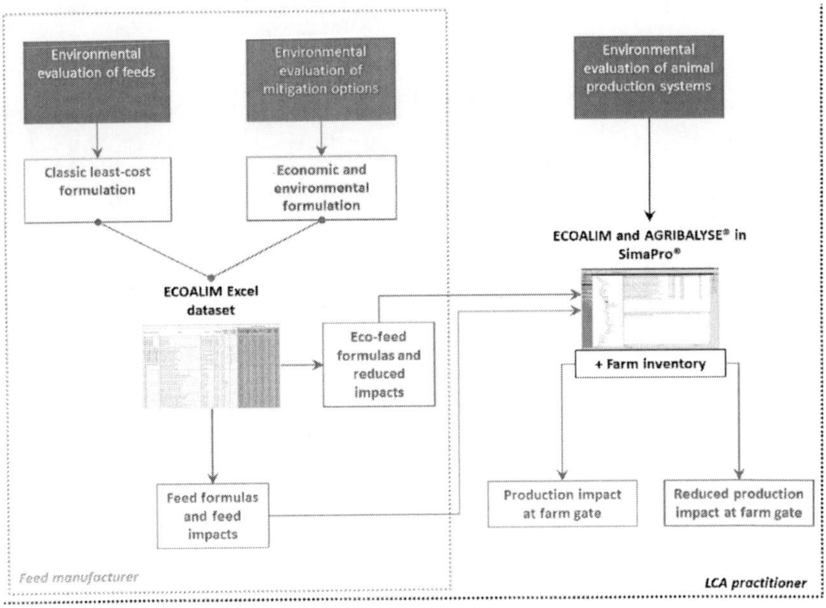

Figure 6. Diagram of potential uses of the ECOALIM dataset.

The blue pathway corresponds to the methodology classically used when estimating environmental impacts of feeds and animal products at the farm gate. The red pathway illustrates the methodology used when evaluating strategies to mitigate environmental impacts of feeds and animal products at the field gate.

Both feed manufacturers and LCA practitioners can formulate feeds according to the classic least-cost approach and calculate the environmental impacts of the resulting feeds using the ECOALIM dataset in Excel. To investigate mitigation options, they can also formulate "eco-feeds" based on both economic and environmental approaches by using the ECOALIM dataset in Excel in formulation software. In an initial application, eco-feeds formulated for pig production had impacts 6–15% (assuming current availability of feed ingredients) and 12–26% (assuming improved availability of some feed ingredients) lower than those of the least-cost formulation, depending on the impact considered [54]. The ECOALIM dataset in the GRIBALYSE® database in SimaPro® can be used by LCA practitioners in more detailed farm LCIs to assess environmental impacts of animal products at the farm gate using either least-cost feed or eco-feed formulas. For these applications, the ECOALIM dataset offers standardised, updated and reviewed data. It covers a range of feed ingredients that was not previously available in AGRIBALYSE® and is, to date, the only dataset that relies on reliable foreground data representative for France, collected from multiple surveys following a common framework. To our knowledge, ECOALIM is the first dataset available for feed ingredient LCIs and impacts that provides all feed ingredients used in conventional livestock production in France and specific LCIs for systematic cover cropping, systematic organic fertilisation, and introduction of a protein crop. It also applies to other European countries for several imported feed ingredients, as well as for French feed ingredients which are exported. This is a major added-value of the dataset because it provides reliable data to non-experts in LCA who may find it very difficult to gather properly data from various databases and publications [20].

Incorporation of the ECOALIM dataset into the AGRIBALYSE® database allows future updating of the LCIs and the addition of new feed ingredients, such as fish meal and oil for aquaculture production. Furthermore, integration of the AGRIBALYSE® database into SimaPro® software ensures appropriate maintenance of the data and wider distribution among LCA practitioners.

4. CONCLUSION

The ECOALIM dataset provides life cycle impacts of feed ingredients used in French livestock production based on a standardised methodology and homogeneous foreground and background data. This approach avoids double-counting of environmental burdens. The approach was developed to be consistent with the AGRIBALYSE® database, which will allow future use of the ECOALIM dataset for feed formulation based on environmental impacts and environmental labelling of animal products. The ECOALIM dataset relies on representative and recent data that cover a wide range of production practices in France. Built with organisations such as feed manufacturers, raw material producers and R&D institutes, this dataset contains feed ingredients necessary to formulate complete feeds and several French feed ingredients for which no data were available before (i.e. sorghum, flaxseed, and soya bean). The ECOALIM dataset will allow feed formulas to be developed which incorporate nutritional, economic and environmental constraints. This new approach to designing diets needs further research to evaluate its potential as an option for mitigating livestock-related impacts. It would be interesting to conduct a similar study in other countries with the same degree of accuracy in LCI data of feed ingredients. The framework developed in the AGRIBALYSE® database and ECOALIM dataset is adequate to further develop such databases.

ACKNOWLEDGMENTS

The dataset described in this article was developed in the ECOALIM project, funded by the French Ministry of Agriculture (CASDAR partenariat et innovation-2012) and the French Environment & Energy Management Agency (ADEME REACCTIF-2012, 1260C0061). The authors are very grateful to Sarah Willmann, Morgane Magnin, and Marie Laustriat for helping to conceive and design the inventories; Paul Ponchant for reviewing the data; and to all the animal-nutrition organisations for sharing their data.

REFERENCES

1. Gerber PJ, Steinfeld H, Henderson B, Mottet A, Opio C, Dijkman J, et al. Tackling climate change through livestock: a global assessment of emissions and mitigation opportunities. Report. Food and Agriculture Organization of the United Nations, 2013.
2. Billen G, Lassaletta L, Garnier J. A vast range of opportunities for feeding the world in 2050: trade-off between diet, N contamination and international trade. Environmental Research Letters. 2015;10(2).
3. FAO. World livestock 2011-Livestock in food security. Rome, Italy: Food and Agriculture Organization of the United Nations (FAO), 2011.
4. Steinfeld H, Gerber P, Wassenaar T, Castel V, Rosales M, Haan Cd. Livestock's long shadow: environmental issues and options. Rome Italy: Food and Agriculture Organization of the United Nations (FAO), 2006 978-92-5-105571-7.
5. Environmental Management—Life Cycle Assessment—Principles and Framework. EN ISO 14040, (2006).
6. van der Werf HMG, Petit J. Evaluation of the environmental impact of agriculture at the farm level: a comparison and analysis of 12 indicator-based methods. Agricult Ecosyst Environ. 2002;93(1/3):131–45.
7. de Vries M, de Boer IJM. Comparing environmental impacts for livestock products: A review of life cycle assessments. Livest Sci. 2010;128(1–3):1–11.
8. Basset-Mens C, van der Werf HMG. Scenario-based environmental assessment of farming systems: the case of pig production in France. Agr Ecosyst Environ. 2005;105(1–2):127–44.
9. Dourmad JY, Ryschawy J, Trousson T, Bonneau M, Gonzalez J, Houwers HWJ, et al. Evaluating environmental impacts of contrasting pig farming systems with life cycle assessment. Animal. 2014;8(12):2027–37. pmid:25170767
10. Boggia A, Paolotti L, Castellini C. Environmental impact evaluation of conventional, organic and organic-plus poultry production systems using life cycle assessment. World's Poultry Science Journal. 2010;66(01):95–114.
11. Prudêncio da Silva V, van der Werf HMG, Soares SR, Corson MS. Environmental impacts of French and Brazilian broiler chicken production scenarios: An LCA approach. J Environ Manag. 2014;133:222–31.
12. Leinonen I, Williams AG, Wiseman J, Guy J, Kyriazakis I. Predicting the environmental impacts of chicken systems in the United Kingdom through a life cycle assessment: Broiler production systems. Poultry Science. 2012;91(1):8–25. pmid:22184424

13. Leinonen I, Williams AG, Wiseman J, Guy J, Kyriazakis I. Predicting the environmental impacts of chicken systems in the United Kingdom through a life cycle assessment: Egg production systems. Poultry Science. 2012;91(1):26–40. pmid:22184425
14. MacLeod M, Gerber P, Mottet A, Tempio G, Falcucci A, Opio C, et al. Greenhouse gas emissions from pig and chicken supply chains—a global life cycle assessment. Rome, Italy: Food and Agriculture Organization of the United Nations (FAO), 2013.
15. FAO. Greenhouse Gas Emissions from the Dairy Sector: A Life Cycle Assessment. Rome, Italy: Food and Agriculture Organization of the United Nations (FAO), 2010.
16. Nguyen TLT, Hermansen JE, Mogensen L. Environmental consequences of different beef production systems in the EU. J Cleaner Prod. 2010;18(8):756–66.
17. Garcia-Launay F, van der Werf HMG, Nguyen TTH, Le Tutour L, Dourmad JY. Evaluation of the environmental implications of the incorporation of feed-use amino acids in pig. production using Life Cycle Assessment. Livest Sci. 2014;161:158–75.
18. Nguyen TTH, Bouvarel I, Ponchant P, van der Werf HMG. Using environmental constraints to formulate low-impact poultry feeds. J Cleaner Prod. 2012;28:215–24.
19. Prudêncio da Silva V, van der Werf HMG, Spies A, Soares SR. Variability in environmental impacts of Brazilian soybean according to crop production and transport scenarios. J Environ Manage. 2010;91(9):1831–9. pmid:20452717
20. Bertoluci G, Masset G, Gomy C, Mottet J, Darmon N. How to Build a Standardized Country-Specific Environmental Food Database for Nutritional Epidemiology Studies. Plos One. 2016;11(4).
21. Durlinger B, Tyszler M, Scholten J, Broekema R, Blonk H. Agri-footprint; A life cycle inventory database covering food and feed production and processing. In: Schenck R, Huizen D, editors. Proceedings of the 9th International Conference on Life Cycle Assessment in the Agri-Food Sector; San Francisco, USA.2014. p. 310–7.
22. Nemecek T, Kägi T. Life cycle inventories of Swiss and European agricultural production systems. Final report ecoinvent report v2.0, n°15. Agroscope Reckenholz-Taenikon Research station ART. Swiss Centre for Life Cycle Inventories. Zurich and Dübendorf, Switzerland: 2007.
23. Colomb V, van der werf HMG, Gac A, Mousset J, Tailleur A, Wilfart A. How to reconcile eco-design and eco-labelling in LCI database construction? AGRIBALYSE experience and links with database harmonization initiatives. 10th International Conference on Life Cycle Assessment of Food 2016, LCA Food 2016; Dublin, Ireland.2016.
24. Koch P, Salou T. AGRIBALYSE(R): Methodological report—Version 1.2. Angers. France: ADEME, 2015.
25. EMEP/EEA. Air pollutant emission inventory guidebook—Technical Report N°12. Copenhagen, Denmark: European Environment Agency (EEA), 2013.
26. Agreste. Enquête sur les pratiques culturales 2006. http://agreste.agriculture.gouv.fr/publications/chiffres-et-donnees/article/enquete-sur-les-pratiques.
27. Agreste. Statistique Agricole Annuelle 2005–2012. http://www.agreste.agriculture.gouv.fr/enquetes/statistique-agricole-annuelle-saa/.
28. UNIP. Statistiques France protéagineux 2005–2009. http://www.unip.fr/marches-et-reglementations/statistiques-france/surfaces-rendements-et-productions.html.
29. Roeder M, Thornley P, Campbell G, Bows-Larkin A. Emissions associated with meeting the future global wheat demand: A case study of UK production under climate change constraints. Environmental Science & Policy. 2014;39:13–24.
30. Feed Grains: Yearbook Tables [Internet]. USDA Economic Research Service. 2014. http://www.ers.usda.gov/data-products/feed-grains-database/feed-grains-yearbook-tables.aspx.
31. Farm Financial and Crop Production Practices [Internet]. USDA Economic Research Service. 2014. http://www.ers.usda.gov/data-products/arms-farm-financial-and-crop-production-practices/tailored-reports-farm-structure-and-finance.aspx#P369773ec09fe4390a25fcbff02a2728a_7_66iT0R0T0R0x2.

32. Quick Stats [Internet]. USDA National Agricultural Statistics Service. 2010. http://quickstats.nass.usda.gov/results/3C1C82DB-0009-3F07-91DE-1D1907917ABD.
33. State Statistics Service of Ukraine. Agriculture of Ukraine—Statistical Yearbook. Kiev, Ukraine: 2012.
34. FAOStat Production statistics [Internet]. Food and Agriculture Organization of the United Nations (FAO). 2013. http://faostat.fao.org/site/291/default.aspx.
35. Boissy J, Aubin J, Drissi A, van der Werf HMG, Bell GJ, Kaushik SJ. Environmental impacts of plant-based salmonid diets at feed and farm scales. Aquaculture. 2011;321(1–2):61–70.
36. Weidema BP, Bauer C, Hischier R, Mutel C, Nemecek T, Reinhard J, et al. The ecoinvent database: Overview and methodology, Data quality guideline for the ecoinvent database version 3, www.ecoinvent.org. 2013.
37. IPCC. Emissions from livestock and manure management. IPCC guidelines for National Greenhouse Gas Inventories. 4(10): IPCC; 2006. p. 87
38. Foster GR. Revised Universal Soil loss equation—Version 2 (RUSLE 2). Report. Washington, D.C.: USDA, Agricultural Research Service; 2005.
39. COMIFER. Lessivage des nitrates en systèmes de cultures annuelles. Diagnostic de risque et propositions de gestion de l'interculture. COMIFER, editor. Puteaux, France2001.
40. Tailleur A, Cohan JP, Laurent F, Lellahi A. A simple model to assess nitrate leaching from annual crops for life cycle assessment at different spatial scales. In: Corson MS, van der Werf HMG, editors. Proceedings of the 8th International Conference on Life Cycle Assessment in the agri-food sector (LCA Food 2012), 1–4 October 2012; Saint-Malo, France: INRA Rennes, France; 2012. p. 903–4.
41. Freiermuth R. Modell zur Berechnung der Schwermetallflüsse in der Landwirtschaftlichen Okobilanz. Zurich, Switzerland: Agroscope FAL Reckenholz, 2006.
42. SOGREAH. Bilan des flux de contaminants entrant sur les sols agricoles de France métropolitaine. Angers, France: ADEME, 2007.
43. JRC. Characterisation factors of the ILCD recommanded life cycle impact assessment methods. database and supporting information. European Commission, Joint Research Centre, Institute for Environment and Sustainability, 2012.
44. SimaPro Database Manual [Internet]. 2015. https://www.pre-sustainability.com/download/DatabaseManualMethods.pdf.
45. R Development Core Team. R: A language and environment for statistical computing. Vienna, Austria: R Foundation for Statistical Computing; 2015.
46. Meul M, Ginneberge C, Van Middelaar CE, de Boer IJM, Fremaut D, Haesaert G. Carbon footprint of five pig diets using three land use change accounting methods. Livest Sci. 2012;149(3):215–23. http://dx.doi.org/10.1016/j.livsci.2012.07.012.
47. Marinussen M, Kool A. Environmental impacts of synthetic amino acid production. Netherlands: Blonk Milieu Advies BV, 2010.
48. Dalgaard R, Schmidt J, Halberg N, Christensen P, Thrane M, Pengue WA. LCA of soybean meal. The International Journal of Life Cycle Assessment. 2007;13(3):240–54.
49. Schmidt JH. Life cycle assessment of rapeseed oil and palm oil: Ph.D. thesis, Part 3: life cycle inventory of rapeseed oil and palm oil. Denmark: Aalborg University; 2007.
50. Monteiro ANTR, Garcia-Launay F, Brossard L, Wilfart A, Dourmad JY. Effect of feeding strategy on environmental impacts of pig fattening in different contexts of production: evaluation through Life Cycle Assessment. Journal of Animal Science. 2016;in press.
51. Cozannet P, Lessire M, Gady C, Metayer JP, Primot Y, Skiba F, et al. Energy value of wheat dried distillers grains with solubles in roosters, broilers, layers, and turkeys. Poultry Science. 2010;89(10):2230–41. pmid:20852114
52. Cozannet P, Primot Y, Gady C, Metayer JP, Lessire M, Skiba F, et al. Energy value of wheat distillers grains with solubles for growing pigs and adult sows. Journal of Animal Science. 2010;88(7):2382–92. pmid:20228235

53. Zanten HHE, Mollenhorst H, Vries JW, Middelaar CE, Kernebeek HRJ, Boer IJM. Assessing environmental consequences of using co-products in animal feed. The International Journal of Life Cycle Assessment. 2013;19(1):79–88.
54. Garcia-Launay F, Wilfart A, Dusart L, Nzally C, Gaudré D, Espagnol S. Multi-objective formulation is an efficient methodology to reduce environmental impacts of pig feeds. 10th International Conference on Life Cycle Assessment of Food 2016, LCA Food 2016; 19–21 October 2016; Dublin, Ireland2016.

CHAPTER 17

A Reproductive Management Program for an Urban Population of Eastern Grey Kangaroos (*Macropus giganteus*)

Andrew Tribe [1,5], Jon Hanger [2], Ian J. McDonald [1], Jo Loader [2], Ben J. Nottidge [3], Jeff J. McKee [4] and Clive J. C. Phillips [5,*]

[1] *School of Agriculture and Food Sciences, University of Queensland, Gatton, Queensland 4343, Australia; E-Mails: a.tribe@uq.edu.au (A.T.)*

[2] *Endeavour Veterinary Ecology Pty Ltd, Toorbul, Queensland 4510, Australia;*

E-Mails: jonhanger@hotmail.com (J.H.); jo@endeavourvet.com.au (J.L.)

[3] *GreenLeaf Ecology, Mooloolah, Queensland 4553, Australia*

[4] *Ecosure, West Burleigh, Queensland 4219, Australia*

[5] *Centre for Animal Welfare and Ethics, University of Queensland, Gatton, Queensland 4343, Australia*

SUMMARY

We designed a programme to control free-ranging kangaroos on a Queensland golf course, using contraceptive implants in females and vasectomisation or testicle removal in males. This reduced the numbers of pouch young to about one half of pre-intervention levels and controlled the population over a 2–4 year period. However, the necessary darting caused a mortality rate of 5–10% of captured animals, mainly due to complications before and after anaesthesia. It is concluded that population control is possible but careful management of kangaroos around the time of anaesthesia induction and recovery is important in such programmes to minimise losses.

ABSTRACT

Traditionally, culling has been the expedient, most common, and in many cases, the only tool used to control free-ranging kangaroo populations. We applied a reproductive control program to a population of eastern grey kangaroos confined to a golf course in South East Queensland. The program aimed to reduce fecundity sufficiently for the population to decrease over time so that overgrazing of the fairways and the frequency of human–animal conflict situations were minimised. In 2003, 92% of the female kangaroos above 5 kg bodyweight were implanted with the GnRH agonist deslorelin after darting with a dissociative anaesthetic. In 2007, 86% of the females above 5 kg were implanted with deslorelin and also 87% of the males above 5 kg were sterilised by either orchidectomy or vasectomy. In 2005, 2008 and 2009, the population was censused to assess the effect of each treatment. The 2003 deslorelin program resulted in effective zero population growth for approximately 2.5 years. The combined deslorelin–surgery program in 2007 reduced the birth rate from 0.3 to 0.06%/year for 16 months, resulting in a 27% population reduction by November 2009. The results were consistent with implants conferring contraception to 100% of implanted females for at least 12 months. The iatrogenic mortality rates for each program were 10.5% and 4.9%, respectively, with 50% of all mortalities due to darting-related injuries, exertional myopathy/hyperthermia or recovery misadventure. The short term sexual and agonistic behaviour of the males was assessed for the 2007 program: no significant changes were seen in adult males given the vasectomy procedure, while sexual behaviours' were decreased in adult males given the orchidectomy procedure. It is concluded that female reproduction was effectively controlled by implantation with deslorrelin and male reproductive behaviour was reduced by orchidectomy, which together achieved population control.

Keywords: behaviour; Deslorelin; orchidectomy; vasectomy; welfare

1. INTRODUCTION

Loss and fragmentation of wildlife habitats and the ecological consequences that follow are major challenges for nature conservation. While loss of biodiversity is one of the most concerning trends, local overabundance, particularly in isolated habitat remnants, may lead to human–animal conflicts and undesirable ecological effects requiring intervention. It is important that management of overabundant species is effective, ecologically sound, humane and socially acceptable.

Since European settlement, the eastern grey kangaroo (*Macropus giganteus*) is one Australian species that is said to have benefited from habitat changes, increasing in population size and range along the eastern coast and for some distance inland [1,2]. Within Australian society, the kangaroo is perceived both as an iconic national symbol and but also invasive in many areas, hence opinion regarding management options varies from full protection to unrestricted culling [3]. However, because of perceived animal welfare and ethical issues and increased availability of non-lethal options, culling is now under greater public scrutiny and socio-political restraint. Non-lethal management options, including translocation and partial or complete reproductive suppression of a population, have greater appeal and hence are the subjects of ongoing investigation [4-8].

Although non-lethal management options may be more socio-politically acceptable, there are nevertheless potential risks to animal welfare associated with capture, anaesthesia and surgical procedures that must be mitigated if such methods are to be widely accepted. Furthermore, behavioural changes associated with pharmacological or surgical intervention may affect the dominance hierarchy and social structure within a population [9-12]. This may reduce group cohesion, which increases their vulnerability in a semi-porous containment area, such as a golf course. Reduced sexual activity may in particular reduce group cohesion.

Herbert *et al.* [5] demonstrated that the gonadotrophin-releasing hormone (GnRH) agonist deslorelin is an effective contraceptive in female eastern grey kangaroos. In their study, the mean contraceptive period for sustained-release implants containing 10 mg of deslorelin was approximately 18 months. Nave *et al.* [8] and Couslon *et al.* [6] demonstrated periods of contraception of approximately 27–48 months in female eastern grey kangaroos after treatment with levonorgestrel implants.

Surgical sterilisation results in permanent reproductive impairment, but is more invasive, labour intensive and requires specialist training. Male vasectomy and distal oviductal transection of females are two methods that have been used to control fecundity in the Kangaroo Island koala population successfully [13]. Castration has not, to our knowledge, been used for suppression of reproduction in wild kangaroo populations, but orchidectomy and vasectomy are routine, safe procedures used for sterilising male captive macropods in zoological collections [14] and to our knowledge have been used in one urban population of kangaroos in Canberra [15]. At present, the behavioural and social effects of a reproductive management program involving a variety of fertility control techniques, such as hormonal contraception and surgical sterilisation, in wild kangaroos have not been documented.

The aim of the reproductive management program reported here was to achieve zero or negative growth in a kangaroo population using a combination of fertility control methods, namely the use of deslorelin implants in females and surgical sterilisation of males using orchidectomy or vasectomy. The mean yearly percentage reduction in recruitment, the effective population growth rates and the cost per percent reduction in recruitment achieved using deslorelin alone are reported and compared with a combination of deslorelin and surgical sterilisation. In addition, the iatrogenic morbidity and mortality rates associated with capture and restraint and the percentage reduction in agonistic and sexual behaviours associated with vasectomy and orchidectomy have been summarised.

2. METHODS

2.1. Study Site

This study was conducted on a 100 hectare golf course situated on the Gold Coast, Queensland, Australia (Figure 1). In 2003, the course was bounded on its northern and western sides by high density residential areas, on its eastern side by a marine inlet, and on its southern side by a grassland area under preparation for residential development. Since that time, the adjacent block on the southern side has been developed as a high-density residential precinct. The southern and western boundaries of the golf course were contained by a 2 m high chainmesh fence, porous in places. The course contained a number of artificial freshwater lakes and waterways and large and small "roughs" vegetated by stands of pine (*Pinus spp.*) or mixed native and exotic vegetation.

In 2003, the managers of the golf course were concerned that the current kangaroo population size could increase the risk of human injury and cause damage to fairways and greens from overgrazing. Consequently, they elected to explore management options. Culling was not considered for ethical, public safety and public relations reasons.

Figure 1. Map showing Pines golf course and surrounding urban areas (Google Earth, 2009).

2.2. The Eastern Grey Kangaroo Population

Notwithstanding the issues with fence integrity, the eastern grey kangaroo population was mainly confined to the golf course. The kangaroos followed a typical crepuscular activity pattern, and during the middle of the day rested in variably-sized mobs in some of the roughs. Anecdotally, the causes of premature mortality in the population were motor vehicle strike on adjacent roads and fox predation on young animals. The kangaroos were habituated to the presence of golfers and their buggies. For the purposes of this study, "flight distance" was defined as the minimum human to animal distance at which kangaroos took flight or displayed avoidance behaviour. We observed that for most kangaroos on the golf course flight distances were in the order of 5–10 m, although this distance increased to 10–35 m over the capture period.

2.3. Census

An initial count of the kangaroo population was undertaken in 2003 by Wildcare Population Health Services. The estimated population at the time was believed to be 194 kangaroos. In 2007, two methods were used to count kangaroos and record basic demographic data:

Method 1: All kangaroos were observed with binoculars to determine sex, age group (adult, sub-adult) and presence of pouch young. This method was the slower of the two and increased the likelihood of double counting due to milling.

Method 2: Kangaroos were counted rapidly irrespective of age and sex class and excluded pouch young.

Kangaroo numbers were assessed from a moving buggy, which stopped if the numbers of kangaroos were too many to count instantaneously. Population estimates were based on the number counted by method 2 plus the number of pouch young observed by method 1, multiplied by a correction factor of 1.2 to allow for undetectable pouch young. This correction factor was based on our estimate that approximately one sixth of pouch young were too small to be detected during the census.

2.4. Capture Restraint and Reproductive Control

The first of two reproductive control programs occurred between August and September 2003. As many as possible of the adult and juvenile kangaroos more than 5 kg were captured by projectile anaesthesia, clinically assessed and tagged. Only females weighing more than 5 kg were implanted with deslorelin-containing slow-release implants (Suprelorin 12®, 9.4 mg deslorelin, Peptech Animal Health). Weight was established by experienced wildlife veterinarians after translocation of each animal. A follow-up census was conducted in November 2005.

The second reproductive control program occurred between May and July 2007. All female kangaroos weighing more than 5 kg were implanted with Suprelorin and all adult and sub-adult male kangaroos more than 5 kg were surgically sterilised. Censuses were conducted in November 2008 and November 2009. A summary of the program timeframes is contained in Table 1.

During the 2003 program, kangaroos demonstrated increasing flight distances as darting progressed. After 3 days of continuous capture the animals' flight distances were beyond the safe range of the dart gun and the program was deferred for 2–3 weeks to allow the kangaroos to settle. Thereafter, each capture session occurred over only 2 days with breaks of 2–3 weeks between sessions. The 2003 program was completed in 5 capture days over a total elapsed time of 2 months. The 2007 program was completed in 6 capture days over a total elapsed time of 2 months.

Table 1. Population management program timeframes.

Time	Activity
March 2003	Initial population estimate
August–September 2003	Capture and tag all kangaroos. Suprelorin implantation of females more than 5 kg only
November 2005	Population census
May–July 2007	Capture and tag all kangaroos. Surgical sterilisation of all males more than 5 kg. Suprelorin implantation of all females above 5 kg
March–September 2007	Behavioural observations
November 2008	Population census
November 2009	Population census

2.5. Kangaroo Capture and Veterinary Procedures

2.5.1. Capture and Anaesthesia

Kangaroos were captured by projectile anaesthesia using Tel-Inject or Dan-Inject re-useable darts fired from either a Taipan 2000 (Montech Industries) or JM Special 25 (Dan-Inject) dart rifle. The anaesthetic agent was Zoletil (tiletamine + zolazepam) (Virbac (Australia), NSW, Australia) delivered at 5–15 mg/kg body weight over darting distances of 5–35 m. Surgeries were performed using standard techniques in a temporary facility observing aseptic techniques. Additional anaesthetic was administered if required. Two darting teams were in operation, consisting of one veterinarian, one veterinary nurse, and one field biologist (spotter), along with kangaroo transport teams consisting of three groups of two wildlife carers which followed each darted kangaroo in a buggy until it was fully immobilised and therefore safe to move. Darting teams returned to the central management area to check on progress with each animal. A veterinary procedures team, consisting of one veterinarian, two veterinary nurses and three volunteer wildlife carers assisted with animal handling and general duties. On arrival, the kangaroo was weighed and placed on the surgery table for a full physical examination and prepared for surgery. A data sheet was then commenced and accompanied the animal to the veterinary procedures area in the golf cart, taking up to 5 minutes.

In 2003, all darted kangaroos were tagged with swivel-type plastic numbered ear tags (Swing-free Tags, Stockbrands, WA, Australia) and marked with temporary stock marking paint to prevent accidental re-darting. Prior to surgery, each kangaroo was given a physical examination to check body condition and gross evidence of any injury or illness, and a cloacal temperature measurement was taken every 15 minutes. Each captured kangaroo was given an injection of Vitamin E + selenium (Ilium Selvite E, Troy Laboratories, Smithfield, NSW, Australia) at a dose rate of 1 mL/30 kg to assist in the prevention of exertional myopathy [16].

All females over 5 kg were injected in the interscapular subcutis with a single 10 mg deslorelin implant (Suprelorin) (Table 2). The following procedures were undertaken on the females:

- checked for pouch young (PY) and if found, the joeys were then sexed and weighed,
- patch of hair shaved and cleaned above the shoulders at the site for the hormonal implant,
- numbered red ear tag placed in the right ear and a red arm band on the right arm.
- The following procedures were undertaken on all the males:
- intravenous injection of Alfaxan® to provide a deeper and more stable level of anaesthesia during surgery. Alfaxan®was also useful because of its shorter period of action than Zoletil® particularly when given intravenously,
- scrotum was shaved, cleaned and prepared for surgery,
- castrated males were also given an injection of lignocaine directly into the testes as a local anaesthetic,
- one numbered red ear tag was placed in their left ear,
- one numbered blue ear tag was placed in the right ear and blue arm band was placed on the right arm if the animal was being castrated,
- alternatively, one numbered green ear tag was placed in the right ear and a blue arm band on the left arm if the animal was being vasectomised,
- injection of engemycin, benacillin antibiotics to help prevent post-operative infection and an injection of metacam for additional for pain relief and a consequently less stressful recovery.

Table 2. Summary of census and veterinary procedure data.

Demographic	Procedure	Number of Kangaroos	
		2003 Program	2007 Program
All kangaroos (incl. pouch young)	Estimated number present at time of management intervention [1]	194	286
All kangaroos	Kangaroos darted	124	187
Females > 5 kg	Deslorelin implant	80	107
Males > 50 kg	Vasectomy	0	19
Males < 50 kg	Vasectomy	0	7
Males > 50 kg	Orchidectomy	0	5
Males > 5 and < 50 kg	Orchidectomy	0	52

[1] Includes pouch young that were captured but not treated.

2.5.2. Processing

Male kangaroos over 5 kg bodyweight were surgically sterilised. Most male kangaroos over 50 kg bodyweight were vasectomised to avoid hormonal changes in dominant males, and most males under 50 kg bodyweight were castrated. In addition, for the purposes of comparison, seven sub-adult males <50 kg were vasectomised and five adult males >50 kg were castrated to determine whether any changes in behaviour could be observed which might be attributed to the sterilisation method for the two classes (sub adult and adult) over the period of this study.

Surgical cases or animals with significant dart wounds were injected with an intramuscular dose of Benacillin (Troy Laboratories, Smithfield, NSW, Australia) at a rate of 1 mL per 10 kg of bodyweight and an intramuscular dose of Engemycin (Intervet Australia, Bendigo, Victoria, Australia) at a dose rate of 10 mg/kg of the active component (oxytetracyline 100 mg/mL). Age, sex, weight, ear tag numbers, anaesthetic details, clinical assessment and procedures were recorded for each animal. When the surgery was completed, each animal was carried to the recovery area and allowed to revive from the anaesthetic. At this stage, the data sheet was given to the veterinary nurse at the shaded recovery area to complete as the animal moved into sternal recumbency and finally left the recovery area of its own accord. During the recovery phase, the animal was spray painted carefully on its back with non-toxic paint, colour coded to indicate the processing session. This was done to further identify which kangaroo had already been sterilised or implanted, to facilitate their behavioural data collection and prevent them from being recaptured during the next day or session.

2.6. Behavioural Observations

Behavioural observations were conducted for 6 weeks prior to the start of the 2007 program, for 8 weeks spanning the capture sessions, and for 6 weeks afterwards. Observations were made on two days each week with the aid of binoculars from a golf cart and occurred over 1 hour sessions: once at sunrise (0700–0800), and once just before sunset (1600–1700).

Preliminary observations identified the relevant behaviours to be recorded, using social and agonistic behaviours previously described [17,18]. Sexual behaviours observed in males were 'following and sniffing females' and penis erection, and agonistic behaviours were fighting with other males, chest beating (male kangaroos beating chest repeatedly with fore paws), high walking and grunting noises. These were recorded each time they were performed by either the focal kangaroo or kangaroos within a group, and had to occur for at least 5 seconds to be recorded. Two bouts were recorded if there was a break of 15 or more seconds between behaviours.

Two observational techniques were used to detect behaviours:

a) Focal animal sampling: one adult and one sub-adult male kangaroo were chosen at random during each observation session. While it would have been preferable to observe the same target kangaroos over the whole period of the study, this was not possible due to the difficulty in locating and identifying particular individuals across the 100 hectare site.

b) Continuous sampling: the mean frequencies of agonistic and sexual behaviours for all visible sub-adult and adult male and all mature female kangaroos in a mob were recorded. Mobs of at least 15 kangaroos were chosen for behavioural observations to increase the detection of interactive behaviours.

Observations of the kangaroos' reactions to darting were recorded during capture sessions for administration of reproductive control measures. A "darting reaction score" of 1–4 was recorded for each kangaroo darting event (1: kangaroo showed a minimal reaction to the dart with no flight response; 2: kangaroo jumped at the time of impact, and/or showed ongoing irritation or annoyance to the presence of the dart; 3: the kangaroo demonstrated an immediate flight response; 4: the kangaroo demonstrated rapid or immediate collapse or bone fracture).

2.7. Statistical Analysis of Behavioural Observations

The rates of sexual and agonistic behaviour before, during and after the program were compared in both the focal and continuous sampling groups using a chi-squared analysis, including a Bonferroni adjustment for multiple comparisons. The 'following and sniffing females' sexual behaviour observed in adult males during the continuous sampling was analysed using a Kruskal-Wallis test. Both tests were analysed with the Minitab™ statistical analysis program.

3. RESULTS

3.1. Reproduction and Demographic Effects

In early 2003, prior to any active population management, there were approximately 130 kangaroos present on the golf course, not including pouch young. Figure 2(a) shows estimates of projected population growth based on a birth rate of 0.3 of total population with two different mortality rates. This model assumes resources are not limiting and there is no immigration or emigration. Recruitment (birth rate) of 0.3 was based on the number of pouch young present as a proportion of the total kangaroo population at the time of the first management program and was based on an approximate pouch life of joeys of 12 months.

Figure 2(b) shows predicted population growth with annual mortality rates of 10% and 20% for different implant responses; assuming the implant was effective in either all or 50% of animals for 12 months.

In 2003, during the first control program there were approximately 194 kangaroos present, including pouch young, of which 181 (93%) were captured (Table 3). In November 2005, 26 months after the completion of the 2003 program, the estimated kangaroo population including pouch young was 174, including 37 pouch young, suggesting that the population growth was effectively zero over that period. This data is consistent with either of the two middle curves in Figure 2(b): 20% mortality and 50% implant efficacy, or 10% mortality with 100% implant efficacy. At the time of the second program in mid-2007, the population had grown to 286, of which 79 were pouch young (Table 4). This data is most consistent with the curve in Figure 2 representing 10% mortality with 100% implant efficacy.

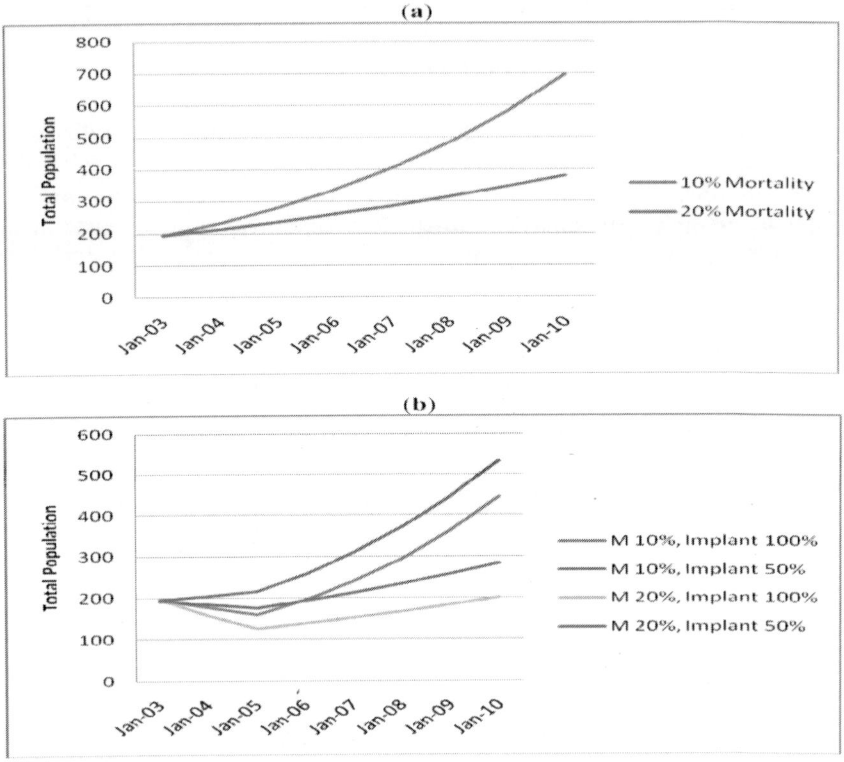

Figure 2. (a) Predicted kangaroo population growth with no intervention; (b) Predicted kangaroo population growth with intervention. M is annual mortality rate.

Table 3. Demographic and management data for the 2003 program.

	Above 5 kg		Pouch young	TOTAL
	Males	Females	Unspecified	
Estimated animals present	47	87	60	**194 (100%)**
Total captured	44	80	57	**181 (93%)**
Estimated not captured	3	7	3	**13 (7%)**
Females implanted with deslorelin	-	80	-	**80 (92% of all females)**
Females with pouch young	-	60	-	**60**

Table 4. Demographic and management data for 2007 management intervention.

	Above 5 kg		Pouch Young	Total
	Males	Females	Unspecified	
Estimated animal present	82	125	79	**286 (100%)**
Total captured	72	115	79	**266 (93%)**
Estimated not captured	10	10	0	**20 (7%)**
Male orchidectomy	53	-	-	**53**
Males vasectomy	19	-	-	**19**
Females implanted with deslorelin	-	107	-	**107**
Females with pouch young	-	79	-	**79**
Mortality	11		2	**13**
Number with 2003 tags	40			**40**

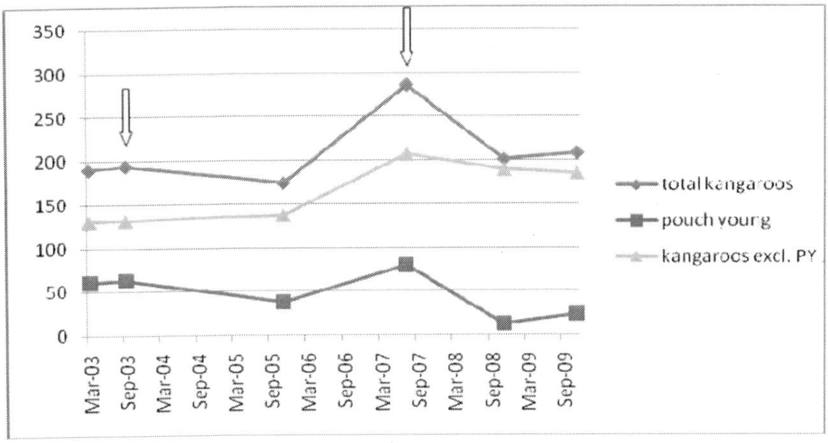

Figure 3. Graphical representation of changes in the total number of kangaroos, the number of pouch young and the number of kangaroos excluding pouch young (PY). The timing of management programs is indicated by arrows.

By late 2008, the population had declined to 201, with 12 pouch young and was only marginally greater in November 2009. The estimated number of pouch young decreased from 63 in 2003 to 37 in 2005 then increased to 79 in 2007, finally decreasing to 12 in November 2008. One year later, there were an estimated 23 pouch young, based on the 19 actually observed during the census.

These changes in total kangaroo population and observed pouch young are shown in Figure 3.

The data is consistent with deslorelin implantation of female kangaroos causing a reduction in the average birth rate from 0.3, expressed as a proportion of total population, including pouch young, to close to zero by one year, and a return to 0.2 by 18 months. In comparison, the deslorelin implantation of females plus surgical sterilisation of males caused a more prolonged reduction in birth rate from 0.3 to 0.06 at 16 months and 0.1 at 28 months. The data is confounded to an extent by some immigration to, and emigration from the golf course of up to 20% of the population per year, based on observation of tagged *versus* untagged animals at the censuses [19]).

3.2. Behaviour

3.2.1. Agonistic Behaviours in Males

Agonistic behaviours tended to decrease post-treatment in castrated sub-adult males (<50 kg males) and were not seen at all in the vasectomised sub-adult males (P = 0.001) (Table 5a). Grunting was observed in untreated but not vasectomised or castrated sub-adult males. Grunting behaviour tended to decrease in castrated adult males compared to untreated and vasectomised adult males (P = 0.06).

Table 5. (**a**) Agonistic behaviours in sub-adult and adult male kangaroos during focal and scanning animal sampling observations (analysed by Chi square test); (**b**) Occurrence of fighting behaviour seen in sub-adult and adult male kangaroos during continuous sampling observation sessions; (**c**) Sexually-related behaviours in sub-adult and adult male kangaroos during focal animal sampling observation sessions (analysed by Chi square test).

(a) Agonistic behaviour

Age Group	Behaviour		Treatment			P Value		
			U*	V*	O*	U* vs. O*	U* vs. V*	O* vs. V*
Sub-adults	Fighting	Present	10	0	3	0.08	0.001	0.20
		Absent	18	29	25			
	Grunting	Present	8	0	0	0.007	0.006	N/A
		Absent	20	29	28			
Adults	Fighting	Present	5	8	2	0.68	0.76	0.21
		Absent	23	21	22			
	High walking	Present	5	4	1	0.33	0.97	0.55
		Absent	23	25	23			
	Chest beating	Present	3	6	1	0.76	0.66	0.21
		Absent	25	23	23			
	Grunting	Present	8	7	1	0.06	0.97	0.12
		Absent	20	22	23			

U*: Untreated, V*: Vasectomy, O*: Orchidectomy

Table 5. *Cont.*

(b) Fighting behaviour

Age Group		Treatment			P Value		
		U*	V*	O*	U* *vs.* O*	U* *vs.* V*	O* *vs.* V*
Sub-adults	Present	6	0	0	0.01	0.01	0.99
	Absent	4	10	10			
Adults	Present	4	2	1	0.32	0.69	0.90
	Absent	6	8	9			

U*: Untreated, V*: Vasectomy, O*: Orchidectomy

(c) Sexually-related behaviour

Age Group	Behaviour		Treatment			P Value		
			U*	V*	O*	U* *vs.* O*	U* *vs.* V*	O* *vs.* V*
Sub-adults	Following & sniffing	Present	8	0	1	0.03	0.006	0.664
		Absent	20	29	27			
	Erection	Present	10	0	1	0.007	0.001	0.68
		Absent	18	28	27			
Adults	Following & sniffing	Present	26	21	6	<0.001	0.122	0.002
		Absent	2	8	18			
	Erection	Present	18	19	4	0.002	0.99	0.001
		Absent	10	10	20			

U*: Untreated, V*: Vasectomy, O*: Orchidectomy

Fighting behaviour was reduced by vasectomisation and castration in sub-adults but not adults (Table 5b).

3.2.2. Sexual Behaviours in Males

Both castrated and vasectomised sub-adult males had a reduced 'following and sniffing females' behaviour and had fewer erections of the penis compared to the untreated sub-adult males, but in the adult males this was only evident for the castrated and not the vasectomised animals (Table 5c). There was no 'following and sniffing females' behaviour observed in castrated adult male kangaroos, whereas there were 4.5 and 4.0 occurrences observed per hour for the vasectomised and untreated males, respectively ($P < 0.001$, Kruskal-Wallis test) during the continuous sampling.

3.2.3. Female Behaviour

Females that had been implanted spent longer in locomotory behaviour than those that had not (implanted 1.75 *vs.* unimplanted 1.35 $\log_n$ bouts/hour, SED 0.06, $P = 0.04$) and less time scanning (implanted 1.70 *vs.* unimplanted 2.85 $\log_n$bouts/hour, SED 0.03, $P < 0.001$) and had fewer grooming bouts (implanted 2 *vs.* unimplanted 5 median bouts/hour, $P < 0.001$). There were no significant ($P < 0.05$) differences in feeding, standing or lying behaviour.

3.2.4. Animal Welfare Impact of Management Interventions

3.2.4.1. Response to Darting

Individual responses of kangaroos to darting were recorded only in 2007. Of 187 kangaroos darted, 152 (81%) were given a "darting reaction score". Most females received a score 1 or 2 and most males received a score 2 or 3 (Figure 4). Of the five kangaroos with the maximum darting reaction score of 4, four kangaroos (3%) collapsed immediately, presumably due to deposition of the anaesthetic agent directly into a vein or bone marrow, and one kangaroo suffered a fracture to the tibia, requiring euthanasia.

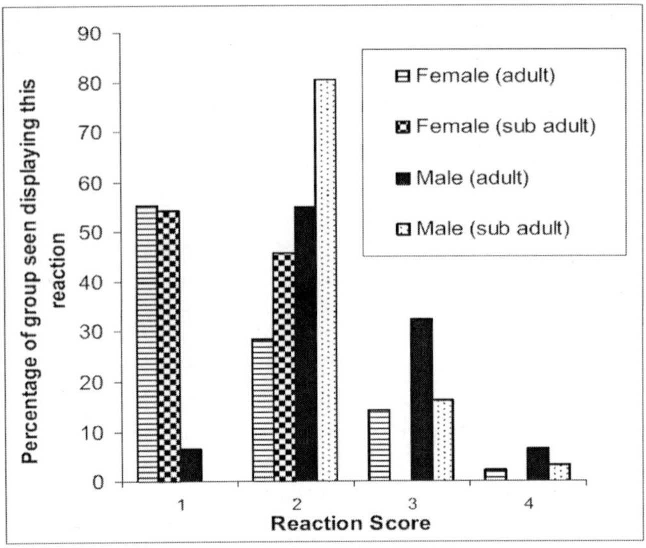

Figure 4. The kangaroos' immediate physical reactions to being darted during the reproductive management interventions, 2007. 1: kangaroo showed a minimal reaction to the dart with no flight response; 2: kangaroo jumped at the time of impact, and/or showed ongoing irritation or annoyance to the presence of the dart; 3: the kangaroo demonstrated an immediate flight response; 4: the kangaroo demonstrated rapid or immediate collapse or bone fracture.

3.2.4.2. Mortality

During both programs, there were darting or anaesthetic mortalities or incidents that required euthanasia. A summary of capture-associated mortalities is presented in Table 6. The animal capture rate (darted animals plus pouch young captured) for both projects was the same (93% of total estimated population), whereas the total mortality rate (as a % of total estimated population) during the 2007 project was less than one half (4.5%) that of the 2003 project (9.8%). Expressed as a percentage of animals captured, the respective mortality rates for 2003 and 2007 were 10.5% and 4.9%.

Five deaths that occurred during recovery from anaesthesia in 2003 were attributed to hyperthermia/myopathy primarily due to high ambient temperatures (26°C or greater). Four animals were found dead up to 2 days after processing and the cause of death was investigated but unclear. Dart-related injuries included fracture of the femur, one spinal injury that required euthanasia and one pouch young that was struck by a dart and died.

In the 2007 management intervention, two kangaroos died as a result of injuries caused by collision with trees during anaesthetic induction after darting. There were three drowning deaths or near-drowning incidents requiring euthanasia that occurred from anaesthesia when kangaroos were ambulatory, but still ataxic.

Table 6. Summary of capture-associated mortalities from the 2003 and 2007 reproductive management plans.

Cause of death/reason for euthanasia	2003 (19 mortalities)		2007 (13 mortalities)		% of all deaths
	Died	Euthanased	Died	Euthanased	
Dart-related injury/fracture	0	5	0	1	19%
Trauma occurring during anaesthetic induction	0	0	0	2	6%
Hyperthermia/myopathy	5	0	1	0	19%
Death after anaesthetic recovery (unspecified cause)	4	0	3	0	22%
Anaesthetic death	0	0	1	0	3%
Drowning/near drowning	0	0	1	2	9%
Predation	2	0	0	0	6%
Untreatable disease	0	1	0	0	3%
Joey death (unviable orphan)	2	0	1	1	13%
Totals	13	6	7	6	100%

3.2.4.3. Non-Lethal Injuries

Non-lethal injuries were generally limited to mild bleeding or haematoma formation at the dart wound requiring minimal veterinary treatment. Virtually all kangaroos experienced a prolonged recovery from anaesthesia (up to 4 hours), during which time they demonstrated significant ataxia and were observed to stand and collapse multiple times, before regaining full coordination and strength.

Spearman's rank correlation indicated a significant positive correlation between the dose rate and the time taken for a full recovery and release from the recovery area back onto the golf course. On average, mature females and males took 154 and 167 minutes, respectively, and sub-adult females and males took 182 and 177 minutes, respectively, to show a full recovery.

In both the 2003 and 2007 management programs, one pouch joey was inadvertently struck by a dart, became anaesthetised and was ejected from the mother's pouch. In 2007, two pouch joeys were abandoned by their mothers during recovery from anaesthesia. All of these joeys were transferred to wildlife carers for hand-rearing and were successfully rehabilitated.

3.2.4.4. Financial Costs and Resource Effort

Each program required 15–20 personnel, including veterinarians, veterinary nurses and wildlife carers. As described in the Method, personnel were divided into three teams: the darting team, consisting of one veterinarian, one veterinary nurse, and one field biologist; the kangaroo transport teams consisting of three groups of two wildlife carers; and the veterinary procedures team, consisting of one veterinarian, two veterinary nurses and three wildlife carers. Additional volunteers (wildlife carers) assisted with animal handling and general duties. The total cost of the 2003 program was AU$18,900, and the 2007 program was AU$49,009.

4. DISCUSSION

4.1. Efficacy of Population Management Program

Based on our population censuses after each program, the reproductive control measures resulted in acceptable reduction of population fecundity, and therefore population growth. Although the census of November 2005 indicated a similar number of adults and sub-adult kangaroos present as were present prior to the 2003 management intervention, the number of pouch young was approximately one half. This result was consistent with our prediction of population number, based upon 100% efficacy of implants and approximately 15% natural attrition (deaths or emigration). At the time of the second management intervention, in mid-2007, the population had increased markedly, with an estimated 207 kangaroos (adults and sub-adults) present, plus approximately 79 pouch young. This is consistent with the expected return of normal fecundity following the loss of effect of the deslorelin implants, as estimated by Herbert and co-workers [5] to be approximately 18 months. However, a significant confounding factor was the potential for movement of kangaroos to and from the site due to numerous breaches in the perimeter fence. At the time of the 2003 management intervention and up to the November 2005 census, there was still a significant population of kangaroos inhabiting or using remnant grassland outside of the golf course, and ample anecdotal evidence of frequent use of breaches in the fence by kangaroos by the course staff and residents. Only 40 of 187 kangaroos darted during the 2007 management intervention had ear tags present from the 2003 management intervention, indicating that a significant proportion of the mature kangaroos present in 2007 were not processed in 2003, and therefore may have immigrated to the site since. In addition, a significant number (estimated to be 20 or more) of kangaroos with ear tags had been killed by motor vehicle strike on roads adjacent to the golf course over the intervening years. By comparison, only 18 of 95 adult kangaroos counted in the November 2008 census were untagged. Given that approximately 20 kangaroos were not captured in the mid-2007 management intervention, the figures are consistent with minimal immigration to the site over that time. Furthermore, urban development had almost entirely destroyed any remnant habitat for kangaroos outside of the golf course by mid-2007.

Our data suggest that deslorelin treatment of female kangaroos alone is sufficient to offer a temporary solution to population growth in eastern grey kangaroos. However, repeat treatments may be required every two years or so, depending upon rates of natural attrition. The addition of surgical (and therefore permanent) sterilisation procedures in male kangaroos was intended to increase the duration of reproductive suppression within the population beyond that provided by deslorelin implantation (of females) alone. Return of fecundity would then rely on the sexual maturation of untreated males (pouch young at the time of the 2007 management intervention), as well as loss of effect of implants in females. In a closed population, this combination is expected to result in a significant drop in the total kangaroo population, followed by a slow increase as normal fecundity returns. However, it is likely that the presence of even a few intact sexually mature males will obviate the effect of permanent sterilisation of other males, as females return to normal reproductive function. The use of levonorgestrel implants in female kangaroos is reported to give a longer period of reproductive suppression compared with deslorelin [6]. Currently, these implants are significantly more expensive than deslorelin implants (AU$250 *vs.* 60/implant), a factor which will influence cost/% reduction in fecundity/annum. Surgical sterilisation of females, such as by tubal ligation/transection or ovario-hysterectomy, is more invasive than in males and was not considered for financial, ethical and logistic reasons in this program. However, it does offer permanent reproductive control, and is worthy of consideration if financial considerations and other disadvantages are able to be addressed or mitigated. The current programs cost approximately AU$500/year/% reduction in birthrate for GnRH program alone and AU$1050/year/% reduction in birth rate for the deslorelin plus surgery program.

4.2. Animal Welfare Considerations

Anaesthesia and darting of wild animals invariably carry some risk [17]. These risks are associated with:

1) Injury caused by the dart
2) Injury or misadventure that occurs during induction or recovery from anaesthesia
3) Anaesthesia and other veterinary procedures
4) Predation as a result of impairment of appropriate flight or defence responses [18]

In our program, the mortality rate associated with dart injuries was significantly lower in the second management intervention (8% of all deaths in 2007) than in the first (25% of all deaths in 2003). With respect to the total number of animals actually darted, the figures are 0.5% (2007) and 4% (2003). This difference we attribute primarily to the greater accuracy and finer pressure adjustments of the Dan-Inject dart gun compared with the Montech Taipan 2000. In this program, the target kangaroos were used to human presence, their flight distances were short, and therefore darting distances were significantly shorter than could be expected of kangaroos that were not familiar with humans. This factor contributed significantly to the safety and efficiency of darting capture. We suspected an upward drift in flight distance over time. This could be due to the animals that were easier to dart being targeted first. Alternatively, it could arise from a learnt response to the stress of the darting events. Regular excursions to the site with simulated dart firing prior to a darting event would be one possibility to potentially reduce fear responses.

Injury and mortality associated with anaesthetic induction and recovery were significant. Rapid induction of anaesthesia with minimal excitement is desirable as it reduces the risk of injury and misadventure during the period between darting and recumbency and reduces the likelihood of an animal becoming lost when, or if, it takes flight in response to darting. An accurate weight estimate is essential to calibrate the dose. Tiletamine/zolazepam (Zoletil®, Virbac (Australia), NSW) is a preferred anaesthetic agent for darting of kangaroos because of its short average induction time [17,20,21]. However, it has some significant disadvantages, mainly associated with its prolonged and sometimes violent recovery phase. In addition, in our experience, the risk of serious or fatal hyperthermia in kangaroos anaesthetised with Zoletil is significant at ambient temperatures above approximately 24 °C, because of factors such as tachypnoea and exertion or muscular activity during induction and recovery. We therefore recommend that darting of wild kangaroos not be performed when ambient temperatures are expected to exceed 24 °C at any time between anaesthetic induction and complete recovery. Another factor contributing to poor animal welfare outcomes, and sometimes death of kangaroos, was physical injury that occurred during anaesthetic induction and recovery. Those occurring during anaesthetic recovery may be minimised by confining recovering kangaroos to hessian or shade cloth pens supported at each corner by posts protected by polystyrene padding. These provided some support if the kangaroo knocked into the sides during recovery, whilst minimising injuries. We also avoided excessive disturbance or noise during this phase and used removable eye masks to decrease the amount of light when kangaroos were recovering. Avoiding injuries during anaesthetic induction is somewhat more difficult, but may be minimised by a quiet and calm approach to darting, and avoiding disturbance of chasing darted kangaroos during the induction phase of anaesthesia. Various anaesthetic combinations, including the use of the reversible agents medetomidine and xylazine, were tried in our program, but resulted in unacceptably longer and less predictable anaesthetic induction times when used at recommended dose rates [16]. Consequently, we used Zoletil as our primary agent, despite the issues associated with recovery from anaesthetic [22,23]. In our study, recovery time was longer following higher anaesthetic dose rates (Figure 5). Further development of safe and rapid-induction anaesthetic regimes with superior recovery characteristics to Zoletil may significantly reduce morbidity and mortality associated with anaesthetic recovery. Other important considerationsrelatedto high-volume macropod anaesthesia include human safety issues, costs (of anaesthetic drugs and reversal agents) and the limitations of dart volume.

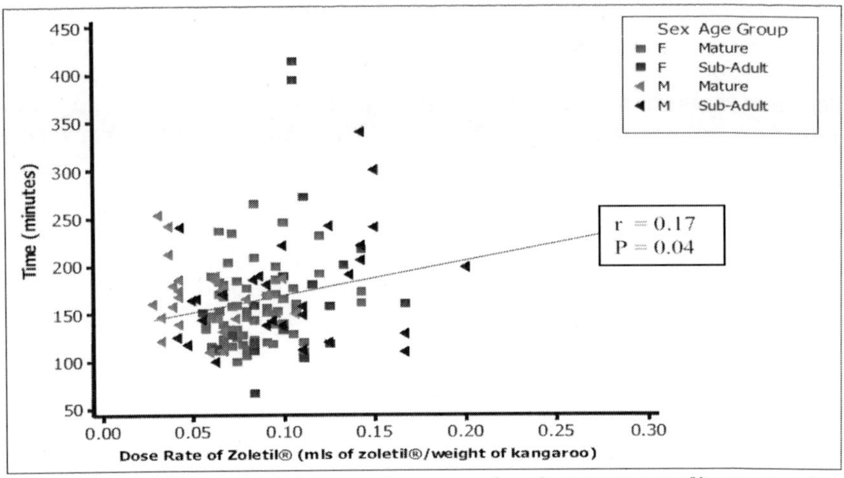

* Kangaroo was able to maintain balance and move away from the recovery area of its own accord.

Figure 5. Time taken to full recovery* from anaesthetic (Zoletil®) during the 2007 intervention at the Pines golf course.

4.2.1. Animal Reaction to Darting

Most kangaroos displayed only a minimal reaction to the impact of the dart, suggesting that darting generally did not inflict significant pain. As this program involved a population of kangaroos highly habituated to close human presence, it is possible that different reactions would be observed if kangaroos were not as habituated to humans or vehicles. The most serious welfare impacts associated with darting and anaesthesia were those associated with recovery from anaesthesia, during which time kangaroos were prone to injury and hyperthermia, which in some instances led to death.

4.2.2. Short Term Behavioural Responses of Kangaroos to Reproductive Procedures

The results demonstrate that some agonistic behaviours, such as fighting and grunting, still occurred in the sub-adult male kangaroos given the orchidectomy procedure 6–8 weeks post surgery. However, compared with the males given the vasectomy procedure and with observations of untreated sub-adult males during the pre-treatment period, the frequency of these behaviours was reduced.

Surgically-sterilised adult male kangaroos did not show a significant difference in agonistic behaviours during the post-treatment period of observation, neither was there any difference between these groups when comparing pre- and post-treatment observations. This may be due to the relatively short time between surgical procedures and the last set of observations, and persistence of significant levels of testosterone in castrated males. This finding is similar to the results of a study of rock hyrax [24], in which adult males given the orchidectomy procedure still displayed agonistic behaviours when approached by other males, but they were never seen to initiate them. Such behavioural inertia is to be expected given the significant time and learning required for the development of the behaviour. We expect that longer-term observations of our population of kangaroos may show a decrease in initiated agonistic behaviours in castrated male kangaroos, although our current (short-term) data do not demonstrate this.

In contrast to our observations, other studies [25,26] suggest that the display of sexual behaviours is infrequent in sub-adult male eastern grey kangaroos and that only dominant males will 'follow and sniff' the females. We observed a decrease in the sexual behaviours of surgically sterilised sub-adult males in the post-treatment period, when compared with the control (pre-treatment) period. The reason is unclear, but it may be a function of the relatively short period of observation post-treatment or low sampling effort. In the adult males given the orchidectomy procedure, sexual behaviours decreased significantly during the post-treatment observation period, and were almost completely absent by the end of the observation period. The plasma levels of testosterone in the male eastern grey kangaroos following this procedure have not been reported in the literature, although it decreases after just 5 days in tammar wallabies (*M. Eugenii*) [27].

Staker [28] found that macropodid males post-vasectomy did not have decreased agonistic behaviours, which is consistent with the fact that vasectomy does not significantly alter the level of circulating androgens in the body. This is consistent with our observations and those of Bolitho *et al.* [29], who reported similar results in male eastern grey kangaroos given a vasectomy.

In the females, the longer time spent walking may indicate a less cohesive social group, reflecting reduced sexual activity in both males and females. The need for an effective perimeter fence is evident, particularly if there is a resident population outside the target population, as in this study. In the females there was also less time spent scanning, possibly for potential mates, and less time self-grooming, perhaps reflecting less tension in the group as a result of reduced sexual activity. These changes were not observed in previous research using deslorelin in kangaroos [30].

Our study targeted the entire population on the golf course, and approximately 90% of animals were actually treated. This appeared effective in controlling population for several years. If a smaller proportion had been treated, or there had been greater incursion into the course for example as a result of a drought, the duration of reproductive control would diminish.

5. CONCLUSIONS

Population management plans involving non-lethal reproductive manipulation are increasingly required to manage landlocked wildlife populations. The development and assessment of effective and humane management techniques are essential if wildlife managers are to appropriately respond to community expectations for compassionate and ethical solutions. Our study demonstrates that although medium-scale kangaroo reproductive management programs are, or can be, effective and achievable, there are nevertheless significant risks to animal welfare that must be mitigated if such programs are to be utilised widely. Key issues affecting animal welfare are generally associated with induction of, and recovery from anaesthesia, during which time kangaroos are susceptible to misadventure and injury. These risks can be mitigated to some degree by minimisation of disturbance and rapid induction of anaesthesia, appropriate confinement during recovery, use of well-trained and experienced personnel, and avoiding darting during warmer months. In spite of the best preparation and precautions we suggest that mortality rates in kangaroos of 5–10% may be expected until better anaesthetic regimes for field darting and chemical restraint become available.

ACKNOWLEDGEMENTS AND APPROVALS

We acknowledge, with thanks, the assistance of the staff and management of "The Pines Golf Course", Sanctuary Cove, the generous support of volunteers from Wildcare Australia, and staff and volunteers of Australia Zoo, Wildlife Warriors Worldwide Ltd.

Permission to access the site was provided by the management of the golf course. The eastern grey kangaroo is protected under the Queensland *Nature Conservation Act 1992*. Under the provisions of this

Act a Damage Mitigation Permit (WIMP04453007) for the reproductive management program itself and a Scientific Purposes Permit (WISP04452507) for the behavioural observations during and after the management interventions were acquired. Animal experimentation ethics approval (SAS/030/07) was provided by the University of Queensland Animal Ethics Committee. All veterinary procedures were conducted by registered veterinary surgeons with wildlife expertise.

REFERENCES

1. Archer, M.; Grigg, G.; Flannery, T. *The Kangaroo*; Kevin Weldon: McMahons Points, NSW, Australia, 1985.
2. Dawson, T.J. *Kangaroos, Biology of the Largest Marsupials*; University of New South Wales Press Ltd.: Sydney, Australia, 1995.
3. Tribe, A. Managing Kangaroos as a Sustainable Resource. In Proceedings of Annual Conference of the Australian Veterinary Association, Sydney, Australia, 12–17 May 1991; p. 66.
4. Adderton Herbert, C. Long acting contraceptives: A new tool to manage overabundant kangaroo populations in nature reserves and urban areas. *Aust. Mammal.* **2004**, *26*, 67–74.
5. Herbert, C.A.; Trigg, T.E.; Cooper, D.W. Fertility control in female eastern grey kangaroos using the GnRH agonist deslorelin. 1. Effects on reproduction. *Wildl. Res.* **2006**, *33*, 41–46.
6. Coulson, G.; Nave, C.D.; Shaw, D.; Renfree, M.B. Long-term efficacy of levonorgestrel implants for fertility control of eastern grey kangaroos (*Macropus giganteus*). *Wildl. Res.* **2008**, *35*, 520–524.
7. Kitchener, A.L.; Harman, A.; Kay, D.J.; McCartney, C.A.; Mate, K.E.; Rodger, J.C. Immunocontraception of eastern grey kangaroos (*Macropus giganteus*) with recombinant brushtail possum (*Trichosurus vulpecula*) ZP3 protein. *J. Rep. Immunol.* **2009**, *79*, 156–162.
8. Nave, C.D.; Coulson, G.; Poiani, A.; Shaw, G.; Renfree, M.B. Fertility control in the eastern grey kangaroo using levonorgestrel implants. *J. Wildl. Manag.* **2002**, *66*, 470–477.
9. Brooks, R.P.; Fleming, M.W.; Kennelly, J.J. Beaver. Colony response to fertility control: evaluating a concept. *J. Wildl. Manag.* **1980**, *44*, 568–575.
10. Davidson, J.M. Characteristics of sex behaviour in male rats following castration. *Anim. Behav.* **1966**, *14*, 266–272.
11. Ramsey, D. Effects of fertility control on behavior and disease transmission in brushtail Possums. *J. Wildl. Manag.***2007**, *71*, 109–116.
12. Zamaratskaia, G.; Rydhmer, L.; Andersson, H.K.; Chen, G.; Lowagie, S.; Andersson, K.; Lundstrom, K. Long-term effect of vaccination against gonadotropin-releasing hormone, using Improvac (TM), on hormonal profile and behaviour of male pigs. *Anim. Rep. Sci.* **2008**, *108*, 37–48.
13. Asquith, K.L.; Kitchener, A.L.; Kaya, D.J. Immunisation of the male tammar wallaby (*Macropus eugenii*) with spermatozoa elicits epididymal antigen-specific antibody secretion and compromised fertilisation rate. *J. Rep. Immunol.* **2006**, *69*, 127–147.
14. Kitchener, A.L.; Edds, L.M.; Molinia, F.C.; Kay, D.J. Porcine Zonae Pellucidae Immunization of Tammar Wallabies (*Macropus eugenii*): Fertility and Immune Responses. *Reprod. Fertil. Dev.* **2002**, *14*, 215–223.
15. Coulson, G. Management of overabundant macropods—Are there conservation benefits? In *Managing Marsupial Abundance for Conservation Benefits*; Austin, A., Ed.; Cooperative Research Centre for the Conservation and Management of Marsupials: Sydney, Australia, 1998.

16. Vogelnest, L.; Portas, T. Macropods. In *Medicine of Australian Mammals*; Vogelnest, L., Woods, R., Eds.; CSIRO Publishing: Collingwood, Australia, 2008; pp. 133–225.
17. Coulson, G. Repertoires of social behaviour in captive and free-ranging grey kangaroos, Macropus giganteus and Macropus fuliginosus (Marsupialia: Macropodidae). *J. Zool.* **1997**, *242*, 119–130.
18. Coulson, G. Repertoires of social behaviour in the Macropodoidea. In *Kangaroos, Wallabies and Rat-Kangaroos*; Grigg, G.C., Jarman, P.J., Hume, I.D., Eds.; Surrey Beatty & Sons: Sydney, Australia, 1989; pp. 457–473.
19. Mcdonald, I.; University of Queensland, Gatton, Australia. Personal observation, 2008.
20. Vogelnest, L. Chemical Restraint of Australian Native Fauna. In *Wildlife in Australia: Healthcare and Management*; Proceedings 327; Post Graduate Foundation in Veterinary Science, University of Sydney: Sydney, Australia, 1999; pp. 149–187.
21. Arnold, G.W.; Steven, D.; Weeldenburg, J.; Brown, O.E. The use of alpha-chloralose for the repeated capture of western grey kangaroos, *Macropus-fuliginosus*. *Austr. Wildl. Res.* **1986**, *13*, 527–533.
22. Mauthe von Degerfeld, M. Personal experiences in the use of Zoletil for anaesthesia of the red necked wallaby (*Macropus rufogriseus*). *Vet. Res. Comm.* **2005**, *29*, 297–300.
23. Pitt, J.; Lariviere, S.; Messier, F. Efficacy of Zoletil for field immobilization of raccoons. *Wildl. Soc. Bull.* **2006**, *34*, 1045–1048.
24. Manharth, A.; Harris-Gerber, L. Surgical castration and the effect on aggression in rock hyrax (*Procavia capensis*). *J. Zoo Wildl. Med.* **2002**, *33*, 80–82.
25. Ganslosser, U. Agonistic behavior in Macropodoids—A Review. In *Kangaroos, Wallabies and Rat Kangaroos*; Grigg, G., Jarman, P., Hume, I., Eds.; Surrey, Beatty and Sons: Sydney, Australia, 1989; Volume 2.
26. Hohn, M.; Kronschnabel, M.; Gansloßer, U. Similarities and differences in activites and agonistic behavior of male eastern grey kangaroos (*Macropus giganteus*) in captivity and the wild. *Zoo Biol.* **2000**, *19*, 529–539.
27. Catling, P.C.; Sutherland, R.L. Effect of gonadectomy, season and the presence of female tammar wallabies (*Macropus Eugenii*) on concentration of testosterone, luteinizing hormone and follicle stimulating hormone in the plasma of male tammar wallabies. *J. Endocrinol.* **1980**, *86*, 25–33.
28. Staker, L. *The Complete Guide to the Care of Macropods: A Comprehensive Guide to the Handrearing, Rehabilitation and Captive Management of Kangaroo Species*; Matilda Publishing: Townsville, Australia, 2006.
29. Bolitho, E.; Coulson, G.; Bricknell, S. Body Size, Dominance and Mating Success in Grey Kangaroos. In Proceedings of the 6th International Behaviour Ecology Congress, Canberra, Australia, 29 September–4 October 1996.
30. Woodward, R.; Herbertstein, M.E.; Herbert, C.A. Fertility control in female eastern grey kangaroos using the GnRH agonist deslorelin. 2. Effects on behaviour. *Wildlife Res.* **2006**, *33*, 47–55.

CHAPTER 18

Characterization of broiler poultry production system in Rwanda

Francis Mbuza[1,2], Rosine Manishimwe[2], Janvier Mahoro[2], Thomas Simbankabo[1] and Kizito Nishimwe[3]

1 Centre for Livestock Research and Development (CLRD), Butare, Rwanda

2 School of Animal Sciences and Veterinary Medicine, Uppsala, Sweden

3 School of Food Science and Technology, College of Agriculture, Animal Sciences and Veterinary Medicine (CAVM), University of Rwanda, Butare, Rwanda

ABSTRACT

A study was conducted on 37 randomly selected broiler poultry farmers in Rwanda to characterize the production system using pre-tested semi-structured questionnaires. The data were processed in SPSS and presented as means, percentages and ranges in tables and text. All respondents kept Cobb breed and young stock was mainly (73 %) imported from abroad. The majority of respondents were males (68 %) and most farmers had attended only primary level of education (40.5 %). Most of the farms were in the peri-urban (48.6 %) and urban (37.8 %) areas and hired male youth (62.2 %) mainly aged 19–35 years. The majority of respondents (68 %) kept less than 500 birds per batch. Recordkeeping was well practiced (91.9 %) and (62.6 %) had permanent poultry houses and all farmers used deep litter system. Purchased feedstuffs were reportedly (92 %) mixed at farm level as the main feed resource. Maize bran was reported (97.06 %) the main, basal feedstuff. The mortality rates of chick and growers were 12.3 and 9.4 %, respectively. The slaughter age was reportedly 60 days with an average dressing percentage of 75.67 %. The main challenges reported were scarcity and unaffordability of quality feeds (59.5 %), lack of market access (45 %) and lack of credit (21 %). The farmers had various views on improving broiler production in Rwanda ranging from establishing feed processing industries 62.2 %, improving marketing facilities 35.1 %, increasing availability of day-old chick and access credit 27 %, to intensification of farmer training 16.2 %.

Keywords: Broiler poultry, Broiler sector, Performance parameters, Challenges, Rwanda

1. INTRODUCTION

Livestock rearing is an important subsidiary occupation that supplements the income of smallholder farm families and rural households in most tropical developing countries. As indicated by Assa (2012), poultry production is the fastest growing component of global meat production, with developing and transitional countries assuming a leading role. The poultry sector all over the world is continuously growing due to increasing of human population, increase in purchasing power and urbanization. However poultry production in Africa is lagging behind that of other continent such us Asia (FAOSTAT 2014).

Poultry possess greater efficiency in converting feed into egg and meat compared to other livestock enterprises. Commercial poultry farms play an important role in meeting national protein supply (Jabir and Hague 2010) through the supply of eggs in addition to poultry meat. Commercial broiler production on the other hand provides mainly poultry meat as birds are raised solely for meat.

But in some African countries, broiler production is becoming important. For example in South Africa, broiler meat production occupies 35 % of all animal products (South Africa Department of Agriculture and Forestry and Fisheries 2012).

In comparison to other eastern Africa countries, poultry meat production in Rwanda is still very low. The total poultry population in 2012 was estimated (NISR 2013) to be 4688 million birds.

Rwanda is a country characterized by a high population density, where the population growth is estimated at 2.9 % per year (Minagri 2012). The poultry sector of Rwanda is one of the cluster areas for value chain development. Poultry are the most predominant in terms of numbers with an estimated population of 4.8 million birds (FAOSTAT 2014). Over 80 % of the small-scale farmers in Rwanda rear chicken, majority of which are indigenous (FAO 2009; NISR 2011).

In order to further alleviate poverty and insure food security in the long run, the efficient use of the available land through poultry and small livestock production could be one of the options for emphasis. The agricultural sector contributes over 40 % of the national Gross Domestic Product and almost 90 % of the export earnings. Livestock including poultry contributes 6 % of the GDP and about 30 % of the agricultural GDP. It therefore calls for an urgent need to increase broiler production at both household and commercial holdings in Rwanda. The Rwandan vision regarding the poultry industry (Minagri 2012) is to make the poultry industry a flagship of the Rwandan livestock in order to increase protein intake and address food insecurity in the country.

In developing, tropical countries like Rwanda, a number of factors are likely to constrain optimum broiler poultry production. In study on 13 factors likely to influence broiler production in Bangladesh (Ali and Hossain 2008), it was found that education, land possession, annual family income, training exposure, broiler farming experience, broiler farm size, capital in broiler farming and extension contact had significant positive relationships with broiler performance.

Due to the lack of grazing land in Rwanda, the scope for development of livestock industries based on large animals is limited. Broiler production is one of viable enterprises under such conditions. Broiler farming is likely to become popular with farmers due to short life cycle and low capital investment (Raha 2007). Chicken meat is acceptable by all sectors of consumers, irrespective of sex and religion.

Broiler can be produced within the least possible time compared to other meat-producing animals. However, broiler production in Rwanda is still low due to some problems during its development, as broiler farmers face a great many input problems. Successful broiler rearing depends on many inputs—such as availability of quality chicks, supply of quality feed, medicine and vaccine and other support services. Each of these factors is vital for profitable broiler farming.

In a recent study sponsored by the ministry of agriculture (Menagerie) of Rwanda (Minagri 2012), the poultry industry of Rwanda was assessed to develop the strategy and investment plan to strengthen the poultry industry in Rwanda within the period 2012–2017. The study concluded that there was great potential for improved poultry production in Rwanda for the local market and possible export to the accessible markets of Democratic Republic of Congo, Burundi and Congo Brazzaville. Whereas this study dwelt on value chains emphasizing marketing, processing and price, it ran short of assessing the production practices of emerging layer and broiler farmers. Unfortunately, little is known about the current situation of the broiler production industry in Rwanda. This study therefore aimed at characterizing the broiler poultry production systems in Rwanda to determine the current status of management, production and marketing practices, identifying the challenges faced by farmers so as to propose improvement and interventions.

2. MATERIAL AND METHODS

The study was done in the four provinces of Rwanda (eastern, northern, southern and western) in addition to Kigali City in the period of 2014–2015. The 37 respondents were chosen through a multistage sampling procedures at province, district and sector levels.

Due to time, financial resources and availability of technical staffs, the sampling rates at different stage were that all provinces were surveyed and 50 % of districts selected from each province except in Kigali City where all the districts were selected. Within the selected district, 10 % of the sectors were selected except in Kigali City where three sectors were selected from each district (Table 1).

Table 1 Selected district and sectors in the study areas

Province	Total districts	Selected districts (50 %)	Total sectors	Selected sectors 10 %
East	7	3.5 ≈ 4	95	10
West	7	3.5 ≈ 4	96	10
North	5	2.5 ≈ 3	89	9
South	8	4	101	10
Kigali City	3	3 (100 %)	35	9

All broiler poultry farmers in each selected sector were interviewed ending up with a total of 37 respondents. The local government staff at sector level guided a numerators to locate the broiler poultry farmers.

A pre-tested semi-structured questionnaire was used by pre-trained academic staff to gather the necessary data from identified broiler poultry farmers. The data were entered in the computer using the software SPSS 16.0 (2009) and processed.

3. RESULTS AND DISCUSSION

3.1 Distribution of respondents among the surveyed districts

The 37 respondents comprised of female 51.3 % and males 68 % as owners and managers of the broiler poultry farms. The education level of the owners was generally basic as 40.5 % of the respondents had only attend primary education while 31.5 % had attend secondary education.

Most farmers were from eastern and western provinces (10 farmers each), followed by Kigali City (9 farmers), and northern and southern provinces with 4 farmers each. Among the districts, the largest number of respondents were from Rubavu in western province which is known to have a high population of poultry farmers whereas Kayonza and Rwamagana in eastern province had only one farmer each. The large population of broiler poultry farmers in western province can be attributed to its proximity to Democratic Republic of Congo (DRC) where a ready market of broiler chicken prevails. Kigali City also had a high population of broiler farmers attributable to existence of super markets, hotels and big restaurants.

Most of the respondents (48.6 %) were from the peri-urban areas while 37.8 and 13.5 % were from the urban and rural areas, respectively. The location of broiler poultry farmers was mainly cross to feeder roads as 43.2 % were within 1 km, 37.8 % were within 1–2 km with only 20 % above 2 km from the feeder road. These results indicate that broiler farmers were aware of the importance of feeder road and urbanization for easy access to farm inputs and marketing of farm produce.

3.2 Gender and education levels of managers

In most cases, managers of broiler poultry farms (75.7 %) were males and most of them had attained primary and secondary level of education. This shows some potential for improved broiler production as level of literacy and numeracy is required for effective management and production of broilers poultry. It is logical that large training exposure with high education and experience makes a farmer better able to do his job properly (Ali and Hossain 2008).

In addition, husbands (78.4 %) were responsible for management of funds accruing form the farm enterprise. The male age group of 19–35 years was mostly responsible for daily farm tasks such as cleaning 64.9 %, treatment 48.6 %, purchasing 48.6 %, selling 45.8 % and feeding 64.9 %. Overall, women of all age groups were rarely involved in broiler poultry farm activities.

In a study in western Kenya (Okitoi et al. 2007), ownership of rural poultry was found to be shared among the family members but was predominantly by women (63 %) and children (18 %). On the other hand, decision-making regarding selling, consumption and gifts to guests in rural poultry reflects plurality. All family members provided labour to a rural poultry production enterprise. Men and children mainly did construction of poultry sheds as women did cleaning, feeding and treatment of rural poultry. Women and children did most of the daily routines in rural poultry management. Men did occasional jobs that were cash requiring such as purchase of inputs and treatment of poultry using conventional drugs. Women did occasional sale of eggs. Women dominated the access and control of food and gifts to guests while men dominated cash and cultural benefits arising from poultry.

In conclusion, ownership of rural poultry and access to benefits was not exclusively the domain of women. Decision-making by women in the rural poultry production system was limited to non-cash-related decisions while cash-related decisions were made mostly by men. The family labour input into the rural poultry production system was a plurality.

3.3 Breeding

Most of broiler poultry farmers (73 %) chose the rearing stock based on the performance (growth late), followed by survival traits, e.g. resistance to disease (Table 2) as cited by neighbouring farmers. The cost of the day-old chicks was also reported to be considered in breed choice criteria.

Table 2 Criteria of choice of breeding stock

Selection criteria	Percentage of farmers who use the criteria
Performance (carcass weight)	73 %
Cost	32.4 %
Resistance	13.5 %

Cobb breed was predominately the preferred breed by all broiler poultry farmers in the study. This breed has very high growth performance potential especial when fed on appropriate rations with proper amino acid regimes (Vieira et al. 2012).

The challenges therefore is whether the broiler farmers in Rwanda have the knowledge and capacity to ensure optimal management of such breeds. Improved exotic chickens produce higher number of eggs and more meat than the indigenous chicken breeds, but the tropical climate is a great challenge. They are not adapted to adverse environmental conditions, such as high temperature, disease and shortage of feed.

Importation (94.6 %) was the commonest mode of securing day-old chicks and most of the respondents (75 %) reported importation of day-old chicks from neighbouring Uganda. Whereas importation ensures access to high-performing genotypes from parents stocks abroad, it can be very costly to farmers as day-old chicks become expensive and often reach the farmers when they very tied. In this study, the average cost of day-old chicks was 800–1000 Ruff equivalent (1.14–1.42 USD per chick) at date of the study.

The average flock size per farm varied from province to province and district to district. Within a province, there was great variation between the minimum and maximum flock size which is a feature indicative of an evolution process from small substance farms to medium semi-commercial and commercial production farms.

It is noteworthy that most broiler poultry farmers (68 %) kept less than 500 birds per batch (Table 3) and only a few (22 %) kept more than 1000 birds per batch. This could be attributed to minimal availability of space, funds, day-old chicks or other resources to enable perches of a large number of chicks in a single batch. With such small flocks, it is doubtable weather farmers can make sufficient profit margin for social economic developments. In Bangladesh (Kawsar et al. 2013), profitability of small-scale broiler farming was found to depend on several factors, of which FCR had the highest impact (−0.711) and profitability increased with the size of the farm.

Table 3 Number of reared birds per batch

Range of the number/batch reared birds	Frequency	Percentage
Less than 100 birds	8	22 %
100–500 birds	17	46 %
600–1000 birds	4	11 %
More than 1000 birds	8	22 %
Total	37	100 %

These results are similar to those of a study done in Nigeria (Emaikwu et al. 2011), where it was found that majority of the respondents (83 %) had broilers birds ranging from 100 to 500 followed by 13 % in the range of 501 to 1000 broilers while 4 % had a flock size in the range of 1001 to 1500 broilers.

3.4 Farmer training in broiler poultry management

Broiler poultry farmers lacked extension services as only 35 % of the respondents reported having previous training in poultry management. Extension services are not only important to facilitate adoption of an innovation but also a necessary tool for success due to ease of implementation of innovation for profitability (Ali and Hossain 2008).

3.5 Categories of labour on broiler poultry farms

Family labour was found to play a significantly low role on the surveyed broiler farms as majority (62.2 %) of the farms used hired labourers on full time or occasionally, only 21.6 % were using family workers and 16.2 % were using hired and family labours concurrently.

This resembles the situation cited in Kaduna state, Nigeria (Emaikwu et al. 2011) were the majority of the respondents (87 %) employed both family and hired labour in their production process while 13 % of them employed purely family labourers. None of the respondents in Nigeria exclusively employed hired labour.

3.6 Keeping of farm records

Record keeping was a common practice as broiler poultry (91.9 %) farmers reported keeping records especially those of feed usage. The records were most reportedly kept in farm record books (56.8 %), soft copies (17 %) and the rest on loose papers, wall and human memory.

3.7 Housing

Permanent poultry promises were gaining ground as 63 % of broiler poultry farmers had constructed permanent structures (concrete floor, brick walls and iron sheet roofing). It was also noted that most of the poultry houses (91.9 %) were correctly oriented in the east-west direction and their distance was always less than 100 m from homestead. Most farmers however were not well conversant with the required levels of ventilation in poultry houses as only 5.4 % had poultry houses with the required ventilation levels of over 40 % of the wall area.

3.8 Stocking rate in broiler poultry pens

Whereas the recommended average area per chick less than 4 weeks is 0.01–0.03 and 0.07 m^2 for growers (Arbor 2009), the results showed that most broiler farms (46.8 %) were understocking. The situation was the same for other age groups of more than 4 weeks as most farmers (91.4 %) were providing space more than 0.05 m^2 per bird. Such excessive space allowance per bird may be wasteful and counterproductive as birds spend more energy meandering and there is loss of efficiency due to space underutilization.

3.9 Type of material used in the deep litter system

Most of all assessed broiler farms (94.6 %) in the study were using the deep litter system and saw dust was the most used material for litter (91.5 %). This results is similar to that found in Ghana (Anang et al. 2013) where the method of rearing the birds was the all-in all-out method. Thus, birds were kept in the same housing from day-old till they were disposed of farmers using the deep litter system because it was cheaper and easier to operate compared to the battery cage system.

3.10 Feedstuffs, feeds and feeding

The results showed that majority of broiler poultry (92 %) farmers just bought feedstuffs mainly from commercial supplies and mixed the feeds at farm level. Only 8 % of broiler farmers purchased premixed commercial feeds. Maize bran was the main basal feedstuff used (97.06 %) in farm made broiler poultry rations followed by rice bran (35.29 %) and wheat bran (17.65 %).

With regard to protein sources, cotton seed cake and fish meal (79.41 %) were the most commonly used followed by soy bean meal (67 %); bone meal was most reported as sources of minerals (70.59 %) followed by lake shell (29.41 %).

It is notable that most of the feedstuffs such as cotton seed cake, fish meal, lake shells and others were being imported since the raw materials are not produced in Rwanda or are produced in very small quantities. Over dependence on importation of feedstuffs is likely to increase the cost of the final feeding rations and thus the need for encouraging the use local available possible substitute of the imported feedstuffs evil leguminous tree forage can substitute imported cotton seed cake.

USAID (2008) cited by Ali and Hossain (2008) reported that high price of feed was a major problem in Bangladesh due to raw materials (maize, wheat and soya meal) being imported from other countries. In Ghana (Anang et al. 2013), feed cost was also reported to be the most important cost item followed by chick cost and labour cost. Together, these accounted for 90 % of the total cost of broiler production.

In Rwanda there is growing competition between livestock and humans for maize and its by-products. However, the maize yield per hectare is too low to meet the needs of the ever growing human and livestock population. This situation is enough to justify the exploration of the usefulness of other carbohydrate-rich feedstuffs as maize replacers in feed formulations. The possible maize replacers included cereal and by-product, roots and tubers, fruits and by-product, etc.

In most developing countries like Rwanda, the most commonly used protein sources for animal nutrition are by-products of oil seed and fish meal, all of which are very expensive and scarce. In a recent study in Zimbabwe (Gadzirayi et al. 2012), it was found that dried forage leaf of Moringa oliefera as protein supplement in broiler diet at 25 % inclusion level produces broiler of similar 274 weight and growth late compare to those fed under conventional commercial feeds. In this 275 way, poultry rations will become more cost effective than hitherto.

3.11 Daily quantity of feed and frequency of feeding for the different age categories

The daily quantity of feed given to broiler birds at different age categories (Table 4) was general at the top end of the recommended range according to Arbor 2009 and the frequency of feeding was less ad libitum feeding.

Table 4 Daily quantity (gs/per bird) and frequency of feeding in broiler farms as compared to the expected standards (Arbor 2009)

	Mean	Standard deviation	Recommended range (Arbor 2009)
Quantity of feed DOC (g/day/chick)	35.2	40.7	12–35 g/chick/day
Frequency of feeding DOC/day	2.7	0.86	Adlib
Quantity of feed 4–8 weeks chicken (g/day/chick)	135.0	47.5	35–135 g/bird/day
Frequency of feeding 4–8 weeks chicken/day	2.5	0.70	Adlib

3.12 Health management on broiler poultry farms

The majority of the respondents (83.8 %) reported read access of veterinary services whenever required and most of them (58 %) preferred private veterinary service providers. Biosecurity measures such as antiseptic foot bath were available at 51.4 % of broiler poultry farms. However, most of the farmers (62.2 %) did not properly clean the poultry premises as they never disinfected the pens before introducing new batches of day-old chick. This could partly explain the reported high mortality rates (14 %) of broiler chicks before 4 weeks while after 4 weeks it drops to 9 %. In Ghana (Anang et al. 2013), a study showed that mortality rate of broiler birds averaged 17 % and the reliability of the source of day-old chicks was found to be important in ensuring low mortality rates.

3.13 Production parameters on broiler poultry farms

The results revealed inefficiency in the production systems as most of the parameters were generally lower than the recommended levels (Table 5). The reported mean age at slaughter of 60 days ± 14.9 is likely to be uneconomic to the farm. The results revealed a dressing percentage (75.67 % ± 16.5) which is inferior to that reported by Kalio and Okafor (2012) in Nigeria, where the dressing percentage was 78.41 % in a semi-intensive rearing system and 81.78 % in an intensive rearing system. The high age at slaughter indicates inefficiency as it leads to unnecessary expenditure on feeds and other management costs, and it delays restocking.

Table 5 Production parameters of broiler poultry on surveyed farms

	Mean	Standard deviation
Age at slaughter (days)	60	14.9
Average live weight of broilers at slaughter (kg)	2.4	0.76
Average carcass weight (kg)	1.98	0.66
Dressing percentage (%)	75.7	16.5

3.14 Broiler marketing

Three modes of marketing were identified. Contract sale was reported as the most common mode of selling for the birds with less than 8 weeks of age (48.2 %). For chickens with more than 8 weeks of age, direct sale at the market place was reported as the most common mode (41.4 %). Direct sale at the farm gate was least practice (24.1 %) for both birds' categories below and above 8 weeks.

4. CHALLENGES REPORTED BY BROILER POULTRY FARMERS

Broiler poultry farmers mentioned many challenges of which lack of quality feeds (59.5 %) and poor access to markets (45.9 %) were most cited. Prevalence of poultry diseases (32.4 %) and lack of training on modern poultry production practices (8.1 %) were also cited as challenges. These may indicate poor service provision from private companies that supply feeds, credits, etc. These results resemble those recently found in Ethiopia (Yemane et al. 2016) where the high price of feed, shortage of land, unavailability of pullets in time, high cost of pullets, feed quality, shortage of water, unavailability of feed in the nearby area, marketing difficulties during selling, health problem, lack of access to credit and inadequate training were the major constraints in small-scale intensive urban poultry production.

In Ghana (Anang et al. 2013), inadequate finance was identified by broiler poultry farmers as the most critical constraint limiting farmers' ability to carry out management practices like feeding and diseases control as well as the purchase of day-old chicks. The broiler poultry farmers proposed suggestions for improving poultry production in Rwanda (Table 6) with the major one being the establishment of feed processing industries.

Table 6 Suggestions from respondents to improve broiler production

Suggestion	Percentage of famers that stated the proposition
Establish feed processing industries	62.2 %
Increase availability of day-old chicks	27 %
Improvement of market facilities	35.1 %
Increase access to credit facilities	27 %
Training	16.2 %

5. RECOMMENDATIONS

The entire poultry value chain of Rwanda require special attention in the areas of research extension and developments to propel it to international standards. Poultry farmers need training in all aspects of production and management such as feeding, breeding, housing, health and entrepreneurship. Special attention should be put on developing the national animal feeds industry using the supply chain approach. Alternative sources of poultry feedstuffs should be identified, evaluated and commercialized. Poultry farmers should also be encourage to form production and marketing farmer groups or cooperatives. Adaptive research is required for testing and evaluating improved technologies from abroad to adapt them to the Rwandan situation. Effective linkages should be developed and strengthen between poultry producers and financial institutions to enable easy access to credit facilities.

A national poultry policy should be in place to improve the organization of production and marketing, allowing increase in stability and security of poultry output throughout the year. In addition, efforts should be taken to ensure safety standards of poultry meat for human consumption. Experts from the government, research institutes, universities, NGOs and other relevant sectors need to work in a collaborative manner in order to allow sustainable production and fight challenges jointly whenever they arise. Corresponding attention to research and development will allow the development of poultry sector in Rwanda. As government funding is limited, poultry industries need to come forward either to establish their research facilities or to provide funds to universities and research institutes in order to undertake research works of national and international importance.

ACKNOWLEDGMENTS

The Kingdom of the Netherlands is grateful acknowledged for the financial support through The Netherlands Initiative for Capacity development in Higher Education (Nuffic project, NICHE RWA173) that enabled this study. In the same vein, the Government of Rwanda is also recognized for the immense support through the College of Agriculture Animal Science and Veterinary Medicine (CAVM) of the University of Rwanda.

REFERENCES

1. Ali M S, Hossain M M, 2008. Factors influencing the performance of farmers in broiler production of Faridpur District in Bangladesh
2. Anang B T, Anthony A A, Cosmos Y, 2013. Profitability of broiler and layer production in the bring Afro region of Ghana. ARPN Journal of Agricultural and Biological Science VOL. 8, NO. 5, ISSN 1990–6145
3. Arbor, A, 2009. Broiler nutrition supplements Aviagen Incorporated Cummings Research Park 5015 Bradford Drive Huntsville, AL 35805 USA https://www.google.com/search?sclient=psy-ab&client=firefox-b
4. Assa M. M. 2012. Poultry production and rural poverty among small-scale farmers in Zumba District of Malawi. Livestock Research for Rural Development. Vol. 24, Article # 177. Retrieved December 17, from http://www.lrrd.org/lrrd24/10/assa24177.htm.
5. Emaikwu K. K, Chikwendu D. O. and Sani A. S, 2011. Determinants of flock size in broiler production in Kaduna State of Nigeria.
6. FAO (2009). Revue du secteur avicole Revue du secteur avicole. Retrieved from. Rome Italy. http://www.fao.org/avianflu/en/poultryproduction.html
7. FAOSTAT, 2014. Livestock production, poultry meat production. http://faostat.fao.org/site/569/default.aspx#ancor
8. Gadzirayi C.T, Masamba B, Mupangwa J.H, S. washaya, 2012. Performance of broiler chicken fed on mature Moringa oliefera leafs meal as protein supplement to soya bean meal.
9. Jabir F., Hague M. T. 2010. Study on production performance of ISA Brown strain at Krishibid Firm Ltd. Trishal, Mymensingh. Bangladesh Research Publications Journal. 3(3): 1039–1044. January–February 2010. Retrieved August 02, 2016
10. Kalio G. A. and Okafor, B. B., 2012. Response of Broilers to Two Management Systems of Housing in Etche Local Government Area of Rivers State, Nigeria. Asian Journal Agriculture and Rural Development, Vol. 2, No. 2, pp. 184–188.
11. Kawsar M.H, Chowdhury S.D. Raha., S.K, Hossain M.M., 2013. An analysis of factors affecting the profitability of small-scale broiler farming in Bangladesh
12. Minagri (2012). Strategic and investment plan to strengthen the poultry industry in Rwanda. Ministry of agricultural (Menagerie) Kigali, Rwanda
13. NISR. (2011). Integrated household living conditions survey .EICV3 thematic report agriculture. Kigali, Rwanda
14. NISR, 2013. Statistical Yearbook, Rwanda. National Institute of Statistics of Rwanda. file:///C:/Users/Rosine/Downloads/Rwanda_Statistical_Year_Book_2013.pdf on the Performance and Egg Quality of Layers, Nigeria. Production and Health, Ladoke Akintola University of Technology, P.M.B 4000. http://www.bdresearchpublications.com/admin/journal/upload/09115/09115.pdf.
15. Okitoi L O, Ondwasy H O, Obali M P , Murekefu F, 2007. Gender issues in poultry production in rural households of Western Kenya Poultry Show, Seminar 2007. World's Poultry Science Association. Bangladesh Branch.
16. Raha S K. (2007) Broiler industry in Bangladesh: some issues. International
17. South Africa Department of Agriculture, forestry and fisheries, 2012. A profile of the South African broiler market value chain. Directorate Marketing- Department of Agriculture, forestry and fisheries South Africa. Private Bag X 15 Arcadia 0007.
18. SPSS 16.0 2009 statistical package for social sciences Version 16 https://en.wikipedia.org/wiki/SPSS
19. USAID (2008) Telling Our Story: online news of USAID on 1st June 2008. Web http://www.usaid.gov/stories/Bangladesh/fp_bangladesh_poultry.html.
20. Vieira L, Taschetto D, Angel C. R, Favero A., Mascharello N. C, Nogueira E, 2012. Performance and carcass characteristics of Cobb × Cobb 500 slow feathering male broilers fed on dietary programs having stepwise increases in ideal protein density, Nigeria.

21. Yemane N, Tamir B, Mengistu A, 2016. Constraints, opportunities and socio-economic factors affecting flock size holding in small scale intensive urban poultry production in Addis Ababa, Ethiopia. Agriculture and Biology Journal of North America, pp. 2151–7525. doi:10.5251/abjna.2016.7.3.146.152

CHAPTER 19

Environmental predictors of bovine *Eimeria* infection in western Kenya

D. N. Makau, G. K. Gitau, G. K. Muchemi, L. F. Thomas, E. A. J. Cook, N. A. Wardrop, E. M. Fèvre, W. A. de Glanville

[1] *Department of Public Health Pharmacology and Toxicology, Faculty of Veterinary Medicine, University of Nairobi, Nairobi PO BOX 29053-00625, Kenya*

[2] *Department of Clinical Studies, Faculty of Veterinary Medicine, University of Nairobi, Nairobi PO BOX 29053-00625, Kenya*

[3] *International Livestock Research Institute, Old Naivasha Road, PO Box 30709-00100, Nairobi, Kenya*

[4] *Centre for Immunity, Infection and Evolution, Institute for Immunology and Infection Research, School of Biological Sciences, Kings Buildings, University of Edinburgh, West Mains Road, Edinburgh EH9 3JT, UK*

[5] *Geography and Environment, University of Southampton, Highfield Campus, University Road, Southampton SO17 1BJ, UK*

[6] *Institute of Infection and Global Health, University of Liverpool, Leahurst Campus, Neston CH64 7TE, UK*

ABSTRACT

Eimeriosis is caused by a protozoan infection affecting most domestic animal species. Outbreaks in cattle are associated with various environmental factors in temperate climates but limited work has been done in tropical settings. The objective of this work was to determine the prevalence and environmental factors associated with bovine *Eimeria* spp.

infection in a mixed farming area of western Kenya. A total of 983 cattle were sampled from 226 cattle-keeping households. Faecal samples were collected directly from the rectum via digital extraction and analysed for the presence of *Eimeria* spp. infection using the MacMaster technique. Individual and household level predictors of infection were explored using mixed effects logistic regression. The prevalence of individual animal *Eimeria* infection was 32.8% (95% CI 29.9–35.9). A positive linear relationship was found between risk of *Eimeria* infection and increasing temperature (OR = 1.4, 95% CI 1.06–1.86) and distance to areas at risk of flooding (OR = 1.49, 95% CI 1.17–1.91). There was weak evidence of non-linear relationship between *Eimeria* infection and the proportion of the area around a household that was classified as swamp (OR = 1.12, 95% CI 0.87–1.44; OR (quadratic term) = 0.85, 95% CI 0.73–1.00), and the sand content of the soil (OR = 1.18, 95% CI 0.91–1.53; OR (quadratic term) = 1.1, 95% CI 0.99–1.23). The risk of animal *Eimeria* spp. infection is influenced by a number of climatic and soil-associated conditions.

Keywords: Eimeria, Bovine, Tropical, Environmental, Extensive production systems

1. INTRODUCTION

Eimeriosis affects a wide range of animal species and is caused by infection with protozoa in the *Eimeria* genus, previously known as *Coccidia*. Eimeriosis in cattle is characterised by diarrhoea, fever, anorexia, weight loss, emaciation and sometimes death, particularly in young animals (Coetzer and Justin 2004). Transmission occurs following ingestion of sporoblasts in the environment. These undergo the first and second stages of schizogony in the small and large intestines, respectively, before gametogony in the colon. Oocysts are then passed in faeces to the environment where sporogony takes place and infection of another host can occur after ingestion of these oocysts with sporoblasts (Urquhart et al. 1996). Outbreaks of the disease have previously been found to be associated with environmental stressors, including low temperatures during cold seasons (Rodríguez et al. 1996; Maas 2007). A number of studies have also looked at management-related factors influencing *Eimeria spp.* infection in cattle. These include housing system, feeding system, watering system, floor type and herd size (Khan et al. 2013). Environmental conditions in cattle sheds, which influence the survival and sporulation of infective oocysts, have also been shown to influence risk of infection (Lassen et al. 2009).

While management factors and very cold weather conditions have been shown to be associated with increased clinical disease in herds in temperate settings, there has been relatively little research on the effect of environmental factors on the epidemiology of eimeriosis in the tropics. In a study done in Zimbabwe to assess the effects of season on gastrointestinal parasitism in cattle, Pfukenyi et al. (2007) reported that high rainfall was significantly associated with high OPG counts of *Eimeria* parasites. In another study in South Africa, Matjila and Penzhorn (2002) reported that the prevalence of *Eimeria* infection was highest in dairy farms that were nearest to the Pienaars River.

Soil composition has also been observed to influence oocyst recovery rate and identification (Lélu et al. 2011), with similar effects observed for a number of other protozoal parasites such as *Cryptosporidium* and *Toxoplasma* (Lélu et al. 2011).

The aim of this study was to determine the prevalence of bovine *Eimeria* spp. infection in cattle in a small-holder farming area of western Kenya and to assess individual and herd level factors associated with infection, with a particular focus on identifying environmental predictors.

2. METHODS

2.1 Study area

All data were collected as part of the People, Animals and their Zoonoses (PAZ) project between August 2010 and July 2012 (Doble and Fevre 2010). The study area was in Busia, Bungoma, Siaya and Kakamega counties in western Kenya (Fig. 1). This is an area with a predominantly small-scale, mixed-farming system, in which livestock production is integrated with agricultural production. The median cattle herd size is five animals of predominantly short-horn zebu breed, with animals typically managed using a mixture of tethered and free grazing systems on communal grazing lands (Bronsvoort et al. 2013).

2.2 Ethical approval

Ethical approval for the animal elements of the study was provided by the Animal Welfare and Ethical Review Body of the Roslin Institute, University of Edinburgh, UK (AWA004).

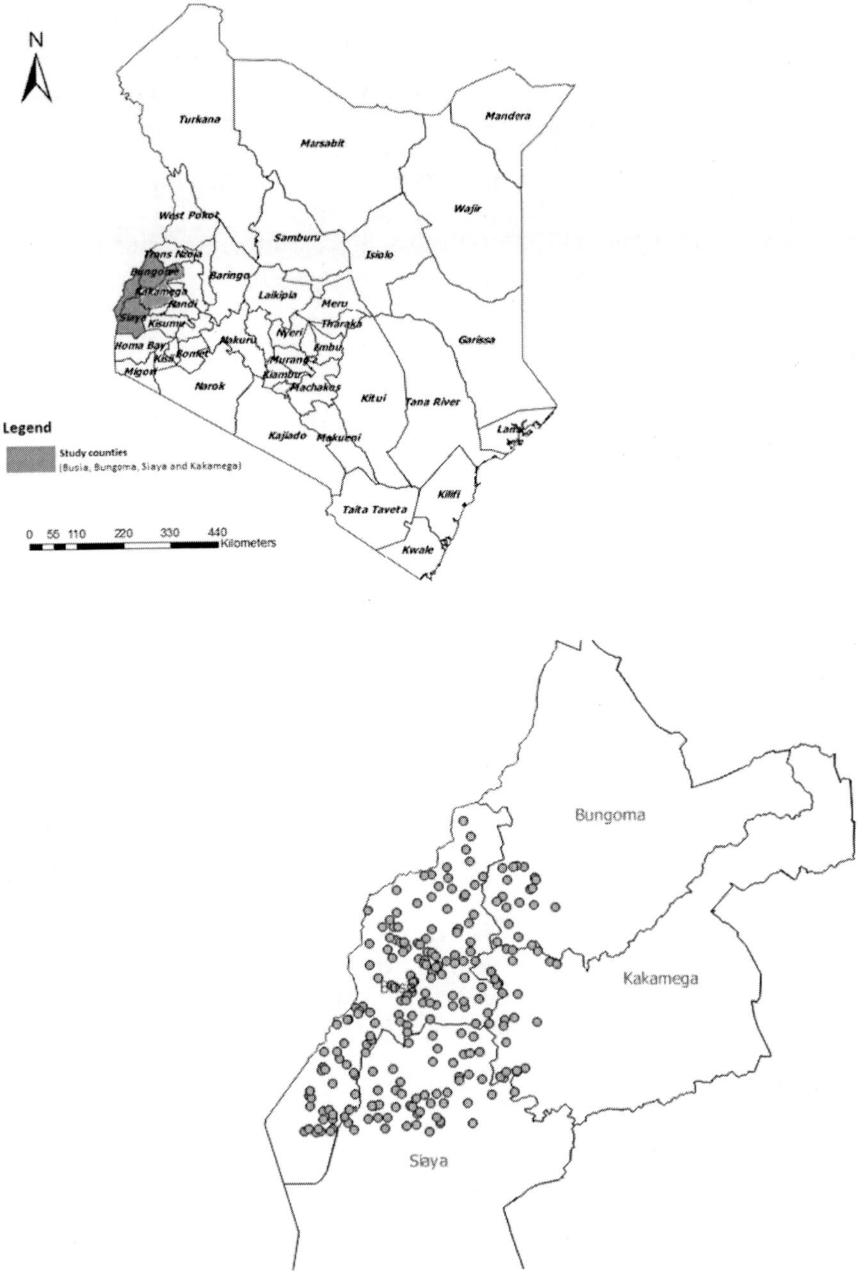

Figure. 1 Study area in western Kenya, including location of sampled households

2.3 Sample collection

The PAZ study was focused on zoonotic pathogens in people and animals and was designed to detect the lowest expected prevalence of 5% in the bovine population with 2% standard error at the 95% confidence interval and a design effect of 3. A multistage sampling approach was adopted, with households randomly selected within sublocations, the smallest administrative unit in Kenya. All sublocations within 45 km of the study field station, located in Busia town, were included. Between 1 and 8 households were selected in each sublocation, with the number selected proportional to the cattle population present in the sublocation. Household selection involved the generation of an appropriate number of random points in each sublocation using a geographic information system (ArcMap 9.1, ESRI, Redlands CA). The closest consenting household that was within 300 m of the actual location of the random point was then selected for inclusion. A back-up point was used when there was no household within 300 m. Both livestock and non-livestock keeping households were included in the PAZ study. In each randomly selected household, a comprehensive household-level questionnaire was administered and data on herd profile, husbandry and management were obtained. All cattle within selected households were sampled, excluding those in the last trimester of pregnancy. Faecal samples were collected directly from the rectum via digital extraction. All faecal samples were analysed using the MacMaster technique (Levecke et al. 2012) to determine the presence or absence of *Eimeria* spp. oocysts. An individual was considered positive if oocysts were observed under microscopic examination.

2.4 Predictors of *Eimeria* infection

Environmental predictors of *Eimeria* spp. infection considered a priori to be important in this setting were the proportion of a 1 km area around a household that was classified as crops and grassland; the proportion of the same area around a household classified as swamp land; household distance from water bodies (Wardrop 2015); distance from areas at risk of flooding; average maximum green vegetation fraction (MGVF) (Broxton et al. 2014); average temperature; average precipitation (Hijmans et al. 2005); and soil sand content (Hengl et al. 2013). Environmental data were derived for the study area from a range of sources, as described by Wardrop et al. (2015). The average and range of values for these environmental variables are presented in Table 1. Sampling for PAZ was carried out continually for 2 years, and to explore rainfall mediated seasonal effects on *Eimeria* infection, we also derived values for the average rainfall measured in the 30 days prior to the sampling visit to a household from the nearest weather station with consistent data (Kisumu, latitude = −0.09, longitude = 34.73, www.wunderground.com). Household geographic position and altitude were derived using a handheld GPS (Garmin eTrex). Household grazing system (zero or tethered grazing vs. free grazing) was extracted from the household questionnaire. Age based on dentition (dichotomised to <18 months and ≥18 months), sex and breed (local zebu vs. pure breed exotic dairy cattle and their crosses) were included as predictors at the individual level. All continuous variables were examined for outliers, and the different scales of each were standardized by subtracting the mean and dividing by the standard deviation. Correlation between continuous variables was assessed using Spearman's rank test, with one of a highly correlated pair (≥0.7) excluded based on an assessment of the biological importance of each of a correlated pair. Correlations were explored using base commands in *R* version 2.1.1. (http://cran.r-project.org/).

Table 1 Summary of environmental determinants analysed in four counties in western Kenya

Factor	Average	Range
Percentage crop or grass land in 1 km area surrounding household	60.68%	35.83–78.10%
Percentage swamp land in 1 km area surrounding household	7.48%	0–36.41%
Soil sand content	42.39%	22–51%
pH of surface water	5.6	5.2–6.3
Temperature	22.1 °C	21.2–22.8 °C
Distance to water body	2826 m	88–9305 m
Distance to areas at risk of flooding	985 m	0–5636 m
Average MODIS-based maximum green vegetation fraction (MGVF).	86%	77–93%
Altitude	1225 m	1127–1405 m
Precipitation	1532 mm	997–1837 mm
Average rainfall in the previous 30 days	1.3 mm	0–6.0 mm

2.5 Covariate relationships with *Eimeria* infection

Potential non-linear relationships between the log odds of *Eimeria* infection and each continuous covariate were explored using the method described by Hosmer et al. (2013). For this, a categorical variable with four levels was created using quartile cut-points. A logistic regression model was then fit replacing the continuous predictor with its four-level categorical derivative, and the resulting coefficients plotted versus the mid-points of the upper three quartiles (with the lowest used as reference, and given a value of 0). Where the relationship was obviously linear, the continuous form of the variable was used for inference. Where a logical parametric shape (e.g. quadratic) could be observed, made biological sense, and resulted in model improvement based on Akaike's information criterion (AIC), the variable was transformed. Where the relationship appeared more complicated, a restricted cubic spline (RCS) with three knots was used. If the resulting fitted shape made biological sense and resulted in lower AIC when compared to the model containing the variable in its linear form, the RCS (defined using default cut-points (Harrell 2001, 2015) was used for inference. Restricted cubic splines were fit using the *rms* package in *R* (Harrell 2015).

Continuous environmental variables transformed according to this approach were included together with categorical predictors of infection (age, sex, breed, herd management) in a full logistic regression model that included household as a random effect. The model was fit using the *lme4* package in *R* (Bates et al. 2015). Model simplification was performed using the purposeful selection procedure described by Hosmer et al. (2013). Variables were removed from the full model in a step-wise manner if they were not significant and not a confounder. The significance threshold was set at the 0.1 alpha level, and confounding was assessed on the basis of a 20% or greater change in the value of the coefficient of any of the remaining variables following removal (Bursac et al. 2008). The final model was assessed for evidence of multicollinearity using variance inflation factors.

2.6 Spatial analysis

A spatial scan statistic using a weighted normal model was implemented in SatScan version 9 (www.satscan.org) to detect spatial clustering in *Eimeria* infection risk in the study area. The statistic was estimated using household level residuals extracted from a null logistic regression model containing only the intercept and household level random effect. The model was weighted using the number of cattle sampled in each household (Alton et al. 2013). A normal model was used with 999 iterations (allowing estimation of *p* values down to 0.001), and a cluster size up to a maximum of 50% of observations. Household level residuals from the final logistic regression model derived from the purposeful selection procedure described above were also examined for evidence of residual spatial autocorrelation using spline correlograms using the *ncf* package (Bjornstad 2016).

3. RESULTS

3.1 Summary statistics

A total 416 households were recruited into the PAZ study. Of these, 226 had cattle and 983 individual animals were included. From these, 93% of animals had faecal samples collected. The study population was 66% female while Zebu breeds and short horn zebu crosses made up 77.4%. A total of 295 (30.0%) of the animals sampled were less than 18 months.

The prevalence of individual animal *Eimeria* infection was 32.8% (95% CI 29.9–35.9%). The median oocysts per gram (OPG) was 200 while the average OPG was 1447.

3.2 Model selection

High levels of correlation were observed between altitude and temperature ($r = 0.97$). Since the effect of altitude on *Eimeria* risk could be expected to be mediated through temperature-associated effects on oocyst sporulation, temperature was retained in the model.

The inclusion of a quadratic term improved the fit (based on AIC, data not shown) of the proportion of the area around a household that was classified as swamp and the sandy content of soil. All other environmental predictors were included in their linear form.

The results of the final multivariable logistic regression model selected using purposeful selection are presented in Table 2. Several environmental variables were found to be associated with the risk of *Eimeria* infection. There was a strong positive effect of both temperature and distance to flooded areas on risk of infection. There was also some evidence of a curvilinear relationship between the risk of *Eimeria* infection and the proportion of the 1 km area around a household that was classified as swamp land as well as with the sand content of the soil. The average marginal predicted probabilities of infection based on a range of values of each of these continuous variables are presented in Fig. 2. There was strong evidence of a large effect of age on risk of *Eimeria* infection, with animals older than 18 months being at substantially reduced risk. There was weaker evidence that male animals were at greater risk of *Eimeria* infection. The distance to a water body had a large confounding effect on the coefficient for temperature and was therefore retained in the model. Variance inflation factors were all below four.

Table 2 Outputs from the multivariable logistic regression model

Predictor	Odds ratio	95% confidence interval	*p* value
Age (>18 months)	0.3	0.21–0.43	<0.001
Male sex	1.35	0.96–1.90	0.088
Soil sand content	1.18	0.91–1.53	0.2
Soil sand content^2	1.1	0.99–1.23	0.07
Swampy land composition	1.12	0.87–1.44	0.38
Swampy land composition^2	0.85	0.73–1.00	0.05
Distance to water bodies	0.85	0.68–1.06	0.14
Distance to flooded areas	1.49	1.17–1.91	0.001
Temperature	1.4	1.06–1.86	0.02

3.3 Spatial clustering

There was no evidence of clustering of high or low values of household level residuals based on a weighted normal spatial scan statistic. The 95% confidence interval for the spline correlogram included zero up to distances of 10000 m, indicating no evidence of residual spatial correlation in household level residuals (Fig. 3).

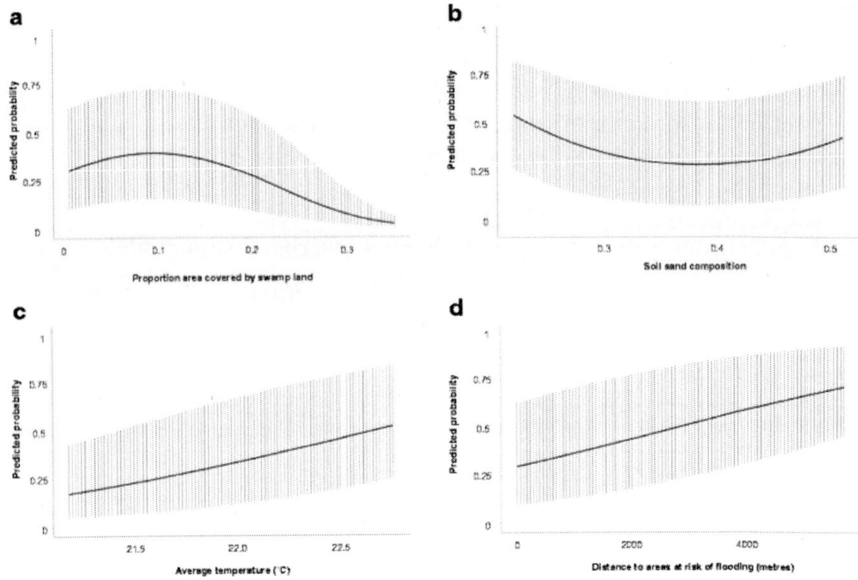

Figure. 2 Predicted average marginal probability of infection with *Eimeria* at a range of standardized values of the identified environmental risk factors

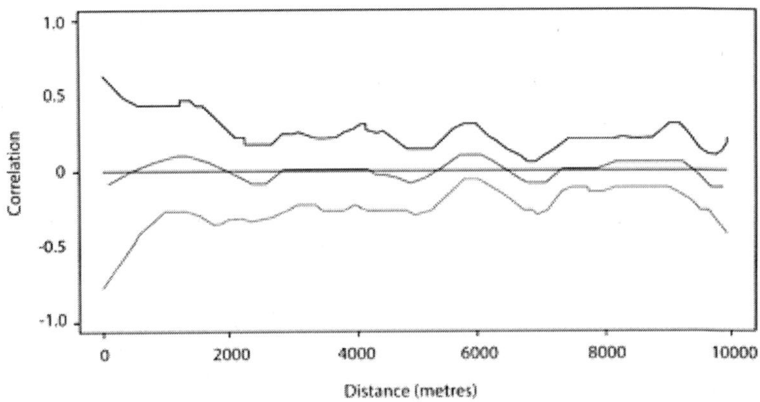

Figure. 3 Spline correlogram comparing values in household level residuals over a range of distances

4. DISCUSSION

This study explores the prevalence of bovine infection by *Eimeria* species and associated risk factors in western Kenya, in a tropical region broadly representative of smallholder livestock production systems in the Lake Victoria basin. The overall prevalence observed was 32.7% (95% CI 29.7–35.7) while the prevalence in calves (<18 months) was 48.2% (95% CI 41.8–54.6). The observed prevalence was lower than what has been documented in other studies in similarly extensive and semi-intensive production systems. Munyua and Ngotho (1990) estimated the prevalence of bovine *Eimeria* infection in Kenya at 67.4%. Pandit (2009) recorded a prevalence of 75.8% in calves in extensive and semi-intensive production systems in Kashmir valley in India while Asfaw et al. (2016) reported a prevalence of 71.7% in calves kept in semi-intensive systems in Ethiopia.

In this study, age was the only individual animal level factor for which there was strong evidence for an effect on *Eimeria* prevalence, with animals older than 18 months being less likely to be infected. This finding has been widely observed (Lassen et al. 2009; Bangoura et al. 2012; Dong et al. 2012; Rehman et al. 2012), and it is likely that an age-related immune effect occurs for this parasite (Fiege et al. 1992). Protective immunity is generally dose dependent and is usually against infection with a homologous species (Rose 1971).

Outdoor husbandry management practices have been documented to reduce the risk of *Eimeria* infection in cattle (McAllister 2007; Rehman et al. 2011). This is expected to be related to differences in the survival of the oocyst in outdoor and indoor environments, as well as the effects of animal density and levels of environmental contamination, which are likely to be highest in closed housing systems (Asfaw et al. 2016; Khan et al. 2013). Animal stress, which tends to be higher in housed animals, may also be important influence on shedding (Priti et al. 2008). In our study, there was no evidence of an association between infection and husbandry practices. However, we observed that very few animals were routinely housed during the day or night: only 0.6% of the study population were reported to be managed under zero grazing systems throughout the whole year.

Various environmental factors were identified as being associated with *Eimeria* infection in cattle in the study area. In particular, there was a significant positive correlation between infection risk and increasing temperature. Warm environments are known to be favourable for the propagation of *Eimeria* parasites, with schizogony and sporogony occurring in warm, humid environments with adequate oxygen (Coetzer and Justin 2004). It is noteworthy that we were able to demonstrate an effect of very small temperature differences on risk of infection, suggesting these small temperature changes affect either the rate of development of oocysts and/or the duration of survival in the environment, which also impact on infection rates in animals.

Lower risk of infection was observed in individual cattle living in households that were closer to areas at risk of flooding. Such an effect could be attributed to the fact that flooded soils may be poorly aerated, with lower oxygen concentrations in the soil discouraging survival of oocysts or affecting the rate of development. For the successful completion of schizogony and sporogony, adequate oxygen concentration in the soil is needed (Coetzer and Justin 2004). Additionally, communal grazing areas that are prone to flooding may be avoided by farmers, reducing the stocking density and potential contamination of such environments for animals that live close to them.

Paradoxically, there was weak evidence of a positive association between *Eimeria* infection and the proportion of land surrounding a household that was covered by swamp, however, this relationship was not linear. The probability of infection increased up to almost 40% when 10% of the area around a household was classified as swamp, before decreasing to almost zero when more than 30% of the area was classified as swamp. This suggests that a small amount of swamp land creates a favourable microclimate for oocyst development, possibly through effects on humidity, but oocsyt survival or development, and animal risk, decreases once a threshold of soil moisture is reached.

Whilst we didn't observe an effect of rainfall in the previous 30 days in this study, the peak prevalence of *Eimeria* infections in cattle in an earlier study in Pakistan was observed in the rainy and post rainy seasons (Rehman et al. 2012). Increased infection risk during rainy season could be expected to occur through contamination of pasture by the parasites spreading from other areas by surface water or through humidity associated effects on oocyst survival and development (Marquardt et al. 1960). The lack of evidence of an effect observed in this study may have been due to the fact that the nearest weather station with reliable data was Kisumu, which is more than 100 km from the study region and closer to Lake Victoria., which tends to have higher levels of rainfall than areas further from the lake (Kenya meteorological service 2015).

There was some weak evidence of a trend towards increased risk of *Eimeria* infection with increasing levels of sandy soil around a household. The relationship was curvilinear, with risk of infection appearing to initially decrease as sand content increases followed by an increase in risk at relatively higher sand contents. Sandy soils have macro-pores which allow water to percolate and are rarely waterlogged (Royal Horticultural Society 2016). Their moisture levels are usually low compared to other soil types, especially clay soils (Bruand et al. 2005). Whilst the evidence is very weak ($p = >0.05$), the shape of the relationship observed may provide further support to that described above for a possible threshold effect of soil moisture on oocyst development and survival. Such an effect warrants further exploration, particularly since land management practices, such as the implementation of drainage, can influence soil moisture content (Skaggs et al. 2006) and may impact on the risk of *Eimeria* infection.

Although eimeriosis can be a severe disease in cattle, subclinical infection is a common finding, including in temperate settings (Cornelissen et al. 1995; Farkas et al. 2007). Importantly, subclinical *Eimeria* infection has been associated with economic losses associated with low productivity and fertility in animals (Dedrickson 2000; Kristjanson et al. 2004; Maas 2007; Lassen 2009). A potential limitation of this study is that individual *Eimeria* species identification was not done, and different species could have different pathogenicity.

The prevalence of *Eimeria* infection in cattle in western Kenya is lower than has been recorded in other parts of the country and the world in similar production systems. While most infections are likely to be subclinical, farmers in western Kenya could consider laboratory diagnosis and treatment of positive cases to reduce production losses associated with infection. The prevalence of *Eimeria* in cattle in western Kenya is influenced by animal age and the temperature. The soil water content in the areas around households also appears to be important. Knowledge of these animal and environmental predictors could assist with the targeting of testing, or observation for clinical disease, to particular animals living in particular environmental conditions.

ACKNOWLEDGEMENTS

We are grateful to all the members of our field and laboratory teams in Busia and Nairobi, in particular James Akoko, Omoto Lazarus, Jenipher Ambaka, Fredrick Opinya, Lorren Alumasa, Daniel Cheruiyot, Alice Kiyong'a, Velma Kivali, George Omondi, Gideon Mwali, John Mwaniki, Hannah Kariuki, Lilian Achola and Maseno Cleophas.

COMPLIANCE WITH ETHICAL STANDARDS

Ethical approval for the animal elements of the study was provided by the Animal Welfare and Ethical Review Body of the Roslin Institute, University of Edinburgh, UK (AWA004).

REFERENCES

1. Alton G.D., Pearl D. L., Bateman K.G., McNab B. and Berke O., 2013. Comparison of covariate adjustment methods using space-time scan statistics for food animal syndromic surveillance. BMC Veterinary Research, 9:231
2. Asfaw D., Deneke Y. and Ibrahim N., 2016. Prevalence and associated risk factors of calf coccidiosis in and around Asela Town, Southeast Ethiopia. Journal of Natural Sciences Research 6(3).
3. Ary Bruand, Christian Hartmann, G. Lesturgez. 2005. Physical properties of tropical sandy soils: A large range of behaviours.Management of tropical sandy soils for sustainable agriculture. confrence proceedings: a holistic approach for sustainable development of problem soils in the tropics. Khon Kaen, Thailand. 2005, 27.11.2005 au 2.12.2005.
4. Bangoura B., Mundt H., Schmäschke R., Westphal B., Daugschies A., 2012. Prevalence of Eimeria bovis and Eimeria zuernii in German cattle herds and factors influencing oocyst excretion. Parasitology Research 110 (2) 875.
5. Bates D., Maechler M., Bolker B., Walker S., 2015. Fitting linear mixedeffects models using lme4. Journal of Statistical Software, 67(1), 1–48.
6. Bjornstad O.N., 2016. R functions for analyzing spatial (cross-) covariance: the nonparametric (cross-) covariance, the spline correlogram, the nonparametric phase coherence function, and related. http://ento.psu.edu/directory/onb1. Accessed 15 August 2016.
7. Bronsvoort B. M., Thumbi S. M., Poole E. J., Kiara H., Auguet O. T., Handel I. G., Jennings A, Conradie I., Mbole-Kariuki M. N., Toye P.G., Hanotte O., Coetzer J.A.W., Woolhouse M.E.J., 2013. Design and descriptive epidemiology of the infectious diseases of East African Livestock (IDEAL) project, a longitudinal calf cohort study in western Kenya. BMC Veterinary Research, 9:171.
8. Broxton, P., X. Zeng,W. Scheftic, and P. Troch, 2014: A MODIS-Based Global 1-kmMaximum green vegetation fraction dataset. Journal of Applied Meteorology and Climatology, 53.
9. Bursac Z., Gauss C. H., Williams D. K. and Hosmer D. W., 2008. Purposeful selection of variables in logistic regression. Source Code for Biology and Medicine 2008, 3:17
10. Coetzer J. A. Wand Justin R. C, 2004. Infectious diseases of livestock. Second edition Vol 3; Oxford University press 319 – 331.
11. Cornelissen A.W., Verstegen R H., van den Brand, Perie N.M., Eysker M., Lam T.J., Pijpers A, 1995. An observational study of Eimeria species in housed cattle on Dutch dairy farms. Veterinary Parasitology 56(1–3): 7–16.
12. Dedrickson B. J., 2000. Coccidiosis in beef calves. Alpharma Animal Health Division Fort Lee, NJ 07024.
13. Doble, L., Fevre E., 2010 Focusing on neglected zoonoses, Veterinary Record, 166(18): 546–546.
14. Dong H., Zhao Q., Han H., Jiang L., Zhu S., Li T., Kong C., and Huang B., 2012.Prevalence of coccidial infection in dairy cattle in Shanghai, China. Shanghai Veterinary Research Institute, Chinese Academy of Agricultural Science, Key Laboratory of Animal Parasitology of Ministry of Agriculture.
15. Farkas R., Szeidemann Z., Majoros G., 2007. Studies on coccidiosis of calves in Hungarian dairy farms. Parasitology Research 101(1) 113–120.
16. Fiege N., Klatte D., Kollmann D. Zahner H., Bürger H. J., 1992. Eimeria bovis in cattle: colostral transfer of antibodies and immune response to experimental infections. Parasitology Research. 78 (1) 32–38.
17. Harrell, F., 2001. Regression modeling strategies, with applications to linear models, survival analysis and logistic regression, Springer International Publisher, Switzerland.ch2

18. Harrell F., 2015. Regression modeling strategies: with applications to linear models, logistic and ordinal regression, and survival analysis, Second Edition, Springer International Publisher, Switzerland.ch4
19. Hengl T, Heuvelink GBM, Kempen B., 2013. Soil property maps of Africa at 1 km. https://www.isric.org: ISRIC: World Soil Information. Accessed 10 May 2016.
20. Hijmans RJ, Cameron SE, Parra JL, Jones PG, Jarvis A., 2005. Very high resolution interpolated climate surfaces for global land areas. International Journal of Climatology. 25:1965–78.
21. Hosmer, Jr., D. W., Lemeshow, S. and Sturdivant, R. X. (2013) Modelbuilding strategies and methods for logistic regression, in applied logistic regression, Third Edition, John Wiley & Sons, Inc., Hoboken, NJ, USA.
22. Kenya meteorological service,2015. Weather and crop review for dekad 26, 11–20 September, 2015 h tt p : / /www.meteo.go.ke/dekad/Dekad26-2015.pdf. Accessed 15 September 2016.
23. Khan,M. N.; Tauseef-ur-Rehman; Sajid,M. S.; Abbas, R. Z.; Zaman, M. A..A. S. M. R. 2013. Determinants influencing prevalence of coccidiosis in Pakistani buffaloes. Journal Pakistan Veterinary Journal., 33(3), 287–290.
24. Kristjanson P., Krishna A., RadneyM., NondoW., 2004. Pathways out of poverty in western Kenya and role of livestock. Working paper14-.http://www.fao.org/ag/pplpi.html. Accessed 13 May 2013.
25. Lassen B., 2009. Diagnosis, epidemiology and control of Bovine coccidiosis in Estonia. Veterinary medicine and zootechnics 48 (70).
26. Lassen B, Viltropa Arvo, Raaperib K, Järvisa T, 2009. Eimeria and Cryptosporidium in Estonian dairy farms in regard to age, species, and diarrhea. Veterinary Parasitology 166, (3–4), 212–219.
27. Lélu, M., Gilot-Fromont, E., Aubert, D., Richaume, A., Afonso, E., Dupuis, E., Gotteland, C., Marnef, F., Poulle, M.-L., Dumètre, A., Thulliez, P., Dardé, M.- L., Villena, I. 2011. Development of sensitive method for Toxoplasma gondii oocyst extraction in soil. – Veterinary Parasitology, 183, 59–67.
28. Levecke B, Rinaldi L, Charlier J, Maurelli MP, Bosco A, Vercruysse J, Cringoli G, 2012. The bias, accuracy and precision of fecal egg count reduction test results in cattle using McMaster, Cornell-Wisconsin and FLOTAC egg counting methods. Veterinary Parasitology 13; 188 (1–2):194–9.
29. Maas J., 2007. Coccidiosis in Cattle. UCD Vet Views. Carlifornia Cattlemen's Magazine.
30. McAllister M., 2007. Bovine neosporosis and coccidiosis. Biológico, São Paulo, 69(2):57–61.
31. Marquardt, W. C., Senger, C. M. And Seghetti, L. 1960, The effect of physical and chemical agents on the oocyst of Eimeria zurnii (Protozoa, Coccidia). The Journal of Protozoology, 7: 186–189.
32. Matjila P., and Penzhorn B., 2002. Occurrence and diversity of bovine coccidia at three localities in South Africa. Veterinary Parasitology 104 (2): 93–102.
33. Munyua W.K., Ngotho J.W., 1990. Prevalence of Eimeria species in cattle in Kenya. Veterinary Parasitology 35 (1–2):163–168.
34. Pandit, B., 2009. Prevalence of coccidiosis in cattle in Kashmir valley. VetScan 4: (1) 33.
35. Pfukenyi D., Mukaratirwa S., Willingham A., and Monrad J., 2007. Epidemiological studies of parasitic gastrointestinal nematodes, cestodes and coccidia infections in cattle in the highveld and lowveld communal grazing areas of Zimbabwe Onderstepoort. Journal of Veterinary Research, 74:129–142.
36. Priti M, Sinha SRP, Sucheta S, Verma SB, Sharma SK, Mandal KG, 2008. Prevalence of bovine coccidiosis at Patna. Journal of Veterinary Parasitology 22:5–12.
37. Rehman TU., Khan MN., Sajid MS., Abbas RZ., Arshad M., Iqbal Z., Iqbal A., 2011. Epidemiology of Eimeria and associated risk factors in cattle of County Toba Tek Singh, Pakistan. Parasitology Research 108(5):1171–1177.
38. Rehman, T., Khan, M.N., Sajid, M. S., Iqbal, Z., Javid,M. T., Riaz,M.,& Ahmad, M. 2012. Epidemiology of Eimeria and associated risk factors in goats of district Toba Tek Singh, Pakistan. Indian Journal of Animal Sciences, 82(3), 282–285.

39. Rodríguez R I. -Vivas, Domínguez J L. -Alpizar, Torres J F. -Acosta, 1996. Epidemiological factors associated to bovine coccidiosis in calves (Bos indicus) in a subhumid tropical climate. Revista Biomédica 7:211–218.
40. Rose E M., 1971. Immunity to coccidiosis: protective effect of transferred serum in Eimeria maxima infections. Parasitology 62 (1)11-25.
41. Skaggs, R.Wayne; Amatya, Devendra M.; Chescheir, G. M.; Blanton, C. D.; Gilliam, J. W., 2006. Effect of drainage and management practices on hydrology of pine plantation. http://www.treesearch.fs.fed.us/pubs/22416. Accessed 15 September 2016.
42. The Royal Horticultural Society 2016. Soil types. https://www.rhs.org.uk. Accessed 15 September 2016.
43. Urquhart G.M, Armour J, Duncun J L, Dunn AM, Jennings FW, 1996. Veterinary parasitology Second edition; Blackwell Science Ltd. 224–231.
44. Wardrop, N A 2015. Landcover classification: western Kenya, 2010. http:// eprints. soton.ac.uk/383135/ Accessed 15 May 2016.
45. Wardrop N. A,., Thomas L. F., Atkinson P. M., GlanvilleW. A., Cook E. A. J.,Wamae C. N., Gabriël S., Dorny P., Harrison L. J. S., Fèvre E. M., 2015. The influence of socio-economic, behavioural and environmental factors on Taenia spp. Transmission in western Kenya: evidence from a cross-sectional survey in Humans and Pigs. PLoS Neglected Tropical Diseases 9 (12).

CHAPTER 20

An investigation of organic sheep and goat production by nomad pastoralists in southern Iran

Hamid R. Ansari-Renani

Animal Science Research Institute (ASRI), P.O. Box 31585-1483 Karaj, Islamic Republic of Iran

ABSTRACT

The nomadic pastoralist system in Baft district in Kerman province is well known in Iran for producing cashmere from Raeini goats. However, there is little information regarding the organic sheep and goat production systems. Interviews and field observations were carried out with 30 Siahjel nomad families of Raen origin in proximity of Baft city to characterize the organic production system in terms of feeding, animal health and veterinary treatments, husbandry management practices, transport, slaughtering and housing. Unimproved rangeland was found to be the main source of sheep and goat nutrition belonging to nomads in southern Iran. Nomad livestock were not fed in stables or in restricted areas but moved and grazed freely in extensive open grazing areas.

As a nomadic traditional feeding management practice, there were no minerals, vitamins, pro-vitamins or GMOs for animal feed. Nomad sheep and goat breeds were considered to be robust, adapted to the environment and disease-tolerant livestock. In the nomadic system, no animal cruelty practices such as tail ducking, dehorning and tethering were allowed. To keep ruminants in groups to meet their social needs, nomad families stayed and kept animals together to support each other in different livestock activities, including shepherding, feeding, milking and health care. Due to natural breeding in nomadic herds, the male breeding stock was kept and grazed separately from does during the breeding season. Nomadic livestock products, i.e. meat, milk, wool, could be labelled organic after making local studies, surveillance, regulation and certification.

Keywords: Nomad, Rangeland, Goats, Sheep, Organic products

1. BACKGROUND

The increasing incidences of residues of pesticides, chemical fertilizers, antibiotics and hormones in livestock products overrecent years are of great concern. For ensuring food safety and protecting human health, the attitude toward organic products is rapidly gaining positive perceptions from people in developing countries including Middle Eastern countries (IFOAM 2005; Jaffee and Howard 2010; Schleenbecker and Hamm 2013; FiBL 2013).

Livestock organic production entails production of highly nutritive quality foods free from all kinds of impurities for sound human health, in which ethological characteristics of animals are respected. Organic livestock producers commit to respect a list of specifications governing animal care, welfare and feeding, obliging them to give their livestock access to pasture (Leroux et al. 2009) which is nutritionally favourable, as products from pasture-fed livestock have been shown to have a nutritionally more desirable composition than products from livestock fed on concentrate diets (Fisher et al. 2000; Aurousseau et al. 2004; Sante-Lhoutellier et al. 2008).

Organic sheep and goat production based on grazing (Rahmann 2002, 2014) could be a valid alternative to animals kept in intensive or industrial systems fed with standard rations of concentrates. Existing similarities between organic agricultural products and extensive farming systems in many developing countries (Ben Kheder 2001; Znaidi 2001) enable many traditional farmers, including nomads, to convert to organic systems.

The geographical and ecological conditions of Iran are well-suited to small ruminant production. The relatively low cost of sheep and goat farming (local breeds - well adapted to their environment plus extensive free communal grazing areas) and the increasing demand for expensive organic products in domestic and regional export markets (Herman and Steidle

2014; Steidle and Herman 2014; Ak and Koyuncu 2002) encourages nomads to shift to organic production.

There is evidence that, besides being usually free from undesirable residues, the products from pastoralists are also appreciated for their high nutritional value and better taste. In Iran, milk and meat produced by nomads is regarded as a local specialty and is much preferred to that of animals raised by large industrial complexes. In addition to these material benefits of nomadic pastoralist products, there are significant immaterial values. Pastoralist breeds are part of the local heritage and contribute to local and regional identity, besides often being essential for traditional rituals. Despite this array of advantages, nomadic pastoralists currently continue to market their products generically and there is no awareness about the taste and health benefits of their animals among consumers, policy-makers and even themselves.

Nomads play an important role in sheep and goat production mainly because they keep 58.5 % of the sheep and 39.7 % of the goat population of Iran. Sheep and goat populations of Iran comprise 53.8 and 25 million head, which rank sixth and fifth in the world, respectively (FAO 2014). Nomadic systems in Iran are characterized by low population densities, movement of livestock between grazing areas (regions and provinces) in different seasons, weak linkages to markets and public services, and based on several multiple co-resident family units (clusters of two to five households staying together). The majority of the nomadic pastoralists do not have permanent settlements and consequently use mobile homes such as tents (Ansari-Renani 2015; Ansari-Renani et al. 2013).

Growing demand for organic sheep and goat products will continue to be the main driver of nomadic livestock systems for domestic and export markets. At present, information regarding Iranian nomadic organic sheep and goat production systems is very limited. The objective of the present study was to describe the characteristics of and to evaluate potentials and conditions under which nomad pastoralists of southern Iran are able to produce different organic livestock products. Attempts were made to address constraints and shortcomings of the sustainable nomadic system in livestock organic production.

1.1 Study area

This study was undertaken in Kerman Province, Baft region in the southern part of Iran (Figure 1). Kerman Province is a highland region with <250 mm annual rain. Summer is hot (up to 35 °C) and dry, and winter is moderate. Baft is a region in the south of Kerman province, 2270 m above sea level at latitude 29° 17′ N and longitude 56° 36′ E. In this area, most nomad households belong to the Siahjel sub-tribe of the Raen tribe. This region has two main livestock breeds: the Raeini cashmere-producing goat and the carpet wool-producing Kermani sheep. Male Raeini goats have an average live weight of 35 kg and females 30 kg. They produce on average 507 g of cashmere of different colours with averages of 56.5 % down yield, 19.5-μm fibre diameter and a staple length of 54.2 mm (Ansari-Renani et al. 2012). Kermani sheep produce on average 2.0 kg of wool with 70.0 % efficiency, a staple length of 150 mm and a fibre diameter of 27 μm.

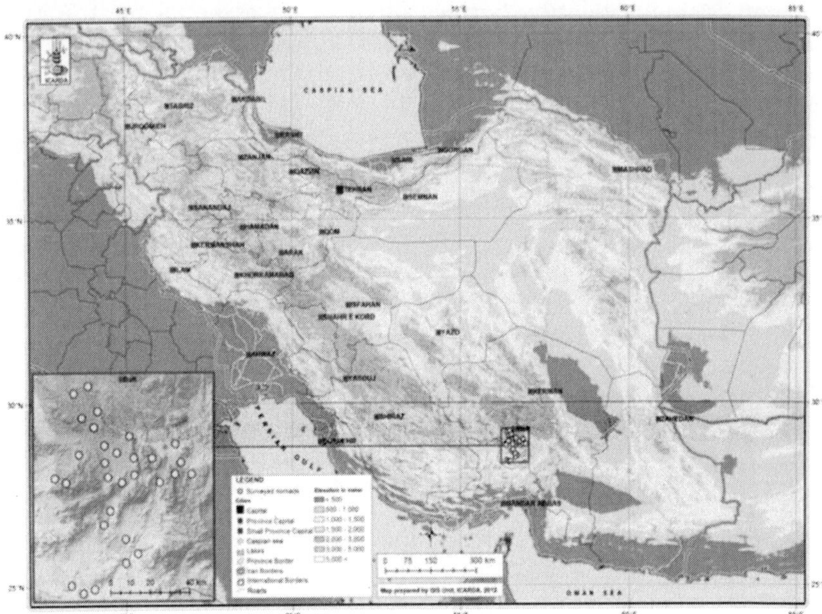

Figure 1 Map of Iran showing the study area and distribution of 30 nomad households chosen within the Baft region in Kerman province

The nomads are completely dependent on livestock as a source of income. Nomad livestock production system is based on mixed herds with 89 % of heads being goats, 8 % sheep and 3 % horses, mules, donkeys and sometimes camels used for transportation (Ansari-Renani et al. 2013). A typical nomad family would run some 250 goats, of which adult female goats (does) constitute 44 %; bucks, castrated adult males, male and female yearlings and male and female kids represent 8 %, 5 %, 7 %, 12 %, 10 % and 14 % of the herd population, respectively.

2. METHODS

2.1 Selection of nomad settlements and information gathering

A total of 30 nomad settlements were chosen at random within ±20 km of Baft city in Kerman Province (Figure 1, Table 1). Information was gathered primarily through in-depth interviews with nomad men and women and field observations. Four periods of six to seven days of fieldwork were conducted within Baft region. Each interview lasted approximately three hours and consisted of about 50 predetermined questions. A structured questionnaire was completed for each individual family of settlement heads, including family composition and labour allocation structure, herd structure and management, housing, stocking rate, nutrition, feeding, watering, health, veterinary treatment, breeding, transportation, management practices, slaughtering, processing and reproduction.

Table 1 Details of the 30 nomad cashmere-producing farms

Nomad family	Place of origin	Size of family (persons)	Number of animals (heads)[a]		
			G	S	C
1	Arzooeih	7	225	40	30
2	Galoogiran	5	142	40	41
3	Janat Abad	6	251	33	35
4	Janat Abad	8	306	-	50
5	Geloo Mahmoudabad	9	323	4	70
6	Khobr	7	280	21	50
6	Khobr	7	280	21	50
7	Khobr	4	200	10	30
8	Soltani-Baft	6	373	150	20
9	Gelook - Baft	9	315	33	15
10	Geloo Anjeer	6	225	25	25
11	Se Chah Dehsard	4	185	-	18
12	Dashtab - Baft	5	179	47	45
13	Sanouheh Dashtab	5	176	40	1
14	Dokoohe - Baft	6	155	17	30

15	Dokoohe - Baft	4	200	-	25
16	Zarab	5	130	-	30
17	Sechah Dahsard	5	356	3	25
18	Gelook - Baft	4	303	53	15
19	Esmailabad - Baft	5	208	58	20
20	Esmail Abad - Baft	3	109	21	13
21	Esmailabad - Baft	3	102	17	1
22	Geloo Mahmoudabad	4	130	24	15
23	Geloodar Kooshki	5	150	-	12
24	Godar Zarab	3	83	-	15
25	Zarab	6	219	20	20
26	Zarab	3	140	50	13
27	Dahaneh Zardan	4	155	-	20
28	Geloo Mahmoudabad	5	120	15	15
29	Dehsalar	5	286	106	30
30	Dehsalar	5	153	15	15

G, goats; S, sheep; C, chicken

The responses to those questions were tallied, and the percentages of the various responses were calculated. Minimum, maximum, standard deviation (SD) and standard error values were measured using SAS package.

3. RESULTS AND DISCUSSION

Compared with conventional and organic systems (Table 2), nomadic sheep and goat production was characterized by natural breeding of locally adapted native breeds, extensive use of rangeland as a source of livestock feed, no use of prophylaxis, minimal allopathical treatments, protection of the environment, improved animal welfare and sustainable animal husbandry practices. In the nomadic system of sheep and goat production, one objective was to achieve animals' wellbeing through animal welfare-oriented husbandry and appropriate use.

Table 2 Characteristics of conventional, organic and nomadic animal husbandry

	Conventional[a]	Organic (834/2007)[a]	Nomadic
Breeds, origin	Highly performing, special breeds and cross-breeds according to product aimed for	Only animals reared in organic farms, diversity of breeds, sometimes rare breeds, natural breeding	Native breed, locally adapted, natural breeding
Keeping (buildings and free runs)	Animal protection laws (requirements for keeping of animal according to species)	Special requirements for keeping animals oriented toward animal welfare (stock density, space, grazing, tiding, etc.)	Animal are kept in rangeland, oriented toward animal welfare
Feeding	According to current foodstuff legislation (permitted food additives such as enzymes and synthetic amino acids)	Foodstuffs produced as much as possible on site, feeding rations according to animal welfare (e.g. minimum use/parts of roughage) only specifically permitted additives, no synthetic amino acids, no genetically modified organisms	Rangeland is considered the main source of livestock nutrition, no synthetics, no GMOs, no pesticides and chemical fertilizers, livestock are not fed in stables or in restricted areas but they move and graze freely in extensive open grazing areas
Management and treatment	Managed breeding, if necessary stable-wide prophylaxis, legally required waiting periods according to drug prescription law	No prophylaxis (exception: legally required vaccination), only three allopathical treatments per year for long-lived animals (more than one year) respectively one treatment for livestock, which is not used more than one year; double the waiting period after use of drugs, minimum 48 h. Restricted interfering with the animal's integrity (no polling, beak trimming, tail clipping, etc.)	No prophylaxis (exception: legally required vaccination), minimal allopathical treatments per year, no tethering, polling and tail clipping
Transport	Animal transport regulation	Animal transport regulation with short transport	Animals are grazed by migration mainly, some transportation by trucks

[a]Adapted from Rahmann (2014)

Curtailing freedom of movement, sensory deprivation and unsocial ways of husbandry; not allowing any contact with animals of the same species; or forcing too close a contact were not permitted in the nomadic farming system.

3.1 Nomad sheep and goat breeds

The livestock breeds kept by the nomads had unique characteristics, as they were subjected to selection criteria that were almost entirely different from those used in 'scientific animal breeding'. They were integral to their respective eco-systems and provided a host of environmental services. Taking droughts and hunger in their stride and acting as insurance, these breeds walk for miles in harsh terrain and seek out scattered, spiky, fibrous plants that survive in areas where crops could never be grown.

In organic farming, the breeding of small ruminants should be done by natural mating. Artificial insemination is allowed, but not embryo transfer, oestrus synchronization, etc. In the nomadic system, animal breeding was only by natural mating, and techniques such as artificial insemination, embryo transfer and oestrus synchronization were not popular among nomads. As a result of natural breeding, bucks and male yearlings were in disproportionate ratio (15 %) compared to adult and yearling females (56 %).

As industrial modes of livestock production are spreading, domestic animal diversity is in rapid decline. According to FAO, one third of all livestock breeds have either perished or are threatened with extinction, due to intensive selection for high production by means of artificial insemination and embryo transfer and spreading of a small number of genetically narrow high-performance breeds around the whole world. In this scenario, the main stewards of livestock genetic diversity are nomadic pastoralists and other 'small-scale livestock keepers' that raise animals under low-input conditions. Termed 'guardians of biological diversity' (FAO 2009), these people conserve biodiversity at the level of livestock breeds, vegetation, eco-systems and landscape.

3.2 Husbandry management practices, transport and slaughtering

Management of livestock among nomads was a social process, and they did their utmost for the wellbeing of their animal and to avoid animal cruelty of any kind. In the nomadic system, there were no tail ducking, dehorning and tethering (Table 3).

Table 3 Percentage of nomad herds moving together and adopting different management practices

Items	Percentage
Management practices	
Dehorning	0
Tethering	0
Shortening of tails	0
Castration of male lambs and kids	100
Families moving together	
Single	13
Two or more	87

In organic systems, castration of male stock is allowed so as to keep traditional animal husbandry practices. In the nomadic areas, breeding management was difficult in mixed flocks of male and female animals, without male castration. All nomad livestock owners castrated the male kids and lambs at a very young age. The surgical technique of cutting the scrotum open with a knife and pulling the testicles was the common method of castration among nomads. They castrated their animals during the cool months of the spring and autumn, to reduce the chances of infection being spread by flies and other insects.

In organic farming, ruminants have to be kept in groups to meet their social needs; however, it is not defined how social needs can be fulfilled in farm conditions. Contrarily, nomads had well-defined social methods for keeping livestock under farming conditions. To keep ruminants in groups, 87 % of nomad families stayed and move together with other families (Table 3). Furthermore, families staying and keeping their livestock together allowed nomads to herd adult and young animals separately as the social needs and feed requirements of different aninal age groups differ and require particular management practices.

The kidding period at the beginning of winter was at a season of lower temperatures and lower feed availability. Hence, animals were supplemented with limited amount of barley, and to avoid losses, at the end of autumn, most nomads migrated to warmer areas in the southern provinces adjacent to the Persian Gulf.

As a result of natural breeding and the high proportion of males to females in the nomadic flock, male breeding stock was kept and grazed separately during the breeding season. Bucks come into rut during the breeding season. Rut is characterized by a decrease in appetite, obsessive interest in the does and a strong heat. Nomads knew that when strong-smelling bucks are not separated from the does during breeding season, their scent will affect the milk. In organic farming, male breeding stock has to be kept in the farm. It is permissible to use conventionally-kept male breeding stock.

In organic farming, the animals have to be slaughtered in abattoirs which fulfil the regulations of organic farming and are certified (Leu 2014). Traditionally, the nomads consumed more milk in their diets than meat. In fact, they often express a dislike for killing and trading animals. Animals were sold to certified abattoirs or butchers directly either for cash needed for income or for culling unwanted livestock. Meat production was almost exclusively for sale. Home slaughtering for own consumption and sale to neighbours or relatives occurred only occasionally. The proportion of nomads buying animals was very low; the reason was that they usually depend on their existing animals to reproduce and increase their herd size.

The transport of livestock is not clearly defined in organic farming, but a stress-reduced loading, transporting and unloading of livestock without the use of allopathic tranquillizer, electrical shockers or similar tools is preferred. Nomad livestock movements between communal grazing areas in different provinces are gradual, and animals are relocated by migration on foot which may take up to three to five days, depending on the distance covered. Sometimes, trucks were used to transport livestock between grazing areas in different Provinces.

3.3 Feeding

In organic farming, sheep and goats have to be fed with 100 % organic feedstuff (EEC Regulation 2007). The statement that livestock has to be fed 'predominantly' with self-produced feedstuff is not specific enough. Fifty percent of organic feeds for ruminants can be purchased from other organic farms. Comparatively, in this study, rangeland was considered the main source (85 %) of sheep and goat nutrition of nomad farms without any use of chemicals such as fertilizers and pesticides (Figure 2) while other sources of feed such as crop stubble composed only a small portion (15 %) (Table 4). The ratio between the high price of feed inputs and lower price of livestock products provides insufficient incentives for the nomads to purchase synthetic chemical inputs for developing intensive production systems.

Figure 2 Rangeland is considered the main source of nomad sheep and goat nutrition of southern Iran

Table 4 Percentage of nomad farms using different chemicals, sources of feed, type of grazing land and feeding method

Items	Percentage
Farms using chemicals	
Fertilizers	0
Pesticides	0
Others	0
Source of feed	
Rangeland	85
Stubbles, agriculture residues, etc.	15
Use of genetically modified organisms (GMO)	0

GMO derivatives	0
Minerals, vitamins and pro-vitamins	0
Source of lamb and kid feed	
Colostrum and maternal milk	100
Powdered milk	0
Type of grazing land	
Open grassland	38
Tree covered	23
Bush/shrub and stone covered	39
Feeding method	
Free-range grazing	100
Stable	0

In organic farming, it is not permitted to use anything produced using genetically modified organisms (GMOs) or derivatives. This includes feed for livestock. There are permissible minerals, vitamins and pro-vitamins for animal feed, and artificially produced vitamins may not be used for ruminants. In the nomadic traditional feeding management practice, all farms did not use minerals, vitamins, pro-vitamins and GMOs for animal feed, as the cost of such inputs is too high (Table 4).

In organic farming, a feeding system which leads to anaemic conditions in sheep and goats is prohibited and considered as animal cruelty. To avoid such conditions, as feeding management practice, 23 % of nomad farms preferred tree-covered grazing areas (Figure 3) which include wild oak trees, as the nutritive value of leaves that are rich in iron, sulphur and copper consumed by animals complements the grass very well. Thirty-eight percent and 39 % of nomad farms used open grassland, bush/shrub and stone-covered rangeland, respectively (Table 4).

In focus group discussions, nomad herders frequently emphasized that the diversity of plant species consumed was responsible for the superior taste and healthiness of sheep and goat milk and meat. Most of these plants also have medicinal value. The local knowledge of the nomad pastoralist communities sees a connection between the dietary composition of livestock feed and the nutritional value of livestock products.

Figure 3 Tree-covered grazing land in southern Iran suitable for grazing livestock

Nomadic sheep and goat breeds were social animals in the true sense, living in a and flock, responding to the voice of their keepers. By means of such breeds of livestock that are co-evolved with their eco-system, nomads were in a position to use the dispersed and extremely bio-diverse natural vegetation of drylands and mountainous areas into a range of high-value delicious organic food including meat and milk, as well as a range of other organic products such as fibre, fertilizer and hides. They did this without leaving any carbon footprint, as their animals forage for themselves and no energy is expended to grow or transport feed to them.

On many organic farms, kids and lambs receive only colostrum milk and subsequently powdered milk. The young stock does not suckle and receive natural milk because the organic milk is very valuable and expensive (Rahmann 2002). Even skimmed powdered cow milk can be used as long as it has an organic label. In the nomadic system there was no early weaning system, contrary to intensive and organic sheep and goat systems of production in which early weaning of lambs and kids is practised and all the milk and milk products are consumed by humans.In all nomad farms, kids and lambs suckled their mothers for 45 days and received only maternal milk (Table 4). When young animals were still nursing, the flock returned to the tent at least once each day to allow the young to nurse.

3.4 Animal health and veterinary treatments

In organic farming, the principle of animal health is prevention rather than curing/treating. In the nomadic system, as a preventive practice, newborn and young animals were often kept together at the tent with the nomad women and children until they were old enough to go out to pasture with the flock.Animals that were diseased were likewise kept at the family tent, effectively isolated from the flock so chances that infection will spread throughout the flock are thus reduced.

Robust, adapted and disease-tolerant livestock ensure fit and healthy animals. Nomad breeds were considered to fulfil these targets. These were indigenous breeds typical of a specific region, adapted to the local environmental conditions and retaining the pattern for centuries.

Many of the nomads' animal management practices had a direct positive impact on the incidence of livestock diseases. Some of the most significant practices include reproduction management, the isolation of diseased animals, grazing and seasonal migration patterns. Nomad herders believed that by herding more than one species, the risk of livestock losses was buffered, whether losses were due to diseases or extreme environmental conditions. They emphasized that the annual migrations that most Raen nomads undertake as well as the extensive grazing patterns of the herd help to lower the incidence of diseases, including internal parasites.

Nomads highlighted that their indigenous technical knowledge and medicinal plants for health care were effective substitutes for allopathic medicines, giving them an advantage in organic livestock production. The indigenous knowledge of nomads may provide an effective option for veterinary care through proper validation, as also the negligible use of agro-chemicals, especially in drylands and hilly nomadic regions provides a favourable environment for organic livestock production.

Nomads knew that infections of the udder can be spread from one animal to another during milking and washed their hands with water between milking animals. Both men and women helped at animal births. Women often explained that they were better than men at this since they have smaller hands, which is good for repositioning the foetus within the birth canal.

3.5 Housing and stocking rate

Tethering of livestock is prohibited in organic farming. Basically, there was no tethering of any kind of livestock among nomads. When nomad livestock returned from grazing, adult and young animals were penned separately near the tent in circular-shaped pens made up of wood, fenced overnight and milked in the morning before being taken out for grazing.

In organic farming, it is obligatory that ruminants should graze on pastures ('free-range') and not fed in stables as long as the animal, weather and pasture conditions are suitable. If grazing is not possible, a permanently accessible open-air run is obligatory. Free-moving stables with permanent access to open-air runs are the principle of ruminant keeping. Only with permanent summer pasture grazing is an outdoor run not necessary, as long as the animals are not tethered.

The nomad livestock were not fed in stables or in restricted areas, but moved and grazed freely in extensive open grazing areas. Nomad families used the northern highland rangelands in spring and summer for grazing and migrated to the warmer southern Persian Gulf provinces in autumn and winter (Figures 4 and 5). The nomadic pastoralists had no fixed homesteads and covered great distances with their livestock following pasture availability throughout the seasons. The transhumant pastoralists followed a regular seasonal movement between set areas. Their movement was vertical where pastures at high altitudes are used in summer and pastures in the lowlands are used in winter or horizontal in the surroundings. Consequently, the livestock density (stocking rate) in Baft varied throughout the year, with the highest number of livestock and people in summer.

Figure 4 Nomads migrate to warmer areas of southern provinces adjacent to the Persian Gulf in the beginning of autumn

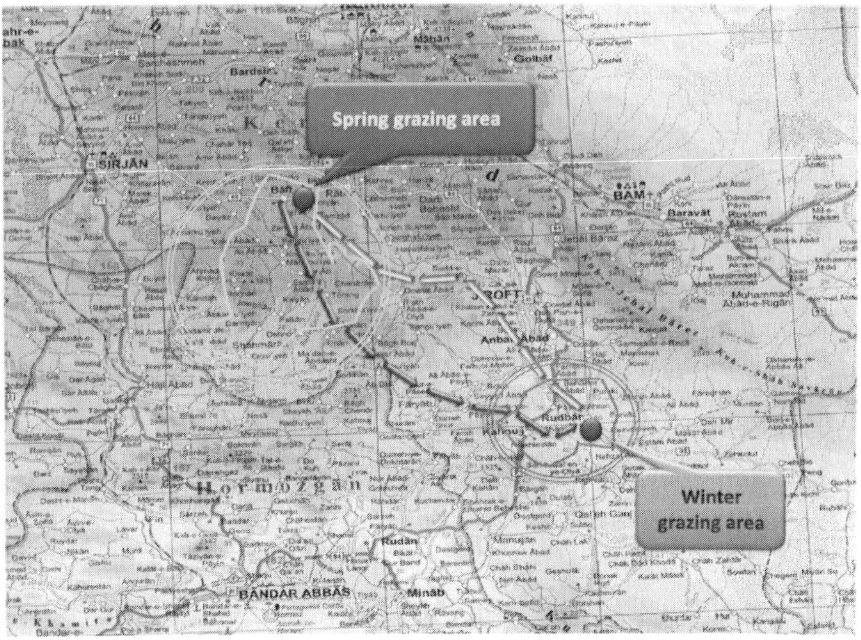

Figure 5 Map of south Iran showing the spring and winter grazing areas in Kerman (Baft region) and Hormozgan (Roodan region) provinces near the Persian Gulf in blue and green colours, respectively

4. CONCLUSION

The ideology behind principles and standards of organic animal husbandry is not new to the nomadic farming system of Iran, whose community have practiced animal welfare and animal natural rearing systems since ancient times. A country rich in indigenous animal genetic resources like Iran is very much suitable for adopting this farming system. Moreover, the nomadic farming system with well-diversified livestock populations in terms of species and breeds is ideal for organic livestock production. Nomad breeds being less susceptible to diseases and stress need less allopathic medicines/antibiotics. Besides, limited external input use, including for animal production and maximum on-farm reliance brings the nomadic system nearer to organic systems. The nomadic sheep and goat production system being largely extensive, animal welfare is not much compromised compared to intensive or conventional type of animal production.

Although the nomadic type of livestockkeeping provides an excellent and 'green' alternative to industrial production, nomad pastoralists need to overcome some challenges and harness strengths and opportunities, while developing their capacity in terms of knowledge, skills, infrastructure, animal feeding, hygiene, sanitation, disease control and assured certified supply chain required for organic livestock production.

Nomad farmers need to be oriented and educated about the organic standards and how to overcome the risks they might face in adoption of organic livestock standards. The livestock advisors should be trained and skilled in providing services in livestock management and permitted therapies in organic rearing systems. Research on the locally adaptable management and disease-preventive measures needs to be emphasized by the government and organic-promoting agencies as well as NGOs. The potential needs to be recognized of Iranian nomad farmers to meet the requirements of organic livestock product demand, not only locally but also globally in the near future. Organic livestock production can be encouraged through research and development efforts, including establishment of model organic livestock farms, processing units, traceability tools and capacity-building measures.

Converting extensive, range-based nomadic system to organic production could become economically attractive, if price premiums could be captured for organic meat and livestock products. Development of business models will definitely attract commercial interests and ensure that vulnerable nomadic communities receive attractive returns for their untapped treasure of organic principles. Systematic studies need to validate the animal husbandry practices of nomads with respect to organic certification, so that revision or improvement can be made wherever necessary. In this way, organic livestock products will have considerable potential for high-value niche markets.

REFERENCES

1. Ak, I., and M. Koyuncu. 2002. Organic meat and milk production by pastoralists from small ruminant in Turkey. EAAP publication, No. 106.
2. Ansari-Renani, H.R. 2015. Cashmere production, harvesting, marketing and processing by nomads of Iran - A review. Pastoralism: Research, Policy and Practice 5: 18.
3. Ansari-Renani, H.R., J.P. Mueller, B. Rischkowsky, S.M. Seyed Momen, and S. Moradi. 2013. Nomadic pastoralism in southern Iran. Pastoralism: Research, Policy and Practice 3: 11.
4. Ansari-Renani, H.R., J.P. Mueller, B. Rischkowsky, S.M. Seyed Momen, O. Alipour, M. Ehsani, and S. Moradi. 2012. Cashmere quality of Raeini goats kept by nomads in Iran. Small Ruminant Research 104: 10–16.
5. Aurousseau, B., D. Bauchart, E. Calichon, D. Micol, and A. Priolo. 2004. Effect of grass or concentrate feeding systems and rate of growth on triglyceride and phospholipid and their fatty acids in the M. longissimus thoracis of lambs. Meat Science 66: 531–541.

6. Ben Kheder, M. 2001. Mediterranean organic agriculture course, organic vegetable production. Bari: Mediterranean Agronomic Institute of Bari.
7. EEC (European Economic Commission). 2007. Regulation (EEC) No. 834/2007. Bruxelles, Belgium.
8. FiBL (Research Institute of Organic Agriculture) and IFOAM (International Federation of Organic Agriculture Movement). 2013. The World of Organic Agriculture 2013. Available: https://www.fibl.org/fileadmin/documents/shop/1606-organic-world-2013.pdf Accessed on 10/06/2014.
9. Fisher, A.V., M. Enser, R.L. Richardson, J.D. Wood, G.R. Nute, E. Kurt, L.A. Sinclair, and R.G. Wilkinson. 2000. Fatty acid composition and eating quality of lamb types derived from four diverse breed X production systems. Meat Science 55: 141–147.
10. Food and Agriculture Organization of the United Nations. 2009. www.FAOSTAT.com.
11. Food and Agriculture Organization of the United Nations. 2014. www.FAOSTAT.com.
12. Herman, G. A., and M. Steidle. 2014. 2nd International Conference on Trade and Market Development of Organic Producers. Iran Organic Association. Tehran, Iran.
13. IFOAM (International Federation of Organic Agriculture Movement). 2005. The IFOAM norms for organic production and processing. Bonn: IFOAM.
14. Jaffee, D., and P.H. Howard. 2010. Corporate co-optation of organic and fair trade standards. Agriculture and Human Values 27(4): 387–399.
15. Leroux, J., M. Fouchet, and A. Haegrlin. 2009. Elevage bio: des cahiers des charges francais a la reglementation europeenne. INRA Productions Animales 22: 151–160.
16. Leu, A. 2014. Achieving food security, climate change adaptation with high yielding organic agriculture in Iran. Second international conference on trade and market development of organic producers. Iran Organic Association. Tehran, Iran.
17. Rahmann, G. 2002. The standards, regulations and legislation required for organic ruminant keeping in the European Union. EAAP Publication 106: 15–26.
18. Rahmann, G. 2014. 2nd International Conference on Trade and Market Development of Organic Producers. Tehran: Iran. Organic sheep and goat farming.
19. Sante-Lhoutellier, V., E. Engel, and P. Gatellier. 2008. Assessment of the influence of diet on lamb meat oxidation. Food Chemistry 109: 573–579.
20. Schleenbecker, R., and U. Hamm. 2013. Consumers' perception of organic product characteristics. A review. Appetite 71(1): 420–429.
21. Steidle, M., and G. Herman. 2014. 2nd International Conference on Trade and Market Development of Organic Producers. Tehran: Iran. How to best design an organic guarantee system for global market access.
22. Znaidi, A. 2001. Sheep and goat organic meat production in the Mediterranean region, DSPU thesis in Mediterranean Organic Agriculture. Bari: Mediterranean Agronomic Institute of Bari.

CHAPTER 21

Live weight, conformation, carcass traits and economic values of ram lambs of Red Maasai and Dorper sheep and their crosses

E. Zonabend König[1,2], J. M. K. Ojango[2], J. Audho[2], T. Mirkena[3], E. Strandberg[1], A.M. Okeyo[2], J. Philipsson[1]

[1] *Department of Animal Breeding and Genetics, Swedish University of Agricultural Sciences, P.O. Box 7023, 750 07 Uppsala, Sweden*

[2] *International Livestock Research Institute, Nairobi, Kenya*

[3] *School of Animal and Range Sciences, College of Agriculture, Hawassa University, Hawassa, Ethiopia*

ABSTRACT

Meat production is the most important trait in the breeding objectives of sheep production in East Africa. The aim of this study was to investigate breed differences in live weight, conformation, carcass traits and economic values for meat production among Red Maasai and Dorper sheep and their crosses. In total, 88 ram lambs, which were reared at the ILRI experimental station, Kapiti plains Estate in Central Kenya, were used for the study. The lambs were slaughtered at Kenya Meat Commission (KMC) at about 1 year of age. Prior to slaughter, the lambs were weighed, measured and assessed by experienced evaluators, and at the abattoir carcass traits were recorded. Large breed differences were found for most traits. Dorper lambs were heavier at delivery for slaughter and had better carcass grade but lower dressing percentage and fat levels than Red Maasai. Crossbreds were generally better than the parental breeds. Evaluators were willing to pay more for the Dorper lambs for slaughter although carcass weights later were shown not to be higher than for Red Maasai. Evaluators undervalued Red Maasai lambs by 8–13 % compared to Dorper lambs according to the prices quoted per kilogramme live or carcass weight by KMC. Live weight was better than any other live measure in predicting carcass weight. Due to the overall higher ranking of the crossbred lambs for meat production, Dorper may be useful as a terminal sire breed for crossing with Red Maasai ewes.

Keywords: Sheep breeds, Evaluation, Lamb growth, Assessments, Carcass merits, Heterosis

1. INTRODUCTION

Small ruminant meat plays an important role in many rural African households, especially in pastoral systems (Juma et al. 2010). In a recent UN-led study on "African Livestock Futures", small ruminant meat production in pastoral systems was shown in different scenarios to have the potential of increasing production by five to six times by 2050 relative to the production levels of year 2000 (Herrero et al. 2014). The relative impact of various small ruminant breeds on the projected increased productivity may however differ.

One popular small ruminant breed raised in the semi-arid and arid lands of Kenya is the Red Maasai sheep, an East African fat-tailed sheep breed. Red Maasai sheep are, however, declining among their traditional keepers, due to widespread crossbreeding with the Dorper breed, initially imported from South Africa for improvement of meat production. Both Red Maasai and Dorper sheep and their crosses are important to the farmers for different reasons (Zonabend König et al. 2016). Red Maasai sheep are appreciated by their keepers for their ability to survive and produce under harsh environments, and the Dorper is valued for its higher growth rate potential and larger mature size, whereas the crosses have good survival rates compared to Dorper and relatively better growth rates than the Red Maasai (Baker et al. 2002). Comparative studies of Red Maasai and Dorper sheep indicate that in harsh or humid environments, differences in performance between the breeds tend to be smaller in magnitude (Baker et al. 2002; Mugambi et al. 2005; Zonabend König et al. 2016).

A large body size as an indicator of good growth rate is an important trait when marketing sheep within different communities in the arid lands of Eastern Africa (Gizaw et al. 2010; Mtimet et al. 2014). Mtimet et al. (2014) showed that in the absence of live weight data, breed and visually assessed size are the most important factors in determining the price for live animals. It is important for farmers to know what the market (middlemen) prefer in order to align their breeding strategy with the market demands (Lambe et al. 2008; Katiku et al. 2013; Ojango et al. 2014).

Comparative studies on carcass traits of indigenous sheep breeds and their crosses with improved breeds within Eastern Africa are rare. Results from carcass studies on indigenous sheep breeds in Ethiopia show some breed differences for both carcass weight and fat deposition (Ermias et al. 2002; Ermias et al. 2006). Crossbred Blackhead Ogaden with Dorper has been reported to have improved carcass weights by 15 % compared to pure Blackhead Ogaden (Tsegay et al. 2013). There are, however, no comparative studies on carcass traits of Red Maasai and Dorper sheep and their crosses.

With the growing importance and significance of the sheep industry in Eastern Africa (FAO 2016), and the use of indigenous breeds and their crosses in pastoral systems, the aims of this study were to (i) investigate potential differences in live weight and carcass traits among ram lambs of the Red Maasai and Dorper breeds and their crosses; (ii) assess the economic values of ram lambs for slaughter for the same breed groups, and (iii) provide information on the accuracy in prediction of carcass traits from live weight and conformation traits, objectively measured or subjectively assessed, prior to delivery for slaughter.

2. MATERIAL AND METHODS

2.1 Experimental animals and recording of traits

Ram lambs of Red Maasai and Dorper breeds and their crosses were raised at the International Livestock Research Institute ranch at Kapiti Plains Estate (Kapiti), Kenya, and slaughtered at the largest abattoir in the country, the Kenya Meat Commission (KMC) on the outskirts of Nairobi.

Animals at Kapiti are reared on natural grasslands in a semi-arid environment with limited supplementation by feeding hay to nursing ewes and their young lambs during periods of extreme droughts. The sheep are maintained as a nucleus flock for selection and improvement of sheep under arid conditions as outlined by (ILRI 2016). All animals are routinely provided with basic vaccinations and are washed in dips with acaricides once a week against ecto-parasites, especially ticks. Animals observed to be suffering from severe haemonchosis were treated with anthelmintics. Lambing is planned to occur twice in a year over a 6-week period in the seasons when pastures are more available. Lambs are weaned in batches at about 3 months of age, with variation in age of the lambs resulting from differences in the day of lambing. The tails of all lambs were docked before the age of 2 months.

Ram lambs used in this study were from two batches: the first batch (batch 1) comprised 27 lambs born in August–September 2011 and the second batch (batch 2) included 61 lambs born in November–December the same year. They comprised 18 (7 and 11 in batches 1 and 2) purebred Red Maasai lambs (RRRR), 24 (3 and 21) lambs of 50 %-crosses (DDRR), 32 (11 and 21) lambs of 75 % Dorper (DDDR) and 14 (6 and 8) purebred Dorper (DDDD) lambs. All breed groups were thus represented in both batches. The DDRR lambs included 13 F1 animals comprising both reciprocal crosses (Dorper males × Red Maasai ewes and Red Maasai males × Dorper ewes) and 11 F2 animals. The backcrosses (DDDR) were included as they constituted a large portion of the crossbreds and would support any effects of crossbreds and are also on high demand by pastoralists in the surrounding area.

Recorded traits are presented in Table 1. Birth weight (BW) was recorded in all lambs (used to compare with later body measures). Prior to delivery for slaughter at the KMC, animals were weighed with a scale and measured using a standard tape to determine their live weight (LW), heart girth (HG) and body length (BL). At the same time, the ram lambs were subjectively assessed by experienced evaluators to determine their measures of weight (LWA), conformation (CA), body score (BSA) and total score (TSA). The first batch was assessed by one butcher and the second batch was assessed by the same butcher, a middleman (trader) and a farmer to allow views of different professions. In addition, the evaluators were asked to quote the prices that they would be willing to pay for each of the lambs if it were to be used for slaughter. The value was quoted in Kenya Shilling (KES) and was considered as the market value of the lamb at the time of slaughter (WPA). It includes primarily the expected carcass value, but may also include the value of offals, hide and skin. Live weight measured on delivery day to the abattoir was referred to as delivery weight (DW). A protocol was developed to be used in the evaluation and grading of the sheep carcasses that was carried out by trained livestock evaluators at the KMC. Dressed carcasses of animals were weighed to obtain a hot carcass weight and then chilled and the cold carcass weight (CWT) was recorded the day after slaughter. Cold carcasses were assessed for their fat level (CFL), muscle formation (CMF) and grade (CG). Routinely, the KMC does not grade carcasses of small ruminants and pays a fixed price of KES 150 (USD 1.79) per kilogramme of DW. Alternatively, KMC offers a flat rate of KES

315 (USD 3.75) per kilogramme hot carcass weight for the small ruminants.

Table 1 Recorded traits with units and abbreviations (Abbr) presented per batch and with average age (in days) of ram lambs at recordings

Trait	Units	Abbr.	Batch	Birth	Evaluation	Delivery	Slaughter
			1	Aug–Sept	317	317	322
			2	Nov–Dec	369	380	384
Birth weight	(kg)	BW		X			
Live weight	(kg)	LW			X		
Heart girth	(cm)	HG			X		
Body length	(cm)	BL			X		
Willingness to pay	(KES)[a]	WPA			X		
Live weight assessment	(1–5)	LWA			X		
Conformation assessment	(1–5)	CA			X		
Body score assessment	(1–5)	BSA			X		
Total score assessment	(1–5)	TSA			X		
Delivery live weight	(kg)	DW				X	
Growth rate to delivery	(g/day)[b]	GR				X	
Cold carcass weight	(kg)	CWT					X
Carcass fat level	(1–3)	CFL					X
Carcass muscle formation	(1–3)	CMF					X
Carcass grade	(1–7)[c]	CG					X
Dressing percentage	(%)[d]	DP					X

[a] 1 USD = 84 KES for both batches

[b] GR = (DW − BW)/age × 1000

[c] Lower value describes a preferred carcass grade

[d] DP = CWT/DW × 100

Average age at slaughter was 364.8 days (359.6 for RRRR, 376.2 for DDRR, 363.8 for DDDR and 354.6 days for DDDD).

2.2 Statistical analysis

All analyses were performed using the software "The R Project for Statistical Computing" (R Core Team 2015). Because the quantile-quantile plots and histograms of residuals for all traits were normally or close to normally distributed, linear models were used. The R packages Companion to Applied Regression (Fox and Weisberg 2011) and the linear and nonlinear mixed effects models (Pinheiro et al. 2015) were used for the general linear and mixed model analyses to adjust for fixed effects, and the package lsmeans (Lenth 2013) was used for estimation of the least squares means to account for unequal size of subclasses. The following three linear models were used for analyses of factors that potentially would influence live animal and carcass traits:

$$y_{ijkl} = \mu + \text{Breed}_i + \text{Batch}_j + b_{1i}\text{Age}_k(\text{Breed}_i) + e_{ijkl} \tag{1}$$

$$\begin{aligned} y_{ijklm} &= \mu + \text{Breed}_i + \text{Batch}_j + b_{1i}\text{Age}_k(\text{Breed}_i) \\ &\quad + b_{2l}\text{LiveMeasure}_l + e_{ijklm} \end{aligned} \tag{2}$$

$$\begin{aligned} y_{ijklmn} &= \mu + \text{Breed}_i + \text{Batch}_j + b_{1i}\text{Age}_k(\text{Breed}_i) \\ &\quad + \text{Evaluator}_l + \text{ID}_m + e_{ijklmn} \end{aligned} \tag{3}$$

where y_{ijkl} in (1) is the trait of interest: BW, LW, HG, LBL, WPA, DW, GR, CWT, CFL, CMF, CG or DP; in (2), y_{ijklm} is the carcass trait of interest: GR, CWT, CMF, CFL, CG, DP or WPA (mainly reflecting carcass value); and in (3), y_{ijklmn} is the subjectively assessed traits: WPA, LWA, CA, BSA or TSA.

In each model, μ is the overall mean for the trait; $Breed_i$ is the effect of the ith breed group (i = RRRR, DDRR, DDDR, DDDD); $Batch_j$ is the effect of jth batch (j = batches 1 and 2); b_{1i} is the regression of the trait on age (in days at recording) nested within breed i ($Age_k(Breed_i)$); b_{2l} is the regression of the trait on $LiveMeasure_l$ (l = an objective measure LW, HG, BL or DW or subjective measure WPA, LWA, CA, BSA or TSA); $Evaluator_l$ is the fixed effect of the lth evaluator (l = 1–3); and ID_m as the random effect of lamb, where ID~ND(0, σ^2_{ID}); e is the random residual error~ND (0, σ^2_e). Adjustment of each trait for age is done to the overall mean age of all the animals at the time of recording the specific trait. For the models (1) and (2), adjusted coefficients of determination (adjusted R^2) were calculated according to (Fox and Weisberg 2011) based on Wherry (1931), as the data set was rather small.

Model (1) was used to estimate differences between breed groups. Based on results from (1), an estimate of heterosis was calculated for the live animal and carcass traits by first dividing the least squares mean of DDRR by the average least squares means for RRRR and DDDD, and given as percent of that average. The heterosis was solved for the expectation of this value: ($\text{propF}_1 \times \text{heterosis} + \text{propF}_2 \times \text{heterosis} \times 0.5$), where propF_1 was 13/24 and propF_2 was 11/24. Associations between traits were analysed by using Pearson correlations of residuals from (1). Model (2) was employed to study how objective live measures of weight and conformation or subjective assessments can explain the variation in carcass traits including assessed economic value of the lamb (WPA). For the assessed traits, records based on the common evaluator between batches 1 and 2 were used. Model (3) was used to study the effect of evaluator on the assessed traits.

3. RESULTS

3.1 Significance of model factors

Results from (1) showed that effects of breed groups and age were significant ($p < 0.05$) or close to significant ($p < 0.1$) for almost all traits (not shown in tables), whereas effects of batch were small despite 2 months difference in age at slaughter. Live animal measures (LW, HG, BL and DW) were shown in (2) to significantly predict GR, CWT, CMF and WPA. Model (3) showed that evaluators significantly ($p < 0.001$) affected all subjectively assessed traits including WPA except for CA.

3.2 Breed group differences

Results in Table 2 show that pure Red Maasai had a 0.7 kg lower birth weight and a 2.1 kg lower live weight (LW) ($p < 0.08$) and weight at delivery (DW) ($p < 0.07$) than pure Dorper. Both crossbred groups exceeded the pure breeds in LW and DW. As regards conformation, Red Maasai had significantly shorter body length (BL) than the other breed groups, whereas heart girth (HG) showed only small differences between the pure breeds. Dorper and the crosses were offered significantly higher prices by the evaluators (WPA), 12 % or KES 519 (US$ 6.18) more than for pure Red Maasai lambs. The latter also produced carcasses with significantly lower scores for carcass grade (CG) compared with Dorper (Table 3). However, carcass weight (CWT) did not differ between the pure breeds, whereas the 50 %-crosses (DDRR) were significantly heavier, 12–13 %, than the pure breeds. The dressing percentage (DP) was significantly higher for Red Maasai (41.6 %) compared with Dorper (39.1 %), whereas the crosses did not significantly differ from any of the parental breeds.

Based on results from the 50 %-crosses (DDRR), positive heterosis effects were realized for all traits, and on an average, the 50 %-crosses were 8–9 % better than the mean of the pure breeds in all traits.

Table 2 Least squares means ± standard error and approximate heterosis (%) for live animal measures by level of breeds based on model (1) for all traits except for WPA that is based on model (3)

Breed	BW	LW	HG	BL	WPA	DW
RRRR	3.40 ± 0.12^{a}	36.07 ± 0.97^{a}	79.61 ± 0.88^{ab}	57.85 ± 0.87^{a}	4321 ± 200^{a}	36.35 ± 1.02^{a}
DDRR	3.79 ± 0.12^{b}	39.47 ± 0.97^{b}	80.99 ± 0.89^{b}	62.36 ± 0.87^{b}	4844 ± 193^{b}	39.89 ± 1.02^{b}
DDDR	4.05 ± 0.09^{b}	38.59 ± 0.77^{b}	79.61 ± 0.70^{ab}	61.83 ± 0.68^{b}	4844 ± 171^{b}	38.93 ± 0.83^{b}
DDDD	4.11 ± 0.14^{b}	38.16 ± 1.13^{ab}	78.25 ± 1.03^{a}	60.98 ± 1.01^{b}	4840 ± 213^{b}	38.48 ± 1.18^{ab}
Heterosis	1.2	8.2	3.4	6.5	7.5	8.6

BW birth weight, *LW* live weight, *HG* heart girth, *BL* body length, *WPA* willingness to pay, *DW* delivery live weight

^{a,b}Means within column with different superscripts are significantly different at $p < 0.05$

Table 3 Least squares means ± standard error and approximate heterosis (%) for growth rate and carcass traits by level of breeds based on model (1)

Breed	GR	CWT	CFL	CMF	CG^{c}	DP
RRRR	92.00 ± 2.74^{a}	15.11 ± 0.49^{a}	2.12 ± 0.11^{b}	2.02 ± 0.18^{a}	5.03 ± 0.33^{b}	41.61 ± 1.10^{b}
DDRR	101.31 ± 2.73^{b}	16.50 ± 0.49^{b}	2.04 ± 0.11^{ab}	2.56 ± 0.18^{b}	4.74 ± 0.33^{ab}	41.35 ± 1.10^{ab}
DDDR	97.53 ± 2.22^{b}	15.80 ± 0.40^{ab}	1.94 ± 0.09^{ab}	2.17 ± 0.14^{a}	4.24 ± 0.27^{a}	40.66 ± 0.89^{ab}
DDDD	96.19 ± 3.15^{ab}	15.03 ± 0.57^{a}	1.81 ± 0.13^{a}	2.25 ± 0.20^{ab}	3.97 ± 0.38^{a}	39.09 ± 1.27^{a}
Heterosis	10.0	12.3	4.9	25.8	6.9	3.2

GR growth rate to delivery, *CWT* cold carcass weight, *CFL* carcass fat level, *CMF* carcass muscle formation, *CG* carcass grade, *DP* dressing percentage

[a,b] Means within column with different superscripts are significantly different at $p < 0.05$

[c] Lower value of CG meaning better carcass grade

Table 4 Adjusted coefficients of determination for growth and carcass traits and for willingness to pay using different factor combinations based on model (2)

Factor combination	GR	CWT	CG	CFL	CMF	DP	WPA
Batch + Age	0.457	0.020	0.251	0.099	0.121	0.000	0.343
Batch + Age + LW	0.951	0.435	0.252	0.111	0.259	0.000	0.522
Batch + Age + LWA	0.477	0.025	0.252	0.092	0.116	0.000	0.459
Batch + Age + WPA	0.606	0.103	0.263	0.101	0.178	0.000	
Breed + Batch + Age(Breed)	0.518	0.120	0.311	0.185	0.199	0.057	0.460
Breed + Batch + Age(Breed) + LW	0.953	0.478	0.308	0.216	0.271	0.067	0.576
Breed + Batch + Age(Breed) + LWA	0.536	0.126	0.305	0.187	0.191	0.049	0.540
Breed + Batch + Age(Breed) + WPA	0.633	0.191	0.305	0.174	0.224	0.050	

LW live weight, *WPA* willingness to pay, *LWA* live weight assessment, *GR* growth rate to delivery, *CWT* cold carcass weight, *CG* carcass grade, *CFL* carcass fat level, *CMF* carcass muscle formation, *DP* dressing percentage

3.3 Influence of live animal traits on carcass traits and willingness to pay

Table 4 shows the amount of variation in growth rate, carcass traits and WPA that was explained by different factor combinations (breed, batch, age, live measures and assessments). LW was the live measurement explaining most of the variation in carcass traits and WPA, accounting for 41–49 % of the variation in CWT and GR, 14 % in CMF and 18 % in WPA. When LW was included in the model, HG, BL or DW did not have an effect on the carcass traits (results not shown). However, LW did not explain any of the variation in DP, CG or CFL. When breed group effects were adjusted for, LW still explained most of the variation, between 12 and 44 % in GR, CWT and WPA. Overall, breed groups accounted for 6–12 % of the observed variation in growth, carcass traits and WPA.

Although much less informative than LW, WPA was the most informative attribute of the variation in carcass traits, 7–8 % of the variation in CWT when data were adjusted for breed, batch and age, but much less informative of the other carcass traits. Subjective assessments of live weight (LWA) and other conformation traits (CA, BSA and TSA) explained no more than 2 % of the variation in growth and carcass traits when breed was adjusted for.

3.4 Residual correlations between weights, conformation and growth traits

Weights, conformation and growth traits are closely related (Table 5). The residual correlation between LW and CWT was 0.64. It was 0.69 between LW and HG, whereas it was only 0.31 between LW and BL. The correlations between CWT and the two conformation measures were 0.47 (HG) and 0.24 (BL). All correlations between LW and live conformation measures with CG and DP were weak and non-significant.

4. DISCUSSION

4.1 Breed differences in live weight and carcass traits

The 2 kg, or 6 %, higher live weight of Dorper lambs at about 1 year of age compared to Red Maasai confirms the superiority of Dorper live weights found in previous studies (Kiriro 1994; Baker 1998; Baker et al. 2002; Mugambi et al. 2005; Zonabend König et al. 2016). However, Zonabend König et al. (2016) indicated even larger differences between the two breeds, especially in semi-arid environments, which also characterize the environment of this study. A reason for the smaller difference in live weight between the two breeds in the current study may be that the Kapiti herd of Red Maasai sheep has undergone a significant improvement through selection for live weight and conformation in the past decades (A.M. Okeyo, personal communication, 2015).

Table 5 Residual correlations between live weight, conformation, growth rate, carcass traits and willingness to pay based on model (1) and the common evaluator between batches 1 and 2 in the upper right triangle

	BW	LW	HG	BL	WPA	DW	GR	CWT	CFL	CMF	CG	DP
BW		0.49	0.40	0.31	0.05	0.44	0.30	0.29	0.07	0.28	−0.26	−0.12
LW	***		0.69	0.31	0.47	0.98	0.95	0.64	0.22	0.32	−0.09	−0.20
HG	***	***		0.27	0.33	0.66	0.64	0.47	0.08	0.15	−0.09	−0.12
BL	**	**	*		0.07	0.29	0.26	0.24	0.09	0.19	−0.16	−0.02
WPA	ns	***	**	ns		0.48	0.50	0.31	0.00	0.21	−0.06	0.04
DW	***	***	***	**	***		0.98	0.61	0.16	0.29	−0.07	−0.17
GR	**	***	***	*	***	***		0.58	0.18	0.26	−0.02	−0.15
CWT	**	***	***	*	**	***	***		0.23	0.42	−0.11	0.32
CFL	ns	*	ns	ns	ns	ns	ns	*		0.09	−0.02	−0.04
CMF	**	**	ns	ns	*	**	*	***	ns		−0.31	0.14
CG	*	ns	ns	ns	ns	ns	ns	ns	ns	**		0.02
DP	ns	ns	ns	ns	ns	ns	ns	**	ns	ns	ns	

Significance levels are shown in the lower left triangle

BW birth weight, *LW* live weight, *HG* heart girth, *BL* body length, *DW* delivery live weight, *GR* growth rate to delivery, *CWT* cold carcass weight, *DP* dressing percentage, *CFL* carcass fat level, *CMF* carcass muscle formation, *CG* carcass grade, *WPA* willingness to pay

Significance levels: ***$p < 0.001$, **$p < 0.01$, *$p < 0.05$, ns = nonsignificant

Despite the difference in live weight, no significant difference was observed in carcass weight between the two pure breeds in the present study, 15.1 vs 15.0 kg cold carcass weight for Red Maasai and Dorper. This result has been obtained due to a higher ($p < 0.05$) dressing percentage of Red Maasai, 41.6 vs. 39.1 % for Dorper. Significantly leaner (CFL) carcasses of Dorper may have contributed to the lower dressing percentage. As a fat-tailed sheep breed, the Red Maasai has fat deposits around the tail and rump contributing to a higher dressing percentage, whereas the fat in Dorper carcasses are more distributed around the internal organs and are easier to separate at slaughter. It may also be speculated whether larger and heavier head, offals and skin of the Dorper lambs contribute to the lower dressing percentage of this breed. Unfortunately, there are no previous studies appropriate to compare with from the region except the Ethiopian study on crosses between Dorper and Blackhead Ogaden, where the Dorper crosses clearly had the highest dressing percentage (Tsegay et al. 2013).

The higher fat levels and, thus, also higher dressing percentage of Red Maasai may be an expression of its well-known adaptation characteristics necessary to withstand harsh climates with intermittent droughts and for reproduction ensuring the animal has energy depots for use when needed.

Carcass grades in the present study were significantly better ($p < 0.01$) for Dorper than for Red Maasai. However, the standards of grading were less rigorous than the case under international standards, and the differences may therefore be treated more lightly. The scores for muscle formation were also slightly higher for Dorper. As this study, which comprised a limited number of purebred lambs, is to the best of our knowledge the first one reported regarding comparison of carcass traits of Red Maasai and Dorper sheep, and since carcass weights unexpectedly did not differ between the breeds, further studies on slaughter lambs of Red Maasai and Dorper lambs are warranted.

4.2 Crossbreeding effects

The crosses were superior to the two parent breeds for most traits. Approximate heterosis effects based on the 50 %-crosses (DDRR) were estimated at 6.5 % for live traits and about 12 % for CWT. These results can be compared with those by Kiriro (1994), where Red Maasai and Dorper crossbred lambs were 4.5 % heavier at birth and 11 % at weaning than the average of the parental breeds. Results of the backcrosses (DDDR) are generally intermediate between pure Dorper and the 50 % crosses. For dressing percentage, effects of heterosis seem marginal. Both crossbred groups have results regarding dressing percentage more similar to Red Maasai than to Dorper. The superior effects of the crosses indicate one reason why crossbreeding between Red Maasai and Dorper has become very common in the study area. A production system where the more resilient and low maintenance Red Maasai ewes are used as dams to faster growing crosses with Dorper may be ideal.

4.3 Economic evaluations

The economic values of the lambs, as expressed by the willingness to pay assessment by three experienced evaluators, one farmer, one middleman and one butcher, showed a significant ($p < 0.01$) difference in favour of Dorper over Red Maasai by 12 % despite that no difference was detected later in carcass weight. This confirms earlier studies of Mtimet et al. (2014) and of Zonabend König et al. (2016), where Dorper sheep were better priced than Red Maasai at farm gate and in local livestock markets. In the latter study of the value of potential breeding ewes, Dorpers were assigned a 9 % higher value in the least market-oriented area, whereas the more commercial farmers were willing to pay almost twice as much for a Dorper ewe as for a Red Maasai. The WPA value may in this case, however, include much more than just carcass value, such as expected breeding potential including appearance of the animal, skin and wool quality.

Crossbreds were given the same WPA value as pure Dorper, about KES 4840 (USD 57.62) in this study. However, in the study of breeding ewes by Zonabend König et al. (2016), the crossbreds were valued at an intermediate price, a little closer to Red Maasai than the price of Dorper. Also, Kosgey et al. (2008) reported that exotic and crossbred sheep fetched higher prices than indigenous animals. The price quoted by KMC was a flat rate of 150 KES per kilogramme live weight or 315 KES per kilogramme hot carcass weight, which corresponds to approximately 325 KES (USD 3.87) per kilogramme cold carcass weight. If these figures are applied on the present data, evaluators were willing to pay the lambs approximately 900–1100 KES (USD 10.72–13.10), or 19–26 %, less than the quoted price per kilogramme live weight by KMC. If comparisons were made with quoted prices per kilogramme carcass weight, without any premium for better carcass grades, the Red Maasai lambs were undervalued by 545 KES (USD 6.49), or 13 %, compared to the Dorper lambs. The price difference between the two quoted systems of KMC amounted to a 540–880 KES (USD 6.43–9.52), or 10–15 %, higher payment according to the live weight quotes, again the highest difference was for Dorper. These differences should cover costs that normally farmers, middlemen or butchers take as they operate at primary or secondary markets compared to the commercial abattoirs. However, Red Maasai lambs seem again to be undervalued compared to Dorper lambs, in this case by 340 KES (USD 4.05), or 8 %. Crossbred lambs were also undervalued compared to pure Dorper, but less so than Red Maasai.

Thus, large differences exist in how evaluators value the breeds, and where Red Maasai lambs were most undervalued, compared to payments according to empirical weights. Large differences were also shown among the assessors' willingness to pay for the same animals. In the present study, the number of evaluators differed between the two batches, but the participating butcher was common to both batches. Thus, the effects of evaluator on the results were minimized by the statistical model used. Due to the apparent differences among evaluators (middlemen), it is important that farmers strive for delivery of lambs directly for slaughter and that payment is based on objective measures of weight. Middlemen often control the market and exploit the farmers by buying sheep at very low prices (Kosgey et al. 2006). Development of farmers' associations would also be beneficial for farmers in order to better inform on the real market prices of the animals.

Carcass grades were better for Dorper and the crosses, and these breed groups would be favoured if grades were considered in the payment scheme. This may be more so given that exports and local high-end markets prefer carcasses of younger animals with better grades. The markets may, however, vary as regards preferences. The leaner carcasses of Dorper may attract customers close to cities, whereas Red Maasai carcasses may be more suited for rural local markets (Katiku et al. 2013). A research report about exports and market structure for live animals in Ethiopia showed that small traders and export abattoirs value lambs for slaughter differently. Besides weight, small traders valued skin condition highly, which was not as relevant for export abattoirs (Hailemariam et al. 2009).

4.4 Live animal measures for prediction of carcass weight

Weighing animals intended for slaughter is the best way to predict carcass weight. Live weight (LW) explained 36 % of the variation in CWT when adjusting also for breed (Table 4) and the residual correlation between LW and CWT was 0.64 (Table 5). If a weighing scale is not available, a conformation measure may be used as a proxy. Heart girth (HG) showed the highest residual correlation with CWT, 0.47, among the conformation measures taken. Subjective assessments of weight hardly explained any variation in CWT. Willingness to pay (WPA) was the best subjective predictor, explaining 7 % if breed, age and batch were adjusted for. Large differences between evaluators were, however, observed for all live assessment traits except for conformation assessment (CA).

As regards correlations with LW among the live measures, again HG showed the highest residual correlation, 0.69, which is adjusted for the fixed factors breed, age and batch. A study on Dorper ram lambs by Fourie et al. (2002) indicated higher values, with a direct correlation between heart girth and live body weight of 0.8. However, it was based on data with high standard errors of the means of studied variables.

A weighing scale would, according to our study, be highly preferred when deciding which lambs should be sent to slaughter. It also helps the farmers to get lambs more correctly priced when selling their lambs live to middlemen or butchers. A tape for measuring heart girth would be a substitute for a scale, but with lower expected accuracy.

4.5 Aspects on a breeding programme

All ram lambs in this study were born and raised at the Kapiti Plains Estate Limited (Kapiti), which is characterized by semi-arid climate. Larger differences in live weight between Red Maasai and Dorper would therefore have been expected following results of previous studies in the same environment (Baker 1998; Baker et al. 2002). However, the earlier studies reflect the population characteristics as they were about two decades ago and much may have changed since then. Kapiti mainly sources rams from within its own flocks, but also from other farms in the country. Excess animals not used for replacement at Kapiti are normally sold to local farmers interested to obtain breeding animals from the ranch. Thus, the genetic material from Kapiti is widespread in the country. Ram lambs in this study may, therefore, be considered representative for lambs raised in the region.

Several important results relating to the design of a breeding programme were obtained from the current study. Firstly, the advantages of Dorper as regards production of crossbred lambs for slaughter have been clearly demonstrated. Thus, the use of Dorper as a terminal sire breed to be crossed with Red Maasai ewes might be beneficial. Secondly, pure Red Maasai lambs seem to be undervalued by live assessments, 340–545 KES (USD 4.05–6.49) compared with Dorper lambs given that payment is only per kilogramme of empirical live or carcass weight. These discrepancies in economic values need to be considered when evaluating Red Maasai vs. Dorper and their crosses for slaughter and in the choice of a breeding strategy. Recording schemes to support both management decisions by the farmers, when selecting lambs for slaughter, as well as providing live weight data for selection purposes, should emphasize the use of as objective weight information as possible.

5. CONCLUSIONS

Large differences in live weight prior to slaughter were found with higher weights of Dorper over pure Red Maasai. Crosses were better than both parental breeds for most traits. Carcasses of Dorper lambs were not heavier than those of pure Red Maasai but had better carcass grades. Red Maasai lambs had a higher dressing percentage and carcass fat level. Red Maasai lambs were undervalued by assessments of experienced evaluators by 8–13 % compared to pure Dorper based on empirical live weight and carcass weight payment schemes practised by the abattoir. Live weight was better than any other live measure in predicting carcass weight, which underpins the use of weighing scales when selecting lambs for slaughter and for breeding purposes. Based on the results of the crossbreds, Dorper might be useful as a terminal sire breed in crossing with Red Maasai ewes.

ACKNOWLEDGMENTS

We sincerely thank the whole team at the ILRI research ranch, Kapiti Plains Estate Limited, for the work with the lambs and for all the support during the collection of data. We are grateful to the evaluators that participated in the study, answered all questions and allowed us to work with them. Sincere thanks also to Kenya Meat Commission for the support during the slaughter phase of the lambs. The Swedish Ministry of Foreign Affairs funded the study and is gratefully acknowledged.

REFERENCES

1. Baker, R.L., 1998. Genetic resistance to endoparasites in sheep and goats. A review of genetic resistance to gastrointestinal nematode parasites in sheep and goats in the tropics and evidence for resistance in some sheep and goat breeds in sub-humid coastal Kenya. Animal Genetic Resources / Resources génétiques animales / Recursos genéticos animales 24, 13–30
2. Baker, R.L., Mugambi, J.M., Audho, J.O., Carles, A.B., Thorpe, W., 2002. Comparison of Red Maasai and Dorper sheep for resistance to gastro-intestinal nematode parasites, productivity and efficiency in humid and a semi-arid environment in Kenya. Proceedings of the 7th World Congress in Genetics Applied to Livestock Production. WCGALP, Montpellier, France, pp. 639–642.
3. R Core Team, 2015. R: A language and environment for statistical computing. R Foundation for Statistical Computing. http://www.R-project.org/.
4. Ermias, E., Yami, A., Rege, J.E.O., 2002. Fat deposition in tropical sheep as adaptive attribute to periodic feed fluctuation. Journal of Animal Breeding and Genetics 119, 235–246.
5. Ermias, E., Yami, A., Rege, J.E.O., 2006. Slaughter characteristics of Menz and Horro sheep. Small Ruminant Research 64, 10–15.
6. FAO. 2016. FAOSTAT database. FAO Statistics Division, Rome. (available at: http://faostat3.fao.org/browse/Q/QA/E) Accessed: 2016-08-09.
7. Fourie, P.J., Neser, F.W.C., Olivier, J.J., van der Westhuizen, C., 2002. Relationship between production performance, visual appraisal and body measurements of young Dorper rams. South African Journal of Animal Science 32, 256–262.
8. Fox, J., Weisberg, S., 2011. An {R} Companion to Applied Regression. Sage, http://cran.r-project.org/web/packages/car/index.html.
9. Gizaw, S., Komen, H., van Arendonk, J.A.M., 2010. Participatory definition of breeding objectives and selection indexes for sheep breeding in traditional systems. Livestock Science 128, 67–74.
10. Hailemariam, T., Getachew, L., Dawit, A., 2009. Market structure and function for live animal and meat exports in some selected area of Ethiopia. In: Abebe, K. (Ed.), Research Report 79. EIAR, Addis Ababa, Ethiopia.

11. Herrero, M., Havlik, P., McIntire, J., Palazzo, A., Valin, H., 2014. African Livestock Futures: Realizing the Potential of Livestock for Food Security, Poverty Reduction and the Environment in Sub-Saharan Africa. Geneva, Switzerland, p. p. 90.
12. ILRI, 2016. Kapiti Sheep Flock. In: (ILRI Animal Genetic Resources (AnGR)), ILRI. (Ed.),. https://ilri-angr.wikispaces.com/ILRI+Kapiti+Sheep.
13. Juma, G.P., Ngigi, M., Baltenweck, I., Drucker, A.G., 2010. Consumer demand for sheep and goat meat in Kenya. Small Ruminant Research 90, 135–138.
14. Katiku, P.N., Kimitei, R.K., Korir, B.K., Muasya, T.K., Chengole, J.M., Ogillo, B.P., Munyasi, J.W., Karimi, S.K., 2013. Value chain assessment of small ruminant production, challenges and opportunities: The case of southern rangelands in Kenya. Livestock Research for Rural Development 25.
15. Kiriro, P.M., 1994. Estimate of genetic and phenotypic parameters for the Dorper, Red Maasai and their crosses. In: Lebbie, S.H.B., Rey, B., Irungu, E.K. (Eds.), Proceedings of the Second Biennial Conference of the African Small Ruminant Research Network AICC. International Livestock Centre for Africa (ILCA) / Technical Centre for Agricultural and Rural Co-operation (CTA), Arusha, Tanzania, pp. 229–234.
16. Kosgey, I.S., Baker, R.L., Udo, H.M.J., Van Arendonk, J.A.M., 2006. Successes and failures of small ruminant breeding programmes in the tropics: a review. Small Ruminant Research 61, 13–28.
17. Kosgey, I.S., Rowlands, G.J., van Arendonk, J.A.M., Baker, R.L., 2008. Small ruminant production in smallholder and pastoral/extensive farming systems in Kenya. Small Ruminant Research 77, 11–24.
18. Lambe, N.R., Navajas, E.A., Schofield, C.P., Fisher, A.V., Simm, G., Roehe, R., Bünger, L., 2008. The use of various live animal measurements to predict carcass and meat quality in two divergent lamb breeds. Meat Science 80, 1138–1149.
19. Lenth, R.V., 2013. Least-squares means. In: Lenth, R. (Ed.). R. Lenth, http://cran.r-project.org/web/packages/lsmeans/index.html
20. Mtimet, N., Baker, D., Audho, J., Oyieng, E., Ojango, J., 2014. Assessing Sheep Traders' Preferences in Kenya: A Best-worst Experiment from Kajiado County. UMK Procedia 1, 63–73.
21. Mugambi, J.M., Audho, J.O., Baker, R.L., 2005. Evaluation of the phenotypic performance of a Red Maasai and Dorper double backcross resource population: natural pasture challenge with gastro-intestinal nematode parasites. Small Ruminant Research 56, 239–251.
22. Ojango, J.M.K., Oyieng, E.P., Audho, J., Okeyo, A.M., 2014. Indigenous sheep to help improve market access and livelihood security among pastoralists in Kenya: Results of a baseline survey. International Livestock Research Institute, Nairobi, Kenya, p. 23.
23. Pinheiro, J., D., B., DebRoy, S., Sarkar, D., 2015. {nlme}: Linear and Nonlinear Mixed Effects Models. In: Team, R.C. (Ed.). R package, http://CRAN.R-project.org/package=nlme.
24. Tsegay, T., Yoseph, M., Mengistu, U., 2013. Comparative evaluation of growth and carcass traits of indigenous and crossbred (Dorper × Indigenous) Ethiopian Sheep. Small Ruminant Research 114, 247–252.
25. Wherry, R.J., 1931. A New Formula for Predicting the Shrinkage of the Coefficient of Multiple Correlation. The Annals of Mathematical Statistics 2, 440–457.
26. Zonabend König, E., Mirkena, T., Strandberg, E., Audho, J., Ojango, J., Malmfors, B., Okeyo, A., Philipsson, J., 2016. Participatory definition of breeding objectives for sheep breeds under pastoral systems—the case of Red Maasai and Dorper sheep in Kenya. Tropical Animal Health and Production 48, 9–20.

CHAPTER 22

Begait Goat Production Systems and Breeding Practices in Western Tigray, North Ethiopia

Hagos Abraham[1,2*], Solomon Gizaw[3], Mengistu Urge[2]

[1]*Department of Animal Science, Bule Hora University, Bule Hora, Ethiopia*

[2]*School of Animal and Range Sciences, Haramaya University, Dire Dawa, Ethiopia*

[3]*LIVES Project, International Livestock Research Institute (ILRI), Addis Ababa, Ethiopia*

ABSTRACT

A study was undertaken to characterize production system, breeding practices and production constraints of Begait goat in Kafta humera district, northwestern lowlands of Ethiopia. Personal observations, focus group discussions and a detailed structured questionnaire were used to collect data from 150 respondents (99 small-scale and 51 large-scale farmers). Mixed crop-livestock production system was found to be the dominant farming system in the study district. The farmers kept a variety of livestock species including goats, sheep, cattle, chickens and donkeys. Small ruminants are the most numerous and are of great importance in the area. Goats are herded with sheep in the open fields by the young boys in the family/hired grazers. They have a number of roles though mainly kept as a source of cash income. Breeding was generally uncontrolled. Size, growth rate and libido were the most frequently reported traits in selecting bucks, whereas does were selected based on size, twining ability and milk yield. Effective population size and rate of inbreeding were calculated by considering random mating and flocks were not mixed. The major constraints hampering goat production in the area were water scarcity, feed and grazing land shortages and insufficient veterinary services among the others. Therefore, addressing these constraints is important to design a successful genetic improvement scheme in the area for goat.

Keywords: Farming System, Husbandry Practice, Marketing, Production Constraints

1. INTRODUCTION

Goats (Capra hircus) are of the most beneficial animals. Ethiopia is one of the top ten goat producing countries in the world. As per the livestock census carried out in 2016, Ethiopia owns approximately 30 million goats; of which 71.57% are females and the rest 28.43% are males. Majority of the national goat population is found in the lowlands. Nearly all (99.99%) of the goats are indigenous types [1] [2], which have become adapted to a wide agro-ecological zones of the country [3] due to natural selection. The environmental adaptation of indigenous breeds facilitates livestock production in a range of agro-ecological conditions and constraints [4].

Goats are deeply embedded in almost every African culture and are true friends to the rural poor [5]. For instance, goats in Ethiopia are important to the subsistence, economic and social livelihoods of their owners [6] [7] [8]. On top of this, Ethiopia generates hard currency through exporting live goats, goat meat and goat skins. At optimum off-take rates, the country has a potential to deliver 1.13 and 2.00 million goats to national and international markets per annum, respectively [7].

Despite the large goat population is found in the country and its contributions to household income and to the country's national economy, productivity per unit of animal is low. Genetic improvement is one way to increase productivity of the goat genetic resources in the country. According to [9], development of animal genetic improvement plans will only be successful when accompanied by good understanding of the production system and when simultaneously addressing constrains. However, there is little or no adequate information on majorities of indigenous goat breeds/types in Ethiopia to design appropriate breeding plans, suggesting a low level of priority for this area of research. Specifically, the study aims at:

1) Characterizing production system and breeding practices of Begait goat;
2) Identifying the major constraints facing Begait goat production in its home tract.

2. MATERIALS AND METHODS

2.1. Description of the Study Area

This research work was done in Kafta and Setit humera districts of western zone of Tigray National Regional State, northwest lowlands of Ethiopia. Setit humera town, a special district, is the administrative center of Kafta humera district and western zone of the region. In this study, the district has been integrated as part of Kafta humera district. The study area covers a total area of 71.80 km^2 of which 54.20% is cropland.

The study area is situated 1372 km away from the administrative center of Addis Ababa city to northwest direction. It is bordered by Sudan to the west, Tahtay Adyabo district to the east, Eritrea to the north and Wolkayit and Tsegedei districts to the south. Geographically, it is located in between the latitudes 13°14'N - 14°27'N and longitudes 36°27'E - 37°32'E. The study area comprises kolla (lowland) and weinadega (midland) agro-climatic zones with an altitude ranging from 560 - 1849 m above sea level. The annual precipitation in the lowlands and midlands is 448.8 and 1102.5 mm, respectively [10] occurring between Jun and September. The mean minimum and maximum temperatures in the lowlands are 25°C and 27.5°C, respectively while the corresponding values in the midlands are 20°C and 25°C. The hottest months are between April and June with temperatures rising up to 42°C.

2.2. Farming System

The agricultural production system in the study area is mixed farming. There are two types of farming systems in the district viz. small-scale and large-scale based farming systems. As a result, farm activities predominantly crop production and livestock rearing are performed by small-scale and large-scale farmers. According to [11], above half (58%) of the total cropland is cultivated by large-scale farmers while small-scale farmers cultivate the rest 42%. Crop production is a rain fed production system without irrigation practice. Sesame and sorghum are the major staple food and cash crops grown in the area. Other crops like cotton, millet, maize and teff are well grown.

Livestock production is similar in both farming systems, and it is a traditional type which is characterized by minimal inputs as is in other areas of Ethiopia. [12] Demonstrated that low input production system is found in all livestock production systems prevailing in the country except in Peri-urban and urban system. The most dominant farm animal species of the area are cattle and small ruminants. Animal feeding is free grazing.

2.3. Sampling and Data Collection

Prior to site selection, livestock experts in the district were contacted for brief discussions. In addition, secondary data were reviewed to better understand the current production systems and the area dominated by Begait goat. Based on this information, four rural kebeles and the special district were selected as study sites. A total of 150 Begait goat owning farmers (99 small-scale and 51 large-scale farmers) were selected at random.

Interviews were conducted at the farmers' residences using structured questionnaire with the assistance of development agents. The questionnaires covered information on key household characteristics, livestock possession, flock structure, purposes of keeping goat, mating system, marketing system and production constraints. To complement the survey work, focus group discussions were held with a total of 8 - 10 key informants per selected study site. In addition, general information about the district was obtained from secondary data sources.

2.4. Data Analysis

The survey data were analyzed using SPSS software (SPSS for windows, release 16.0, 2007). F test was applied when required to test statistical significances. Indices were calculated for all ranking data using the formula: Index = Σ [3 for rank 1 + 2 for rank 2 + 1 for rank 3] given for an individual reason divided by Σ [3 for rank 1 + 2 for rank 2 + 1 for rank 3] for all reasons [13]. By considering the number of breeding males (Nm) and breeding females (Nf), the effective population size (Ne) and rate of inbreeding (ΔF) were also estimated using the equations of $Ne = 4(Nm \times Nf)/Nm + Nf$, respectively [14].

3. RESULTS AND DISCUSSION

3.1. Household Information

Table 1 presents some key characteristics of the respondents across the two farming systems. The overall average age of the sampled household head was 50 ± 10.44 years, implying that the respondents were adults with a good experience in goat farming. The overall mean family size was 4.85 ± 0.89, which is comparable with the national estimate of 4.80 persons per household [15].

Of the total sampled household heads, 98.67% were males. The rest (1.33%) households were female headed. Indeed, it is unlikely to find a female headed household unless she is widowed or divorced. Literacy rate among the household heads was 69.67%. Similar results were reported by [16] in Dale district and [17] in Goma district. The better educational background obtained in this study would be a good opportunity for further animal genetic improvement programs in the study area, since literate communities are more likely adopt and practice new technologies. The study also revealed that most (98%) of the respondents were married. The percentage of unmarried and widowed was 0.67% and 1.33%, respectively. The present findings indicate that goat farming can be performed by every social class of the community regardless of their background characteristics, viewing the significant importance of goat for its owners.

3.2. Livestock Holding and Flock Structure

Table 2 demonstrates flock structure and livestock possession across the two farming systems. Overall, sampled farmer owned 43.67 goats, 42.90 sheep, 10.19 cattle, 9.70 chickens and 0.96 donkeys. There was a slight increase in mean livestock holding of respondents from small-scale to large-scale farming even if the difference is not statistically significant except for cattle. None of the respondents owned mules and horses. This finding supports the fact that mules and horses could not adapt the hot environmental conditions of the area.

The mean flock size of goats in the present study was close to the work of [18] and [19], who reported that 44.0 and 48.5 heads per household in Siti zone of Somali region and Abergelle district of Amhara region, respectively. In contrast, the mean flock size recorded in this study is higher than the previous average flock size of 21.20 Begait goats in the area [20]. This difference might be due to random sampling error or due to changes in goat farming system over time in the area.

Table 1. Background characteristics of the respondents by farming system.

Descriptor variables	Small scale farming (n = 99)	Large scale farming (n = 51)	Overall (N = 150)
	Mean ± SD	Mean ± SD	Mean ± SD
Age(years)	48.24 ± 9.88	53.10 ± 10.85	50.00 ± 10.44
Family size	4.91 ± 0.94	4.75 ± 0.80	4.85 ± 0.89
Sex	Percent	Percent	Percent
Male	98.33	100.00	98.67
Female	1.67	-	1.33
Educational level			
Illiterate	27.30	37.30	30.33
Literate	72.70	62.70	69.67
Marital status			
Married	97.00	100.00	98.00
Unmarried	1.00	-	0.67
Widowed	2.00	-	1.33

N: Number of interviewed households; SD: Standard deviation.

Table 2. Mean livestock holdings and flock structure across the two farming systems.

Livestock species	Small-scale farming	Large-scale farming	Overall
	(mean ± SD)	(mean ± SD)	(mean ± SD)
Goat	42.21 ± 18.42^a	46.43 ± 15.94^a	43.67 ± 17.49
Kids	11.92 ± 5.22^a	13.16 ± 4.33^a	12.34 ± 4.96
Buckling	3.18 ± 1.28^a	3.51 ± 1.07^a	3.29 ± 1.22
Doeling	5.18 ± 2.29^a	5.71 ± 1.88^a	5.36 ± 2.17
Bucks	1.89 ± 0.92^a	1.99 ± 0.81^a	1.94 ± 0.89
Does	19.54 ± 8.41^a	21.47 ± 6.99^a	20.19 ± 7.98
Castrates	0.67 ± 0.47^a	0.76 ± 0.43^a	0.70 ± 0.46
Sheep	41.68 ± 17.86^a	45.41 ± 15.07^a	42.90 ± 17.09
Cattle	9.62 ± 4.68^b	11.29 ± 3.98^a	10.19 ± 4.51
Chicken	9.45 ± 3.69^a	10.16 ± 0.70^a	9.70 ± 3.57
Donkey	0.97 ± 0.28^a	0.94 ± 3.34^a	0.96 ± 0.20

Means with the same letter within the same row are not significantly different ($p > 0.05$); SD: standard deviation.

The number of males and females in the sample population and their ages were often used as an indicator of a traditional management system in Africa [21]. The present finding revealed that the proportion of does represents the largest class followed by suckling kids, while castrates represent the lowest proportion (Table 2). This is in good agreement with previous findings in Ethiopia [17] [22] [23] [24] and elsewhere in Africa [25] [26] [27]. The higher proportion of does than other age groups suggests that adult females stay in a flock for breeding purposes and/or milk production. However, the less number of buckling as compared to doeling is because of marketing and slaughtering of buckling. There was a high level of elimination of the majority of born male kids.

Small ruminants were the predominant livestock species in the area (Table 3) and this shows that their wide acceptability and ease of adaptation in the area. Sheep and goat numbers are growing fastest in the mixed farming systems [28] as subsistence farmers prefer small stock because the risk of losing large ruminants is high [29]. In rural areas of Ethiopia, because of their subsistence and economic reasons, goats and sheep have been described as bank reserve which can be drawn upon when cash money is required. This confirms a study by [30], who reported that keeping livestock, especially small ruminants plays role as safety net that enables households to get quick income to settle urgent financial needs.

3.3. Purposes of Keeping Goat

Knowledge of reasons for keeping animals is a prerequisite for deriving operational breeding goals [6] [31]. Table 4 summarizes reasons of keeping goats and the ranking of these purposes by owners. The most frequently reported reason for keeping goat was cash income generation followed by milk and meat production for home use. This implies that sale of goats to generate cash constitute the primary purpose among the other benefits of keeping goats in the area. This was also observed in similar studies [32] [33] [34] [35], who reported that cash income as the most important reason in goat farming. However, the purpose of keeping goats in rural areas of South Africa is mainly for traditional purposes and meat consumption [36].

3.4. Land Holding and Goat Production

Mixed crop-livestock production system takes place as an important source of livelihood in the area. Huge ruminant animals such as goats are kept and a vast area of land is cultivated. Overall cropland holding of small-scale and large-scale farmers was observed to be 4.10 and 86.90 hectares, respectively. This indicates that there is huge crop residue production in the area, a considerable source of goat feed. Normally, crop residues are grazed or stockpiled for animal feeding. In some cases, the stored residues are sold during the critical period of feed scarcity. Sorghum crop residues are by far the most abundant feed resource in the area.

Table 3. Percentage of households owning livestock species across the two farming systems.

Livestock species	Small-scale farming		Large-scale farming		Overall	
	n = 99	%	n = 51	%	N = 150	%
Goat	99	100.00	51	100.00	150	100.00
Sheep	80	80.80	38	74.50	118	78.67
Cattle	66	66.67	33	64.70	99	66.00
Chicken	69	70.00	30	58.82	99	66.00
Donkey	67	67.67	29	56.86	96	64.00

Table 4. Reasons for keeping goats as ranked by respondents in the area.

Purpose	Rank			Index
	Rank 1	Rank 2	Rank 3	
Cash income	119	17	12	0.448
Meat home consumption	11	10	118	0.190
Milk	20	123	7	0.348
Skin	0	0	3	0.003
Saving value	0	0	10	0.011

Index = Σ [3 for rank 1 + 2 for rank 2 + 1 for rank 3] for particular purpose divided by Σ [3 for rank 1 + 2 for rank 2 + 1 for rank 3] for all purposes. The highest index value indicates the highest importance of the trait.

4. GOAT HUSBANDRY PRACTICES

4.1. Feeds and Feeding Management

The different feed resources reported in the area were natural pasture, browse species, crop residue and crop aftermath. Private grazing land was not common in the area. It was reported that communal grazing was the most abundant feed source for goat in the area though grazing drastically reduces in the dry season. Indigenous browse species are available feed resources for goats mainly in the dry season and the drought periods (March-June). It was indicated that Acacia mellifera, Acacia seyal, Anogeissus leiocarpus, Balanite aegyptica, Dichrostachys cinerea, Sclerocarya birrea, Combretum hertmannianum, Rhus nataliensis, Zizyphus spina-cristi were the predominant plant species widely utilized as goat feed source.

Crop residue and crop aftermath grazing were the other vital feed resources of goat during crop harvesting and after harvesting has occurred. Animals feed on the residue in two ways. The harvested crop residue is stacked near to homestead and fed to selected group of animals in the dry season. Otherwise, standing residue is left for grazing. However, the expansion of cropland, increase in livestock population and area enclosure are negatively affecting feed resource in the area.

4.2. Water Sources and Watering Management

Water resource is pertinent and vital for the subsistent life of livestock and livestock owners. In the study area, river, pond and borehole form the major source of water for domestic use and livestock watering. Indeed, water points are limited and large number of animals watered at the same water points leading to high chances of spreading diseases and land degradation. There were only two government constructed water points. Besides, private water points were very few.

Distance to water points varied with season and was invariably longer during the dry season. This posed challenges to breeders and sometimes limited access to water. A large proportion (45.33%) of the respondents travel 6 - 10 km to water their animals in the dry season. While in the wet season 67.33% of the total respondents watered their animals at a distance of less than 1 km. Shorter watering frequencies were used to water animals in the area. The proportions of the respondents that watered their goats once a day and twice in a day were 82% and 18%, respectively. Watering with more than a day frequency is not reported.

4.3. Housing System

Based on personal observations and interviewed households' information, goats were housed with sheep in nights. None of the respondents had provision for accommodating different age groups. Only sick ones and newly born kids/lambs were retained alone in sheds. Sheds were constructed of locally available materials. Sheds were not permanent as animals migrate from place to place in search of feed and water. However, farmers in urban areas used permanent houses separated from or attached to their homesteads.

4.4. Herding Practice

The common herding management in the area was free grazing. All the interviewed farmers replied that they herded goats with sheep. One or two herdsman and at least one herding dog used to accompany flocks for grazing. There is also the possibility of mixing with other goats and sheep flocks in the nearby vicinity of villages. Flocks of small size of less than 30 were pooled together to a maximum size of 60 to 120 heads for grazing. And then the flocks owned by different farmers were taken together by hired grazers.

4.5. Common Goat Diseases and Health Management

Diseases are a major constraint to the improvement of livestock industry in the tropics as they decrease production and increase the morbidity and mortality [37]. The most commonly prevailing diseases which hamper goat production in the study area are presented in Table 5. Respondents indicated that diseases affect all age groups of goats and mortalities from disease are high. More than half (53%) of the sampled respondents were able to separate between the different diseases. Diseases were named according to their symptoms.

Vaccination services were provided by Bureau of Agriculture and Rural Development of the district at seven clinics. In addition, farmers use modern drugs either from government clinics or open markets and treat their animals themselves. Some farmers also claimed that veterinary officers were called to treat their sick animals. The most extensively used drug in the area was oxytetracycline and locally known as leshlesh. Likewise, branding was applied when animals get sick and for identification purpose.

4.6. Origin, Distribution and Current Status of Begait Goat

Indeed, there is no documented information on the exact origin of Begait goat. According to [20], it is believed that the breed was derived from Nubian type goats. In the area the breed is also known as Barka/Hassan. From individual interviews, 37% of sampled farmers agreed that the breed's name "Barka" is an indicator of the breed's origin that is Barka area from Eritrea while most (63%) of the respondents believed that the breed's origin could be Eritrea and Sudan. From the focus group discussions, it became obvious that Begait goat is primarily kept by Beni Amir ethnic group who lives in Eritrea and Sudan bordering Ethiopia. This observation is supported by [38], who documented that Beni Amir ethnic group occupies the borders between much of Eritrea's Barka valley and eastern Sudan.

Table 5. Local names and scientific equivalents of the common goat diseases in the area.

Scientific name	Local name (*Tigrigna*)
PPR (pest des petits ruminants)	Gulhay
Foot rot	Mojelle
Orf/Contagious Ecthyma/Sore mouth	Af-gumed
Pasteurellosis	Mi'eta
Small pox	Enfrir
Anthrax	Megerem

Despite the exact date of immigration remains unclear, Begait goat is extensively distributed across northwestern lowlands of Ethiopia. It is estimated that >110,000 Begait goats are found in northwestern and western zones of Tigray region. Based upon the results of focus group discussions and individual interviews, population size of the breed is at increasing trend. The possible reason reported for this trend was the involvement of more farmers in Begait goat production as the breed is more prolific and resistant to drought conditions with better milk yield and meat quality. At present, Bureau of Agriculture and Rural Development of Tigray region has established a ranch, at Mai Woyni kebele, western zone of the region, for all Begait animal species aiming at conservation, breeding and improvement.

4.7. Breeding Practice and Selection Criteria

In the study area, mating was random. There was no report of controlled mating. Bucks run with does throughout the year and castration was an uncommon practice in the area. This resulted in indiscriminate and uncontrolled breeding. The productive life span of buck was reported on average 5 years. Most (83%) of the respondents had own buck. Farmers who had no breeding males, purchase buck from local markets or got buck service from their neighbors. This finding is in line with previous findings [14] [39] [40].

The most common way of selecting goats as parents for the coming generations is to use the offspring of a chosen parent (buck and/or doe). Linear index is the best strategy for selecting replacements in the livestock industries [41]. Selection criteria for buck and doe with corresponding index values are presented in Table 6.

The respondents prioritized body size, growth rate and libido as selection criteria for buck with the indices of 0.329, 0.232 and 0.183, respectively. While the most important selection criterion for does were body size (index = 0.330), twining ability (index = 0.219) and milk yield (index = 0.166). Others like mothering ability, kid survival, kidding interval and age at puberty were also reported as criteria but with lower rankings.

4.8. Effective Population Size and Level of in Breeding

The effective population size (Ne) and level of inbreeding (ΔF) were calculated for Begait goat using the averaged breeding males and females under closed breeding practice. Majority of the bucks were originated from their respective flocks which lead to inbreeding. Communal herding in contrast increases genetic diversity by rising Ne [6], which is a common practice in the study area. The obtained results for Ne and ΔF were 7.08 and 0.07, respectively. The rate of inbreeding was equivalent to the maximum acceptable value of 0.063 [42].

4.9. Reproductive Performance

According to respondents, the age at puberty in Begait goat was 7.41 ± 1.85 months for males and 8 ± 1.78 months for females. The present finding is within the range of age at sexual maturity reported for most of the tropical goat breeds under extensive system of management [20] [43].

The reported mean age at first kidding (AFK) for Begait goat was 14.18 ± 1.24 months. This is in agreement with that reported by [44] for Metema goats that have mean AFK of 13.6 ± 2.44 months. This result is also in agreement with reported age at first kidding for Arsi-Bale goats which is 14.88 ± 0.3 months [16] and shorter than for Somali short-eared goats which is 19.9 ± 7.93 months AFK under traditional management conditions.

Begait goat breed, according to the respondents had an average kidding interval of 8.4 ± 1.37 months. This result was shorter than the reported kidding interval for Abergelle and Central Highland goats which were 11.31 ± 2.21 and 10.3 ± 1.42 months, respectively [45]. Higher longevity under adverse conditions is one of the adaptation traits of tropical livestock. The mean reproductive lifespan of does in the flock was reported 11.44 ± 1.17 years and the average number of kids per doe per lifespan was 15.32 ± 1.76.

Table 6. Selection criteria for buck and doe in the study area as ranked by owners.

Class and selection criteria	Rank			Index
	Rank 1	Rank 2	Rank 3	
Buck				
Body size	82	18	14	0.329
Drought resistance	18	18	28	0.131
Growth rate	17	70	18	0.232
Libido	12	31	67	0.183
Coat color	15	8	17	0.087
Horn	6	5	6	0.038
Doe				
Body size	85	14	13	0.330
Age at puberty	8	4	3	0.039
Kidding interval	8	5	6	0.045
Twining ability	8	75	23	0.219
Milk yield	14	16	75	0.166
Kid survival	10	15	8	0.076
Mothering ability	17	20	22	0.126

Index = Σ [3 for rank 1 + 2 for rank 2 + 1 for rank 3] for particular trait divided by Σ [3 for rank 1 + 2 for rank 2 + 1 for rank 3] for all traits.

4.10. Acquisition and Disposal

The famous methods used to acquire goats were new born, buying and inheritance with the index values of 0.390, 0.343 and 0.186, respectively (Table 7). This is in good agreement with the findings of [33] and [46] in Uganda. Whereas disposal was through selling, slaughtering in holidays and death in that order (Table 7).

Goat marketing in the study area was traditional type. All (100%) of the respondents reported that weighing balance is unknown for selling and/or buying animals; rather visual assessment was used to estimate body condition of animals. Hence, selling price was fixed by negotiation between sellers and buyers. This result agreed with many research reports [16] [47] [48] from Ethiopia. The study results also indicated that higher numbers of goats are sold and better price is fetched in holidays although farmers sell goats at any time of the year depending on their need for money. Goat owners sold their goats primarily for purchasing food items, health expenses, farm input expenses (fertilizer and cropland rent), labor expenses (herdsman and laborers), school expenses for children and to pay back credit.

4.11. Production Constraints

Interviews with the sampled farmers revealed key problems constraining goat production in the area (Table 8). Water shortage, feed and grazing land shortages and inadequate veterinary services were ranked first, second and third, respectively. In line with the present findings previous studies [49] [50] indicated that the major problems of small ruminant farming include inadequate supply of water and pasture mainly in the dry season and problems arising from inadequate veterinary services and infrastructure.

Table 7. Ways of acquiring and disposing goat in the study area as ranked by owners.

Way of acquiring and disposing goat	Rank			Index
	Rank 1	Rank 2	Rank 3	
Acquiring (Entered through)				
Born	82	43	19	0.390
Bought	50	69	20	0.343
Gift	10	11	21	0.081
Inheritance	8	27	89	0.186
Disposing (Exited through)				
Sale	91	28	24	0.392
Death	29	24	78	0.237
Slaughter	17	83	32	0.277
Theft	7	11	11	0.060
Predator	6	4	5	0.034

Index = Σ [3 for rank 1 + 2 for rank 2 + 1 for rank 3] for particular method divided by Σ [3 for rank 1 + 2 for rank 2 + 1 for rank 3] for all methods.

Table 8. Ranked goat production constraints in the study area.

Constraints	Rank			Index
	Rank 1	Rank 2	Rank 3	
Feed and grazing area shortages	28	52	19	0.230
Water shortage	49	24	35	0.256
Inadequate veterinary services	31	32	23	0.200
Labor shortage	9	6	20	0.066
Market problem	20	23	24	0.144
Theft	9	7	17	0.064
Predators	4	6	12	0.040

Index = Σ [3 for rank 1 + 2 for rank 2 + 1 for rank 3] for particular constraint divided by Σ [3 for rank 1 + 2 for rank 2 + 1 for rank 3] for all constraints.

5. CONCLUSION

This study provides insight in to agricultural production system, breeding practices and major production constraints encountered in goat farming in the study area, which are preconditions in developing breeding programs. The present study also revealed that there are two types of farming systems in the area namely small-scale and large-scale farming. Goat production is an integral component of the existing farming systems by providing multifunctional roles to its owners. The high economic significance of goat coupled with its large flock size in the study area would suggest the scope of genetic improvement schemes in the area. However, water scarcity, feed and grazing area shortages, poor veterinary services and market linked problems should be addressed.

ACKNOWLEDGEMENTS

The authors would like to acknowledge farmers for their willingness to share their indigenous knowledge concerning the breed while nursing for the first author during his field stay.

REFERENCES

1. Nether Lands-African Business Council (2010) Livestock in Ethiopia and Opportunity Analyses for Dutch Investment. Fact Sheet, NABC.
2. Central Statistical Agency of Federal Democratic Republic of Ethiopia (2016) Agricultural Sample Survey: Report on Livestock and Livestock Characteristics (Private Peasant Holdings). Statistical Bulletin No. 583.
3. Workneh, A. (1992) Preliminary Survey of Indigenous Goat Types and Goat Husbandry Practices in Southern Ethiopia. M.Sc. Thesis, Alemaya University, Alemaya.
4. Köhler-Rollefson, I. (2000) Management of Animal Genetic Diversity at Community Level. GTZ, Eschborn, Germany. http://hdl.handle.net/10568/3679
5. Peacock, C. (2005) Goats: Unlocking Their Potential for Africa's Farmers. Paper presented at the 7th Conference of Ministers Responsible for Animal Resources, Kigali, 31 October-4 November 2005.
6. Jaitner, J., Sowe, J., Secka-Njie, E. and Dempfle, L. (2001) Ownership Pattern and Management Practices of Small Ruminants in The Gambia-Implications for a Breeding Program. Small Ruminant Research, 40, 101-108. https://doi.org/10.1016/S0921-4488(00)00221-2
7. Adane, H. and Girma, A. (2008) Economic Significance of Sheep and Goats. In: Alemu, Y. and Merkel, R.C., Eds., Sheep and Goat Production Handbook for Ethiopia, Ethiopian Sheep and Goat Productivity Improvement Program (ESGPIP), 1-4.
8. Assen, E. and Aklilu H. (2012) Sheep and Goat Production and Utilization in Different Agro Ecological Zones in Tigray, Ethiopia. Livestock Research for Rural Development, 24. http://www.lrrd.org/lrrd24/1/asse24016.htm
9. Baker, R.L. and Gray, G.D. (2003) Appropriate Breeds and Breeding Schemes for Sheep and Goats in the Tropics: The Importance of Characterizing and Utilizing Disease Resistance and Adaptation to Local Stresses. In: R. Sani, G.D. Gray and R.L. Baker, Eds. https://cgspace.cgiar.org/handle/10568/2933
10. Ethiopian Agricultural Research Organization (2002) An Assessment of the Agricultural Production Base, Technological Packages and Innovation and Intervention Strategies for Commercial farmers in Kafta Humera Wereda of Tigray Regional State.
11. Kafta Humera District Livelihood Report (2007) Disaster Prevention and Preparedness Agency, Ethiopia.
12. Institute of Biodiversity Conservation (2004) The State of Ethiopia's Farm Animal Genetic Resources: Country Report. A Contribution to the First Report on the Stat of the World's Animal Genetic Resources. IBC, May 2004, Addis Ababa.
13. Kosgey, I.S. (2004) Breeding Objectives and Breeding Strategies for Small Ruminant in the Tropics. Ph.D. Thesis, Wageningen University, Wageningen.
14. Falconer, D.S. (1989) Introduction to Quantitative Genetics. 3rd Edition, Longman Scientific and Technical, New York.
15. Central Statistical Agency of the Federal Democratic Republic of Ethiopia (2010) Summary and Statistical Report of the 2007 Population and Housing Census Result, Addis Ababa.
16. Endeshaw, A. (2007) Assessment of Production and Marketing System of Goats in Dale District Sidama zone. M.Sc. Thesis, University of Hawssa, Awassa.
17. Shenkute, B., Getahun, L., Azage, T. and Hassen, A. (2010) Small Ruminant Production in Coffee-Based Mixed Crop-Livestock System of Western Ethiopian Highlands: Status and Prospectus for Improvement. Livestock Research for Rural Development, 22. http://www.lrrd.org/lrrd22/10/shen22186.htm
18. Hulunm, G. (2014) On-Farm Phenotypic Characterization and Performance Evaluation of Bati, Borena and Short Eared Somali Goat Populations of Ethiopia. M.Sc. Thesis, Haramaya University, Haramaya.

19. Solomon, A. (2014) Design of Community Based Breeding Programs for Two Indigenous Goat Breeds of Ethiopia. Ph.D. Thesis, BOKU-University of Natural Resources and Life Sciences, Vienna.
20. Farm-Africa (1996) Goat Types of Ethiopia and Eritrea. Physical description and management systems. Published jointly by FARM-Africa and International Livestock Research Institute.
21. Wilson, R.T. (1986) Livestock Production in Central Mali: Long Term Studies on Cattle and Small Ruminants in the Agro-Pastoral System. ILCA Research Report No. 14, 116.
22. Tsedeke, K. (2007) Production and Marketing of Sheep and Goats in Alaba, SNNPR. M.Sc. Thesis, Hawassa University, Awassa.
23. Dhaba, U., Belay, D., Solomon, D. and Taye, T. (2012) Sheep and Goat Production Systems in Ilu Abba Bora Zone of Oromia Regional State, Ethiopia: Feeding and Management Strategies. Global Veterinaria, 9, 421-429.
24. Ahmed, S., Kefelegne, K. and Kefena,E. (2015) Breeding Objective, Selection Criteria and Breeding Practice of Indigenous Goats in Western Ethiopia: Implications for Sustainable Genetic Improvemt. Greener Journal of Agricultural Sciences, 5, 167-176. https://doi.org/10.15580/GJAS.2015.5.072715105
25. Niekerk, W.A. and Pimentel, P.L. (2004) Goat Production in the Smallholder Section in the Boane District in Southern Mozambique. South African Journal of Animal Science, 34, 123-125.
26. Dossa, L.H., Sangaré, M., Buerkert, A. and Schlecht, E. (2015) Production Objectives and Breeding Practices of Urban Goat and Sheep Keepers in West Africa: Regional Analysis and Implications for the Development of Supportive Breeding Programs. SpringerPlus, 4, 281. https://doi.org/10.1186/s40064-015-1075-7
27. Otieno, O.G., Junga, J.O., Badamana, M.S. and Amimo, J.O. (2015) Indigenous Knowledge Used in Breeding and Management of Capra hircus Populations in Kajiado and Makueni Counties, Kenya. Open Journal of Genetics, 5, 111-135. https://doi.org/10.4236/ojgen.2015.53009
28. de Haan, C., Steinfeld, H. and Blackburn, H. (1997) Livestock and the Environment: Finding a Balance. Report of the Commission of the European Communities, the World Bank and the governments of Denmark, France, Germany, The Netherlands, UK and USA.
29. Sölkner, J., Nakimbugwe, H. and Valle Zarate, A. (1998) Analysis of Determinants for Success and Failure of Village Breeding Programmes. Proceedings of the 6th World Congress on Genetics Applied to Livestock Production, Armidale, 273-280.
30. Apori, S.O., Osei, D.Y., Rhule, S.W.A., Doku, C. and Oppong-Anane, K. (2011) Livestock Marketing in the Coastal Savannah Zone of Ghana. Proceedings of the 17th Biennial Conference of the Ghana Society of Animal Production, 20-23 July 2011, 43-45.
31. Rewe, T.O., Indetie, D., Ojango, J.M.K. and Kahi, A.K. (2006) Breeding Objectives for the Boran Breed in Kenya: Model Development and Application to Pasture-Based Production Systems. Animal Science Journal, 77, 163-177. https://doi.org/10.1111/j.1740-0929.2006.00334.x
32. Kugonza, D.R., Bareeba, F.B. and Kirembe, G. (2001) Characterization of Meat Goat Management Systems in Three Districts of Central Uganda. MUARIK Bulletin, 4, 26-29.
33. Semakula, J., Mutetikka, D., Kugonza, R.D. and Mpairwe, D. (2010) Smallholder Goat Breeding Systems in Humid, Sub Humid and Semi Arid Agro-Ecological Zones in Uganda. Global Veterinaria, 4, 283-291.
34. Arse, G., Feyisa, H., Gurmessa U., Merga, M. and Girma, D. (2013) Assessment on Challenges and Opportunities of Goat Farming System in Adami Tulu, Arsi Negelle and Fantale Districts of Oromia Regional State, Ethiopia. African Journal of Agricultural Research, 8, 26-31.
35. Byaruhanga, C., Oluka, J. and Olinga, S. (2015) Socio-Economic Aspects of Goat Production in a Rural Agro-Pastoral System of Uganda. Universal Journal of Agricultural Research, 3, 203-210. https://doi.org/10.13189/ujar.2015.030604
36. Braker, M.J.E., Udo, H.M.J. and Webb, E.C. (2002) Impacts of Intervention Objectives in Goat Production within Subsistence Farming Systems in South Africa. South African Journal of Animal Science, 32, 185-191. https://doi.org/10.4314/sajas.v32i3.3745

37. Mwacharo, J.M. and Drucker, A.G. (2005) Production Objectives and Management Strategies of Livestock-Keepers in South-East Kenya: Implications for a Breeding Programme. Tropical Animal Health and Production, 37, 635-652. http://www.ncbi.nlm.nih.gov/pubmed/16619880 https://doi.org/10.1007/s11250-005-4253-8
38. Chisholm, H. (1911) Beni-Amer. Encyclopædia Britannica 3, 11th Edition, Cambridge University Press, Cambridge.
39. Ndamukong, K.J., Sewell, M.M. and Asanji, M.F. (1989) Management and Productivity of Small Ruminants in Northwest Province of Cameroon. Tropical Animal Health and Production, 21, 109-119. https://doi.org/10.1007/BF02236189
40. Ngere, L.O., Adu, I.F. and Okubanjo, I.O. (1984) The Indigenous Goats of Nigeria. Animal Genetic Resources, 3, 1-9. https://doi.org/10.1017/S1014233900000109
41. Dekkers, J.C. and Gibson, J.P. (1998) Applying Breeding Objectives to Dairy Cattle Improvement. Journal of Dairy Science, 81, 19-35. https://doi.org/10.3168/jds.S0022-0302(98)70151-1
42. Armstrong, J.B. (2006) Inbreeding: Why We Will Not Do It? http://www.parispoodles.com/Inbreeding.html
43. Wilson, R.T. (1991) Small Ruminant Production and the Small Ruminant Genetic Resource in Tropical AFRICA. FAO Animal Production and Health, Rome, Paper No. 88.
44. Tesfaye, T. (2009) Characterization of Goat Production Systems and On-Farm Evaluation of the Growth Performance of Grazing Goats Supplemented with Different Protein Sources in Metema Woreda, Amhara Region, Ethiopia. M.Sc. Thesis, Haramaya University, Haramaya.
45. Belay, D. (2008) Growth, Reproductive Performance and Carcass Characteristics of Abergelle and Central Highland Goat Types under Traditional and Slightly Improved Management in Sekota Woreda. M.Sc. Thesis, Haramaya University, Haramay.
46. Ssewannyana, E., Onyait, O.A., Okwir, W., Ekoi, M., Okello, M., Masaba, J. and Ajibo, G.E. (2004) Characteristics of Rural Goat Production and Marketing in Kumi and Lira Districts, Uganda. Uganda Journal of Agricultural Science, 9, 289-293.
47. Ayele, S., Assegid, W., Jabbar, M.A., Ahmed, M.M. and Belachew, H. (2003) Livestock Marketing in Ethiopia: A Review of Structure, Performance and Development Initiatives. ILRI Socio-Economics and Policy Research Working Paper 52.
48. Gezahegn, A., Mohammad, A.J., Hailemariam, T., Elias, M. and Getahun, K. (2003) Seasonal and Inter-Market Differences in Prices of Small Ruminants in Ethiopia. Journal of Food Products Marketing, 12, 59-77.
49. Egbe-Nwiyi, T.N.C., Igbjekwe, F.N. and Nwosu, C.O. (1999) Gastro-Intestinal Parasitism and Associated Haemotological Changes in Small Ruminants in the Semi Arid Region of North-Eastern Nigeria. Bulletin of Animal Health and Production in Africa, 47, 61-70.
50. Omoike, A. (2006) Prevalence of Diseases among Sheep and Goats in Edo State Nigeria. Journal of Agriculture and Social Research, 6, 23-31.

INDEX

R

S

V

W

Y